AA002343

2012 12th International Conference on Numerical Simulation of Optoelectronic Devices

(NUSOD 2012)

Shanghai, China
28 – 31 August 2012

IEEE Catalog Number: CFP12817-PRT
ISBN: 978-1-4673-1602-6

Copyright © 2012 by the Institute of Electrical and Electronic Engineers, Inc
All Rights Reserved

Copyright and Reprint Permissions: Abstracting is permitted with credit to the source. Libraries are permitted to photocopy beyond the limit of U.S. copyright law for private use of patrons those articles in this volume that carry a code at the bottom of the first page, provided the per-copy fee indicated in the code is paid through Copyright Clearance Center, 222 Rosewood Drive, Danvers, MA 01923.

For other copying, reprint or republication permission, write to IEEE Copyrights Manager, IEEE Service Center, 445 Hoes Lane, Piscataway, NJ 08854. All rights reserved.

***This publication is a representation of what appears in the IEEE Digital Libraries. Some format issues inherent in the e-media version may also appear in this print version.

IEEE Catalog Number: CFP12817-PRT
ISBN 13:   978-1-4673-1602-6
ISSN: 2158-3234

Additional Copies of This Publication Are Available From:

Curran Associates, Inc
57 Morehouse Lane
Red Hook, NY 12571 USA
Phone: (845) 758-0400
Fax: (845) 758-2633
E-mail: curran@proceedings.com
Web: www.proceedings.com

2012 12th International Conference on Numerical Simulation of Optoelectronic Devices (NUSOD 2012)

Shanghai, China
28-31 August 2012

IEEE Catalog Number: CFP12817-POD
ISBN: 978-1-46731-602-6

Table of Contents

Tuesday, 28 August 2012

TuA Novel Devices

TuA1 The Tunable Plasmonic Resonant Absorption in Grating-gate GaN-based HEMTs for THz detection; W. D. Hu, L. Wang, N. Guo, X. S. Chen, and W. Lu; Shanghai Institute of Technical Physics, China (*invited*)　　1

TuA2 The improvement of figure of merit with infrared perfect absorber for plasmonic resonance sensing; G.H. Li (1), X. Wu (2), Y. Jiang (2), C. X. Shao (2), L. J. Huang (1), X. S. Chen (1), W. D. Hu (1), and W. Lu (1); (1) Shanghai Institute of Technical Physics, China; (2) University of Electronic Science and Technology , China　　3

TuA3 Influence of surrounded metallic layers on whispering-gallery modes in circular microresonators; Q. F. Yao, Y. Z. Huang, X. M. Lv, J. D. Lin, and L. X. Zou; State Key Laboratory on Integrated Optoelectronics, Institute of Semiconductors, Chinese Academy of Sciences, China　　5

TuA5 Surface Plasmon Resonance for Sensing; Zhihong Chen (1,2) and Chun Jiang (1); (1) Shanghai Jiao Tong University, (2) Shanghai Normal University; China　　7

TuB Photodetectors

TuB1 Design of Terahertz Quantum-Well Photodetectors; X. G. Guo (1), J. C. Cao (1), and H. C. Liu (2); (1) Shanghai Institute of Microsystem and Information Technology, China; (2) Shanghai Jiao Tong University, China (*invited*)　　9

TuB2 Crosstalk Suppressing Design of GaAs Microlenses Integrated on HgCdTe Infrared Focal Plane Arrays; Z. H. Ye, Y. Li, C. Lin, X. N. Hu, R. J. Ding and
L. He; Shanghai Institute of Technical Physics, China　　11

TuB3 Impact of Duty Cycle and Nano-Grating Height on the Light Absorption of Plasmonics-Based MSM Photodetectors; Farzaneh Fadakar Masouleh(1), Narottam Das (2), Hamid Reza Mashayekhi (1); (1) University of Guilan, Iran; (2) Curtin University, Australia　　13

TuB4 High-efficiency Optical Coupling to Planar Photodiode using Metal Reflector loaded Waveguide Grating Coupler; G. Li, Y. Hashimoto, S. Ebuchi, T. Maruyama, and K. Iiyama; Kanazawa University, Japan　　15

TuB5 New optical coupling structure of high light absorption quantum well Infrared photodetectors; Q. Li, N. Li, X. S. Chen, Z. F. Li, W. Lu; Shanghai Institute of Technical Physics, China 17

TuB6 Modeling on Current-Voltage Characteristics of HgCdTe Photodiodes in Forward Bias Region; Y. Li, Z. H. Ye, C. Lin, X. N. Hu, R. J. Ding and L. He; Shanghai Institute of Technical Physics, China 19

TuP   Posters

TuP02 Simulation of InGaN/GaN light-emitting diodes with a non-local quantum well transport model; C. S. Xia (1), Z. M. Simon Li (1), Y. Sheng (1), L. W. Cheng (2), W. D. Hu (2), and W. Lu(2); (1) Crosslight Software Inc. China Branch, China; (2) Shanghai Institute of Technical Physics, Chinese Academy of Sciences, China 21

TuP03 Simulation of InGaN/GaN light-emitting diodes with Patterned Sapphire Substrate; Yang Sheng (1), Chang Sheng Xia (1), Z. M. Simon Li (1), and Li Wen Cheng (2); (1) Crosslight Software, China; (2) Shanghai Institute of Technical Physics, China 23

TuP04 Two-dimensional Simulation of Mid-infrared Quantum Cascade Lasers: Temperature and Field Dependent Analysis; Ying-Ying Li(1), Z.-M. Simon Li(2), Guo-Ping Ru(1); (1) Fudan University, China; (2) Crosslight Software, Canada 25

TuP05 All-Optical Gate Switches Employing the Quasi-Phase Matched Cascaded Second-Order Nonlinear Effect: Effect of Fabrication Errors; Ryohei Oshige, Yusuke Osawa, and Yutaka Fukuchi; Tokyo University of Science, Japan 27

TuP06 The Hybridization of Plasmons in GaN-based Two-Dimensional Channels; L. Wang, W. D. Hu, X. S. Chen, and W. Lu; Shanghai Institute of Technical Physics, China 29

TuP07 Light Extraction Enhancement Analysis of GaN-based LED with Surface Spherical Crown Array; Xiaomin Wang, Kang Li, Fanmin Kong, Zhenming Zhang; Shandong University, China 31

TuP09 Electro-optical characteristics for AlGaN solar-blind p-i-n photodiode: Experiment and simulation; X. D. Wang , W. D. Hu, X. S. Chen, J. T. Xu, L. Wang, X. Y. Li, and W. Lu; Shanghai Institute of Technical Physics, China 33

TuP10 Numerical simulation of high-efficiency InGaP/GaAs/InGaAs triple-junction solar cells grown on GaAs substrate; J. Liang (1), W. D. Hu (1), X. S. Chen (1), C. S. Xia (2), and L. W. Cheng (2); (1) Shanghai Institute of Technical Physics, China; (2) Crosslight Software China, China 35

TuP11 Structure Design of Refractive Index Sensor Based on LPFG with Double-layer Coatings; Zhengtian Gu, Tao Luo, Kan Gao; University of Shanghai for Science and Technology, China — 37

TuP12 Experimental determination of minority carrier lifetime and recombination mechanisms in MCT photovoltaic detectors; Haoyang Cui, Naiyun Tang, Zhong Tang; Shanghai University of Electric Power, China — 39

TuP13 Polarization-Independent Self-Collimated Beam Splitting in Two-Dimensional Photonic Crystals; M. B. Yucel (1), O. A. Kaya (2), A. Cicek (3), B. Ulug (1); (1) Akdeniz University, Turkey; (2) Inonu University, Turkey; (3) Mehmet Akif Ersoy University, Turkey — 41

TuP14 Portable Terahertz Spectrometer and Imaging System; Sang-Pil Han (1), Namje Kim (1), Han-Cheol Ryu (1), Hyunsung Ko (1), Jeong-Woo Park (1), Min Yong Jeon (2), and Kyung Hyun Park (1); (1) ETRI, Korea; (2) Chungnam National University, Korea — 43

TuP15 Optimization of InSb Infrared Focal Plane Arrays; N. Guo (1), W. D. Hu (1), X. S. Chen (1), Y. Q. Lv (2), X. L. Zhang (2), J. J. Si (2), and W. Lu (1); (1) Shanghai Institute of Technical Physics, China; (2) Luoyang Optoelectronic Institute, China — 45

TuP16 A general transformation designing high gain lens antennas with homogeneous media; Lujun Huang, Xiaoshuang Chen, Bo Ni, Guanhai Li, Zhifeng Li, and Wei Lu; Shanghai Institute of Technical Physics, China — 47

TuP17 The photocurrent of resonant tunneling diode controlled by the charging effects of quantum dots; Daming Zhou; Shanghai Institute of Technical Physics, China — 49

TuP18 A bisection-function technique to characterize heat transport in high-power GaN-based light-emitting-diodes package; Liwen Cheng(1), Yang Sheng(2), ChangshengXia(2),Weida Hu(1) and Wei Lu(1); (1) Shanghai Institute of Technical Physics, China; (2) Crosslight Software, China — 51

TuP19 Running wave condition of extended states in charactering the photocurrent of multiple quantum well and superlattice structured GaAs/AlGaAs solar cells; X. F. Yang, Y. S. Liu, and X. F. Jiang; Changshu Institute of Technology, China — 53

TuP20 A dual-band polarization insensitive metamaterial absorber with split ring resonator; B. Ni, X. S. Chen, L. J. Huang, J. Y. Ding, G. H. Li, and W. Lu; Shanghai Institute of Technical Physics, China — 55

TuP21 Physical model of an Optical Memory Cell with coupling quantum dots; L. Fan, F. M. Guo; East China Normal University, China — 57

TuP22 General design of compact waveguide coupler with homogeneous media; Lujun Huang, Xiaoshuang Chen, Bo Ni, Guanhai Li, Zhifeng Li, and Wei Lu; Shanghai Institute of Technical Physics, China 59

TuP23 Simulation of Resonant Tunneling Structures: Origin of the I-V multi-peak and Plateau-like Behaviour; J. Wen, Q.-C. Weng, L. Li, D.-Y. Xiong; East China Normal University, China 61

TuP24 An Equivalent Circuit Model for the Long-Wavelength Quantum Well Infrared Detectors; L. Li, Q.-C. Weng, J. Wen, D.-Y. Xiong; East China Normal University, China 63

TuP26 Analytic solution of the nonlinear equation; H.-J. Lee, S. Lee, D. H. Woo, T. J. Lee, J. H. Kim; Korea Institute of Science and Technology, South Korea 65

TuP27 An Evaluation of Photoresist Thickness for Semi-ellipsoid Microlens Fabrication before Thermal Reflow Using the Prolate Spheroid Approximation; S.-Y Hung; Nan Kai University of Technology, Taiwan 67

TuP28 Self-Switching Using SOA-Assisted Sagnac Interferometer; V. Ahmadi(1), M. Jamali(1), and M. Razaghi (2); (1) Tarbiat Modares University, Iran; (2) University of Kurdistan, Iran 69

TuP29 Study on the structure characteristics of HgCdTe infrared detector using laser beam-induced current; X. K. Hong, H. Lu, D. B. Zhang; Changshu Institute of Technology, China 71

TuP30 Simulation of a SlowLight MQWs Semiconductor Device Based on a Practical Sample for EIT Phenomenon; M. Shahriari, H. Kaatuzian; Amirkabir University of Technology, Iran N/A

TuP31 Optimization of Detector Arrays and Circuits Targeted for Precision Calculation in Infrared Laser Interferometer; Xiaojie Sun (1,2), Jianwen Hua (1), Zuoxiao Dai (1); (1) Shanghai Institute of Technical Physics, Chinese Academy of Sciences; (2)Graduate University of Chinese Academy of Sciences, China 75

TuP32 The investigation of the transient photovoltage in HgCdTe infrated photovoltaic detectors; Haoyang Cui, Naiyun Tang, Zhong Tang; Shanghai University of Electric Power, China 77

Wednesday, 29 August 2012

WA   Material Properties

WA1 Simulation of carrier dynamics in graphene on a substrate at terahertz and mid-infrared frequencies; N. Sule, K. J. Willis, S. C. Hagness, and I. Knezevic; University of Wisconsin Madison, USA   79

WA3 Hydrogenated Graphene: Structures and Surface Work Function; N. Jiao, Chaoyu He, C. X. Zhang and L. Z. Sun; Xiangtan University, China   81

WA4 Structures, stability and electronic properties of two- or four-segment BN/C nanotubes; Chaoyu He, C. X. Zhang, H. P. Xiao, L. Z. Sun and J. X. Zhong; Xiangtan University, China   83

WA5 Non Linear Piezoelectricity in ZincBlende GaAs and InAs Semiconductors; G. Tse, J. Pal, R. Garg, V. Haxha and M.A. Migliorato; University of Manchester, UK   85

WB   Photonic Crystals

WB1 Design of Silicon Photonic Crystal Integrated Optical Devices; Zhi-Yuan Li, Lin Gan, and Chen Wang; Institute of Physics, China (*invited*)   87

WB2 Nonlinear optics in photonic crystal nanostructures; Chad Husko; University of Sydney, Australia (*invited*)   89

WB3 Optical design of a qubit embedded in photonic crystals for rotation gate operations; H. Nihei (1), and A. Okamoto (2); (1) Health Sciences University of Hokkaido, Japan; (2) Hokkaido University, Japan   91

WC   Nano Structures

WC1 InGaN Nanorod LEDs: A Performance Assessment; B. Witzigmann, M. Deppner, and F. Roemer; University of Kassel, Germany (*invited*)   93

WC2 Influence of polar surface properties on InGaN/GaN core-shell nanorod LED properties; M. Auf der Maur, F. Sacconi, A. Di Carlo; University of Rome Tor Vergata, Italy   95

WC3 Excitonic Properties of GaN/AlN Quantum Dot Single Photon Sources; Stanko Tomic (1) and Nenad Vukmirovic (2); (1) University of Salford, UK; (2) Institute of Physics, Serbia   97

WC5 Strain-induced modulation of mechanical properties and electronic structure of edge-modification graphene nanoribbons; Cheng Zhang, Mingsen Deng and Shaohong Cai; Guizhou University, China   99

WD   Numerical Methods

WD1 Quantum Mechanical Simulations of Nano-Structures and Nano-Devices; Xiang-Wei Jiang (1), Hui-Xiong Deng (1), Shu-Shen Li (1), Jun-Wei Luo (2), Lin-Wang Wang (3); (1) Institute of Semiconductors, China; (2) National Renewable Energy Laboratory, USA; (3) Lawrence Berkeley National Laboratory, USA (*invited*)   101

WD2 Discretization scheme for drift-diffusion equations with strong diffusion enhancement; Th. Koprucki and K. Gaertner; WIAS Berlin, Germany   103

WD3 Beam Propagation Analysis Using Higher-Order Full-Vectorial Finite-Difference Method; C.-H. Du and Y.-P. Chiou; National Taiwan University, Taiwan   105

WD4 Acceleration of 3D numerical simulation of silicon solar cell using thread parallelism; B. Min, S. Suckow, U. Yusufoglu, T. M. Pletzer and H. Kurz; RWTH-Aachen University, Germany   107

Thursday, 30 August 2012

ThA   Solar Cells

ThA1 Device Simulation of Intermediate Band Solar Cells; Katsuhisa Yoshida and Yoshitaka Okada; University of Tokyo, Japan (*invited*)   109

ThA2 Radiative and Non-radiative Processes in Intermediate Band Solar Cells; Stanko Tomic; University of Salford, UK   111

ThA3 Modeling of N-i-P Vs. P-i-N InGaN Solar Cells with Ultrathin GaN Interlayers for Improved Performance; Jeramy Dickerson, Konstantinos Pantzas, Tarik Moudakir, Paul L. Voss, and Abdallah Ougazzaden; Georgia Tech-CNRS, France   113

ThA4 Green functions for photovoltaic response of quantum wir-dot-wire junctions; A. Berbezier, F. Michelini; Institute Materials Microelectronics and Nanosciences of Provence (IM2NP), France   115

ThA5 Effects of Sulfur Incorporation into Absorbers of CIGS Solar Cells Studied by Numerical Analysis; Chia-Hua Huang and Hung-Lung Cheng; National Dong Hwa University, Taiwan   117

ThB   Laser Diodes

ThB1 Self-Consistent Electro-Thermal-Optical Simulation of Thermal Blooming in Broad-Area Lasers; J. Piprek; NUSOD Institute LLC, USA   119

ThB2 Thermal Simulation of GaAs-Based Midinfrared Quantum Cascade Lasers; Y. B. Shi, Z. Aksamija, and I. Knezevic; University of Wisconsin-Madison, USA   121

ThB3 On the Line Form and Natural Linewidth; Simulation and Interpretation of Experiments; M. G. Noppe; Novosibirsk State Technical University, Russia   123

ThB4 Simulation of a Ridge-Type Semiconductor Laser for Separate Confinement of Horizontal Transverse Modes and Carriers; Hiroki Kato, Hazuki Yoshida, and Takahiro Numai; Ritsumeikan University, Japan   125

ThB6 Efficient Simulation Method for DFB Lasers with Large Gain Saturation Effect ; Y. Xi(1), H. Lin(2), and X. Li(2) ; (1) Huazhong University of Science and Technology, China; (2) McMaster University, Canada   127

ThC   Photonics

ThC1 Recent progress in theory of nonlinear pulse propagation in subwavelength waveguides; Shahraam Afshar V., Wenqi Zhang, M. A. Lohe, and Tanya M. Monro; University of Adelaide, Australia (*invited*)   129

ThC2 Spatio-Temporal Pulse Propagation in nonlinear dispersive optical Media; Carsten Bree, Shalva Amiranashvili and Uwe Bandelow; WIAS Berlin, Germany   131

ThC4 Characterization of Subwavelength Grating Waveguides with 3D Finite Element Method; Y. H. Isayama, M. S. Gonçalves, H.E.Hernández-Figueroa; University of Campinas, Brazil   133

ThC5 All-Optical Discrete Fourier Transform for OFDM Demultiplexing and its Sensitivity to Phase Errors; S. Schwarz(1), A. Rahim(2), C. G. Schaeffer(1), J. Bruns(2), and K. Petermann(2); (1) Helmut Schmidt University, Germany;(2) Berlin University of Technology, Germany   135

ThC6 Determination of Resonance Frequencies in Silica Fiber using SRS Gain; Mrinal Sen, Mukul Kumar Das; Indian School of Mines, India   137

Author Index   139

Postdeadline paper summaries are distributed separately at the conference.

12th International Conference on

Numerical Simulation of Optoelectronic Devices

NUSOD 2012

Editors / Chairs:
Joachim Piprek & Wei Lu

Technical Co-Sponsor: The IEEE Photonics Society

Welcome to NUSOD 2012 !

The 12th International NUSOD Conference welcomes researchers from 18 countries who present more than 70 papers, including 8 invited talks. The conference sessions cover a wide range of topics, such as novel materials and devices, nanostructures, photonic crystals, laser diodes, photodetectors, solar cells, as well as numerical methods.

The NUSOD Conference was started at the University of California at Santa Barbara in 2001 and the participation was far beyond expectations - which provided the motivation to make this meeting an annual event, rotating between North America, Europe, and Asia. Subsequent NUSOD Conferences took place in Zurich, Tokyo, Berlin, Singapore, Newark, Nottingham, Gwangju, Atlanta, and Rome and they keep underscoring the continuing need for exchange and collaboration in this diverse field. NUSOD is now firmly established as one of the key conference fixtures in optoelectronics to network and discuss the latest challenges and developments in device simulation and design.

The NUSOD Institute was established in 2004 to serve as organizational umbrella of the NUSOD Conference and to provide related educational and technical services. Its web site (www.nusod.org) also offers a growing directory of software tools. Researchers and software developers are encouraged to list their tools and to make them available to other users.

The local organizers wish to thank the National Natural Science Foundation of China (NSFC) and Chinese Academy of Sciences (CAS) for financial support, as well as Shanghai Institute of Technical Physics (SITP) and Shanghai Institute of Microsystem And Information Technology (SIMIT) for conference organization and ground support.

We wish you a stimulating and enjoyable experience at NUSOD 2012 !

Wei Lu
Chair

Joachim Piprek
Chair

Weida Hu
Secretary

Chairs and Committees

NUSOD 2012 Chairs

Wei Lu, Shanghai Institute of Technical Physics, China
Joachim Piprek, NUSOD Institute, United States

Conference Secretary

Weida Hu, Shanghai Institute of Technical Physics, China

Program Committee

Urs Aeberhard, Research Center Juelich, Germany
Matthias Auf der Maur, University of Rome " Tor Vergata", Italy
Eugene Avrutin, University of York, UK
Uwe Bandelow, Weierstrass Institute, Germany
Prasanta Basu, University of Calcutta, India
Enrico Bellotti, Boston University, USA
Wallace Choy, University of Hong Kong, China
Lukas Chrostowski, University of British Columbia, Canada
Martijn de Sterke, University of Sydney, Australia
Aleksandra Djurisic, University of Hong Kong, China
Julien Javaloyes, Balearic Islands University, Spain
Max Migliorato, University of Manchester, UK
Wei-Choon Ng, Synopsys, USA
Seoung-Hwan Park, Catholic University of Daegu, Korea
Mauro Pereira, Sheffield Hallam University, UK
Angela Thränhardt, University of Chemnitz, Germany
Stanko Tomic, University of Salford, UK
Shyh-Lin Tsao, Cherry Tree Consulting Co., Taiwan
Eoin O'Reilly, Tyndall National Institute, Ireland
Jayanta Sarma, University of Bath, UK
Slawek Sujecki , University of Nottingham, UK
Hans Wenzel, Ferdinand-Braun Institute, Germany
Siu-Fung Yu, Hong Kong Polytechnic University, China

Steering Committee

Aldo Di Carlo, University of Rome "Tor Vergata", Italy
Eric Larkins, University of Nottingham, UK
Simon Li, Crosslight Software, Canada
Joachim Piprek, NUSOD Institute, USA
Berthold Schmidt, Switzerland
Bernd Witzigmann, Univerity of Kassel, Germany

Invited Talks

TuA1 The Tunable Plasmonic Resonant Absorption in Grating-gate GaN-based HEMTs for THz detection;
W. D. Hu, L. Wang, N. Guo, X. S. Chen, and W. Lu; Shanghai Institute of Technical Physics, China

TuB1 Design of Terahertz Quantum-Well Photodetectors;
X. G. Guo (1), J. C. Cao (1), and H. C. Liu (2); (1) Shanghai Institute of Microsystem and Information Technology, China; (2) Shanghai Jiao Tong University, China

WB1 Design of Silicon Photonic Crystal Integrated Optical Devices;
Zhi-Yuan Li, Lin Gan, and Chen Wang; Institute of Physics, China

WB2 Nonlinear optics in photonic crystal nanostructures;
Chad Husko; University of Sydney, Australia

WC1 GaN Nanorod LEDs: A Performance Assessment;
B. Witzigmann, M. Deppner, and F. Roemer; University of Kassel, Germany

WD1 Quantum Mechanical Simulations of Nano-Structures and Nano-Devices; Xiang-Wei Jiang (1), Hui-Xiong Deng (1), Shu-Shen Li (1), Jun-Wei Luo (2), Lin-Wang Wang (3); (1) Institute of Semiconductors, China; (2) National Renewable Energy Laboratory, USA; (3) Lawrence Berkeley National Laboratory, USA

ThA1 Device Simulation of Intermediate Band Solar Cells;
Katsuhisa Yoshida and Yoshitaka Okada; University of Tokyo, Japan

ThC1 Recent progress in theory of nonlinear pulse propagation in subwavelength waveguides;
Shahraam Afshar V., Wenqi Zhang, M. A. Lohe, and Tanya M. Monro; University of Adelaide, Australia

The Tunable Plasmonic Resonant Absorption in Grating-gate GaN-based HEMTs for THz detection

Weida Hu[*], Lin Wang, Nan Guo, Xiaoshuang Chen[*], and Wei Lu

National Lab for Infrared Physics, Shanghai Institute of Technical Physics, Chinese Academy of Sciences, 500 Yu Tian Road, Shanghai, China 200083

Abstract

The plasmonic resonant phenomenon in terahertz wave band for GaN-based high electron mobility transistors is investigated by using finite difference scheme. Strong resonant absorptions can be obtained with large area slit grating-gate serves both as electrodes and coupler. Such kinds of plasmonic resonant detection devices are compatible to the well-developed GaN process, and possibly overcome the difficulty in fabricating ultra-short-gate devices for terahertz applications.

I. INTRODUCTION

Far infrared (terahertz) remote sensing technology can potentially be used in the astronomy and atmosphere detections. The quality of remote sensing image is strongly restricted by device performance of the THz detection. The plasmonic wave resonant detector basing on slit-grating-gate transistor is one of important devices which can realize tunable and room-temperature THz detection. However, the detection frequency, quantum efficiency, and work temperature of the device cannot fulfill the requirements of remote sensing applications.

Furthermore, utilization of terahertz (THz) radiation in detection of biological and chemical agents is anticipated for long time. Since rotational and vibrational spectra of many molecules locate at terahertz domain, specific terahertz absorption patterns allow the identification, quantification of these molecules. Because of these, terahertz spectroscopy and imaging systems can have important applications (i.e. explosive detection, medicine quality control, and nondestructive evaluation) and have grown dramatically in the last decade [1, 2]. These systems require THz detectors with fast response time. And the most common THz detectors available now include bolometers [3], pyroelectric detectors, schottky diodes [4].

Recently, there has been growing interests in semiconductor plasmonic detectors which utilize plasma wave for generating signals [5]. The collective motions of two dimensional electron gases (2DEG) behave like shallow water under the gate. And hydrodynamic nonlinearity can produce photoresponse in the form of constant source-drain voltage (under open circuit condition) or direct current (with fixed drain voltage) under the irradiation of THz electromagnetic field. In the regime of off-resonance, the photoresponse is a smooth function of gate voltage as well as frequency, while the resonant structure in the photoresponse can be achieved if the resonant condition is satisfied. Since the resonant frequency is proportional to the square root of gate voltage, the devices have inherent advantage of tunability by electrical bias. These would eliminate bulky filters, mirrors and other element to realize compact monolithic detector array [6].

In this paper, we review our recent numerical simulation works on the new kinds of tunable plasmonic resonant detectors basing on slit-grating-gate double-channel and single-channel HEMTs. The results of these work will offer a great sustain to realizing of the tunable and room-temperature THz detections.

II. DEVICE DESCRIPTION AND DISCUSSION

The plasma wave frequency follows this simple linear relation: $\omega = sk$, s is plasma wave velocity ($s = (eV_{gs}/m^*)^{0.5}$) and k is plasma wave vector. In previous GaAs material system, typical surface electron concentration is 10^{12}cm^{-2}, this require very short gate length in order to utilize plasma wave for resonant detection [7]. The gate length usually lies in the deep sub-micrometer domain for GaAs material system. Therefore, devices with higher electron density and mobility are important in the reduction of burden during the technique processing. Recently, gallium-nitride (GaN) system has attracted great interest due to their unique properties such as large band gap and strong polarization effect. As compared with other III-V HEMT devices, GaN-HEMTs have received wider recognition as potential devices for high-frequency and high-power microwave applications [8]. The electron density in the channel of AlGaN/GaN HEMT can be an order of magnitude higher than GaAs HEMT. This will reduce limitation of deep-submicron meter gate length devices like GaAs HEMT for THz detection according to the dispersion law of plasma wave. In previous GaAs HEMTs, periodic grating gate has been used to expand the frequency domain in THz detection. This relies on the higher order resonant plasmon modes excitations with the help of spatial dispersion of grating gate. However, the maximum resonant frequency has not yet exceeded 5THz with the length of gate finger around 1 μm due to the low frequency of fundamental plasmon mode and radiative damping of higher order plasmon mode. And in our recent paper, we also perform electromagnetic simulation of the plasmon resonant oscillations of GaN HEMTs with periodic grating gate serves both as coupler and electrodes. The device structure is shown in Fig.1 schematically. The thickness of barrier and buffer layers is consistent with normally ones, and the electron

---

[*] Corresponding author: wdhu@mail.sitp.ac.cn, xschen@mail.sitp.ac.cn.

density and mobility in the channel are around $2\times10^{13}\mathrm{cm}^{-2}$ and $1200\mathrm{cm}^2/\mathrm{Vs}$. The simulation is performed in the finite difference scheme with drude-optical conductivity to describe electromagnetic wave transport and damping along the channel.

Why the new kind of tunable plasmonic resonant detector basing on slit-grating-gate double-channel and single-channel HEMTs are chosen? Firstly, the designs of the double channel and high-Al barrier hetro-structure into the HEMTs[9-12], as shown in Fig. 1, can supply the HEMTs with high density of two-dimensional electron gas, high mobility, and long momentum relaxation time making the detector get into the real room-temperature THz frequency detection. Secondly, a slit grating gate design is used to better couple the incoming THz signal. The slit grating gate provides the near-field interaction with the two-dimensional electron gas plasmonic resonant making the device being tunable under gate voltages[9, 11]. A new theoretical way is also developed[9], which combines the electromagnetic Finite-difference time-domain method and quantum transportation final-element method at the two-dimensional calculation boundary conditions. With the developed Drude model, the high-order resonant Restrahlen band effect, resonant enhancement at the anticrossing regime and low frequency absorption enhancement by the double-channel hybrid effect are studied[12].

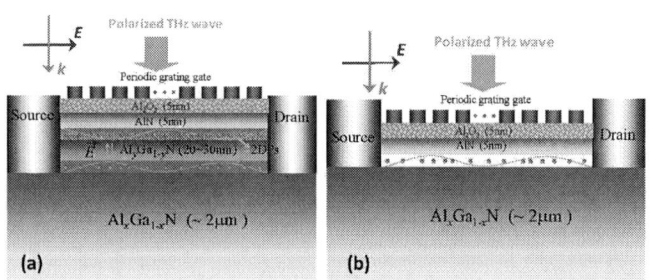

Fig.1. (a) Schematic of double-channel HEMT structure with THz wave incident from top side. (b) Schematic of single-channel HEMT structure with THz wave incident from top side. The current vibration and electron density striction are also shown along the channel.

III. CONCLUSION

The plasmonic resonances of GaN-HEMT are simulated based on the finite difference scheme. There appear two kind plasmonic oscillations along the channel of the device. Due to strong polarization effect and high electron density, superior resonant properties can be obtained even when the gate length reaches more than 1μm. This will expand the applications of plasmonic devices to the high frequency end of THz band.

ACKNOWLEDGEMENTS

The authors acknowledge the support provided by the National Natural Science Foundation of China (61006090), and the Aviation Science Fund (20110190001).

REFERENCES

[1] N. Pala and M. S. Shur, "Plasma wave terahertz electronics, " Electron. Lett., vol. 44, pp. 1391, 2008.

[2] J. Federici, and L. Moeller, "Review of terahertz and subterahertz wireless communications," J. Appl. Phys., vol. 107, pp.111101, 2010.

[3] M. Kroug, S. Cherednichenko, H. Merkel, E. Kollberg, B. Voronov, G.Goltsman, H. W. Huebers, and H. Richter, "NbN hot electron bolometric mixers for terahertz receivers," IEEE Trans. Appl. Superconduct., vol. 11, pp. 962, 2001.

[4] S. Barbieri, J. Alton, H. E. Beere, E. H. Linfield, D. A. Ritchie, S. Withington, G. Scalari, L. Ajili, and J. Faist, "Heterodyne mixing of two far-infrared quantum cascade lasers by use of a point-contact Schottky diode, " Opt. Lett., vol. 29, pp. 1632, 2004.

[5] M. Dyakonov and M. S. Shur, "Detection, Mixing, and Frequency Multiplication of Terahertz Radiation by Two-Dimensional Electronic Fluid", IEEE Trans. Electron Devices, vol .43, pp. 380, 1996.

[6] T. Otsuji, and H. Kitamura, "Numerical analysis for resonance properties of plasma-wave field-effect transistors and their terahertz applications to smart photonic network systems," IEICE Trans. Electron., vol. E84-C, pp. 1470, 2001.

[7] M. Dyakonov and M. S. Shur, "Plasma wave elecronics: Novel terahertz devices using two dimensional electron fluid," IEEE Trans. Electron Devices, vol. 43, pp. 1640, 1996.

[8] O. Ambacher, J. Majewski, C. Miskys, A. Link, M. Hermann, M. Eickhoff, M. Stutzmann, F. Bernardini, V. Fiorentini, V. Tilak, B. Schaff, and L. F. Eastman, "Pyroelectric properties of Al(In)GaN/GaN hetero- and quantum well structures, " J. Phys.: Condens. Matter., vol. 14, pp. 3399, 2002.

[9] L. Wang, X. S. Chen, W. D. Hu, and W. Lu, "Spectrum analysis of 2D plasmon in GaN-based high electron mobility transistors," *IEEE Journal of Selected Topics In Quantum Electronics*, DOI (identifier) 10.1109/JSTQE.2012.2188381, 2012.

[10] X. D. Wang, X. S. Chen, W. D. Hu, and W. Lu, "The study of self-heating and hot-electron effects for AlGaN/GaN double-channel high-electron-mobility-transistors," *IEEE Transactions on Electron Devices*, DOI (identifier) 10.1109/TED.2012.2188634, 2012.

[11] L. Wang, X. S. Chen, W. D. Hu, J. Wang, J. Wang, X. D. Wang and W. Lu, "The plasmonic resonant absorption in GaN double-channel high electron mobility transistors," *Applied Physics Letters*, vol. 99, pp. 063502, 2011.

[12] L. Wang, X. S. Chen, W. D. Hu, and W. Lu, Plasmon resonant excitation in grating-gated AlN barrier transistors at terahertz frequency, *Applied Physics Letters*, vol. 100, pp. 123501, 2012.

The improvement of figure of merit with infrared perfect absorber for plasmonic resonance sensing

G.H. Li[a], X. Wu[b], Y. Jiang[b], C. X. Shao[b], L. J. Huang[a], X. S. Chen[a, *], W. D. Hu[a], and W. Lu[a, *]

[a] National Lab for Infrared Physics, Shanghai Institute of Technical Physics, Chinese Academy of Sciences, 500 Yutian Road, Shanghai, China 200083

[b] Fundamental Science on EHF Laboratory, University of Electronic Science and Technology of China, No.2006, Xiyuan Ave, West Hi-Tech Zone, Chengdu, China 611731

xschen@mail.sitp.ac.cn, luwei@mail.sitp.ac.cn

Abstract-We present an infrared perfect absorber which combines gold nanobars and a photonic microcavity. By adjusting the structural geometry, this device is utilized for refractive index sensing. For proper designed structural parameters, it can yield more than 99% absorbance in the near-infrared frequency regime. Our work directly investigate the effect of geometry on sensing performance and it can sever as a model of coupling between localized surface plasmon within nanoparticles and propagating surface plasmon along planar metal layer for sensing applications with a perfect absorber.

I. INTRODUCTION

When the interface of the metallic structure is large enough, the plasmon, which is attributed to the strong interaction between the conduction electrons and incident electromagnetic field, can propagate in the form of oscillating charges wave, and this is referred to as propagating surface plasmon (PSP). Contrary to the PSP, the localized surface plasmon resonance (LSPR) is within a subwavelength metallic nanostructure. Due to the localized property and the near-field enhancement, surface plasmons have significant applications including near-field optical microscopy [1], surface enhanced Raman spectroscopy (SERS) [2], and sensing devices [3]. Moreover, some special physical mechanisms, such as the classical analog of electromagnetically induced transparency or perfect absorbance [4], are explored for more prospective applications [5].

The LSPR strongly depends on the shape, composition, size, light polarization and surrounding dielectric environment. Particularly, the latter dependence opens a way toward refractive index sensing which can detect small quantities preferably down to single molecules. Recently, different plasmonic structures, such as nanoshells, nanospheres and etc, are proposed to detect the large spectral shift for changes in refractive index [6, 7]

For plasmonic sensors, the sensitivity (S) is commonly defined in terms of the change or shift in a measurable parameter, typically resonant wavelength, for detection of per refractive index. Another factor charactering the performance is the full width at half maximum (FWHM) which is strongly dependent on the shape and size of the nanoparticles. It was mentioned that the linewidth narrowing is potentially helpful to improve the performance of the LSPR sensors. Generally, there defines an overall performance parameter of the plasmonic sensor as figure of merit (FOM), FOM=S/FWHM. However, a second FOM* is proposed by defining FOM*= $[(dI/dn)/I]_{max}$ for a practical way since the intensity change dI is much easier to detect in experiment at a fixed wavelength λ_0 induced by a refractive index change dn rather than that of the wavelength[8, 9].

In this paper, we present a novel plasmonic sensor with the help of an infrared perfect absorber. This device can yield more than 99% absorbance in the simulation. The geometry parameters have great influence on the sensing performance. However, the spacer thickness is of particularly importance to the performance of absorbance efficiency and consequently affects the sensing qualities. The mechanism of LSP and PSP interaction is proposed to demonstrate this phenomenon.

II. SIMULATION MODELS AND DEVICE STRUCTURE

The schematic geometry of the infrared perfect absorber is illustrated in Fig. 1. The large gold planar bottom layer and the top an array of periodic gold nanobars are sandwiched by the microcavity, which is filled with SiO_2. The unit cell of the nanobars is composed of two bars with the same parameters. The test liquids or gases are channeled or diffused over the period gold nanobars. This structure has dimensional parameters of bar length (L), bar width (W), bar thickness (T), bar gap distance (D), the lattice constant in x axis (P1) and in y axis (P2), the dielectric spacer thickness (H), and planar gold layer thickness (G) [10].

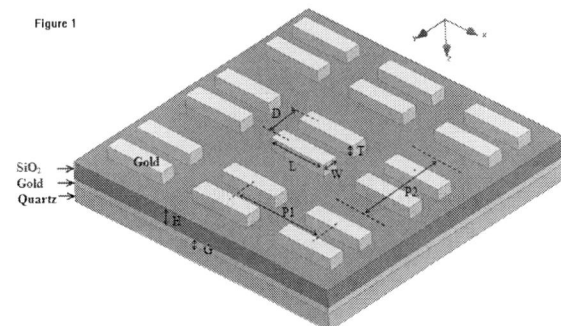

Fig. 1: (colour online) Schematic of the sensing structure and the coordinate setup configuration. The whole structure is placed on the quartz substrate

For the existence of bottom layer, the transmittance of the absorber is almost totally eliminated in the near-infrared (NIR) regime and then the absorbance can be calculated by 1-R (R is the resonant reflectance intensity). Consequently, the perfect absorber can be achieved just by pursuing zero reflectance dip intensity through the following approaches, adopting different nanoparticle shapes, changing the lattice constant or rationally designing the other structure dimensions. Among these choices, optimizing the spacer thickness offers a simple and efficient method to reach the nearly absolute absorbance.

Here, we take advantage of a 3D-FDTD commercial software(from Dongjun Information Technology Co.,Ltd.) to carry out our calculations. With this FDTD method the electric and magnetic fields are temporally separated by half time step, and also are spatially interlaced by half mesh cell. The periodic boundary conditions are imposed in the x and y directions.

III. RESULT AND DISCUSSION

The geometry parameters are systemically optimized and the reflectance dips are gradually moving to the ground of zero intensity. The Fig. 2 shows the reflectance spectrum as a function of key variable spacer thickness.

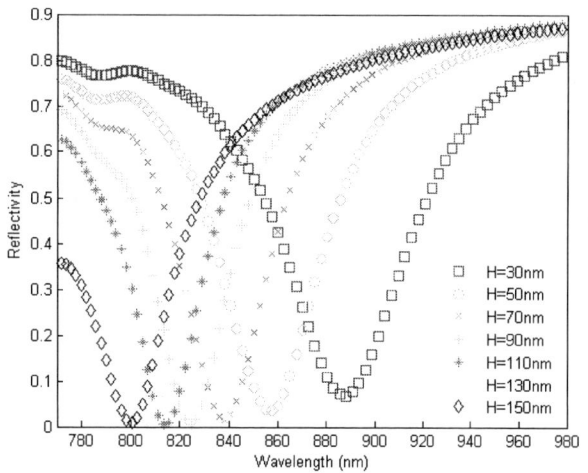

Fig. 2: (colour online) Simulated reflection spectrum of the sensing structure as a function of dielectric spacer thickness from H=30nm to H=150nm for 25%-solution of glucose in water. Within the large thickness range the structure obtains high absorbance efficiency.

As shown in the figure 2, there appear a series of strong reflectance dips. Here we emphasize that the shape of the spectrum is becoming more and more asymmetric as the spacer thickness increases. The intensity of short wavelength part is decreasing to less than half of the long wavelength part. Considering the FWHM values are obtained from the Lorentz fitting for the plasmon resonance, the reflection spectrum are not carried out for thickness more than 150nm. However, it becomes symmetric again for thickness more than about 300nm, but the FWHM is more than twice wider than that shown in figure 2. Based on the above discussion, we will not talk about the thickness more than 150nm.

To have a look at the periodicity effect on the absorbance efficiency, we calculate it and show in table 1. It can be seen that the periodicity of 500nm*500nm achieves the best sensing

performance. The dip intensity is about zero and the FWHM and FOM* all satisfy the demand for sensing quality.

TABLE 1
The periodicity effect on the absorbance efficiency

Periodicity (nm*nm)	Resonant Wavelength (nm)	Resonant Intensity (arb.)	FWHM (nm)	FOM* (arb.)
440*440	792	0.0546	45	30
440*500	786	0.0369	47	42
500*440	851	0.0128	44	155
500*500	824	0.00627	40.8	357
500*600	830	0.00181	*	158
600*500	946	0.124	45	46
600*600	936	0.1518	43	87

* The reflection spectrum does not follow the Lorentz shape.

IV. CONCLUSION

Through the optimization of geometry parameters, we take the inevitable losses to advantage and design a novel plasmonic infrared sensor. It can yield more than 99% absorbance in the near-infrared frequency regime. The FWHM and FOM* are achieved both excellent.

ACKNOWLEDGEMENTS

This work was supported in part by the State Key Program for Basic Research of China grants 2007CB613206, the National Natural Science Foundation of China grants 10725418, 10734090, 10990104, and 60976092, the Fund of Shanghai Science and Technology Foundation grant 09DJ1400203, 09ZR1436100, 10JC1416100, and 10510704700. Also thanks to the Longwen Scholarship for his support..

REFERENCES

[1] T. Kalkbrenner, U. Hakanson, A. Schadle, S. Burger, C.Henkel, and V. Sandoghdar, "Optical microscopy via spectral modifications of a nanoantenna," Phys. Rev. Lett. 95, 200801, 2005.

[2] F. Hao and P. Nordlander, "Plasmonic coupling between a metallic nanosphere and a thin metallic wire," App. Phys. Lett. 89, 103101, 2006.

[3] Jian Ye, Pol Van Dorpe, "Improvement of the figure of merit for gold nanobars array plasmonic sensors." Plasmonics 10, 665-71, 2011.

[4] Liu N, Mesch M, Weiss T, Hentschel M, Giessen H, "Infrared perfect absorber and its application as plasmonic sensor," Nano Lett. 10 2342-48, 2010.

[5] G.H. Li, X.S. Chen L.J. Huang, J. Wang, W.D. Hu, W. Lu, "The localized near-field enhancement of metallic periodic bowtie structure: an oscillating dipoles picture" Phys. B 407, 2223, 2012.

[6] C. Soennichsen, and A.P. Alivisatos, "Gold nanorods as novel nonbleaching plasmon-based orientation sensors for polarized single-particle microscopy" Nano Lett. 5 301-04, 2005.

[7] J.B. Lassiter, J. Aizpurua, L.I. Hernandez, D.W. Brandl , L. Romero, S. Lal, J.H. Hafner, P. Nordlander, and N.J. Halas, "Close Encounters between Two Nanoshells" Nano Lett. 8 1212-18, 2008

[8] L.J. Sherry, S.H. Chang, G.C. Schatz, R.P. Van Duyne, B.J. Wiley, and Y. Xia, "Localized surface plasmon resonance spectroscopy of single silver nanocubes" Nano Lett. 5 2034-38, 2005

[9] J. Becker, A. Trügler, A. Jakab, U. Hohenester, and C. Sönnichsen, "The optimal aspect ratio of gold nanorods for plasmonic bio-sensing " Plasmonics 5 161-67, 2010

[10] G.H. Li, X.S. Chen, O.P. Li，C.X. Shao, Y. Jiang，L.J. Huang, B. Ni, W.D. Hu, and W. Lu "A novel plasmonic resonance sensor based on an infrared perfect absorber" J. Phys. D: Appl. Phys. In press.

Influence of surrounded metallic layers on whispering-gallery modes in circular microresonators

Q. F. Yao, Y. Z. Huang, X. M. Lv, J. D. Lin, and L. X. Zou

State Key Laboratory on Integrated Optoelectronics, Institute of Semiconductors,
Chinese Academy of Sciences, PO Box 912, Beijing 100083, China.
yzhuang@semi.ac.cn

Abstract—**Influences of surrounded metallic layers (Au, Al, Ag, Cu, and Ti) on whispering-gallery modes (WGMs) are numerically investigated by solving eigenvalue equation for multiple-layer two dimensional circular microresonators. For TM modes, metal layer can provide good optical confinement as its thickness is larger than 0.03 μm. For TE modes, an isolation layer should be introduced to reduce the dissipation loss of metallic layers. Al and Ag layers can provide better optical confinement than Au layer, and Ti layer which is usually a layer of p-electrode will result in a large dissipation loss.**

I. INTRODUCTION

Microcavities confined by a metallic layer have recently attracted great interests. The metallic layer can provide strong optical confinement and be used to miniaturize the device size [1]. The introduction of a low refractive index layer between the metal layer and the resonator was investigated for triangle and square microresonators for reducing the dissipation loss [2, 3]. Ti/Pt/Au is usually used as p-electrode for semiconductor lasers. However, copper (Cu) and aluminum (Al) are widely used in microelectronic circuits. So it is interested to use Cu and Al as electrode materials for photonic integrated circuits. In this paper, the influences of different metal layers Au, Ag, Cu, Al, and Ti on mode characteristics are numerically investigated for circular microresonators. We first introduce the analytical solution for the circular resonator with multilayer structure, and then compare mode quality factors for microcircular resonators confined by different metallic layers and dielectric/metallic layers, respectively.

II. ANALYTICAL SOLUTION FOR MICROCIRCULAR RESONATOR WITH MULTILAYER STRUCTURE

The two dimensional microcircular resonator as shown in Fig. 1 is considered, which is consisted of the active layer, the isolation layer, the metallic layer and the external layer. In the cylindrical coordination system (r, φ), Maxwell equations can be reduced to the following Helmholtz equation:

$$\frac{\partial^2 \psi_z}{\partial r^2} + \frac{1}{r}\frac{\partial \psi_z}{\partial r} + \frac{1}{r^2}\frac{\partial \psi_z}{\partial \varphi^2} + k^2 \varepsilon_r \psi_z = 0 \quad (1)$$

where ψ_z is electric field E_z for TM modes or magnetic field H_z for TE modes in the z-direction, ω is the angular frequency, $k = \omega(\mu_0\varepsilon_0)^{1/2}$ is the vacuum wave number, and ε_r is the relative permittivity.

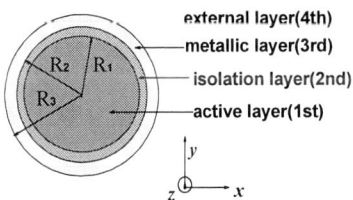

Fig. 1. Schematic diagram of a circular microresonator with multilayer structure.

The solutions of (1) are the combination of Bessel functions:

$$\psi_z(r,\phi) = \begin{cases} J_v(n_a kr)e^{\pm jv\varphi} & 0 \le r \le R_1 \\ \left[A_1 J_v(n_d kr) + B_1 Y_v(n_d kr)\right]e^{\pm jv\varphi} & R_1 \le r \le R_2 \\ \left[A_2 I_v(n_m kr) + B_2 K_v(n_m kr)\right]e^{\pm jv\varphi} & R_2 \le r \le R_3 \\ CH_v^1(n_o kr)e^{\pm jv\varphi} & r \ge R_3 \end{cases} \quad (2)$$

where v and l are azimuthal and radial mode numbers, respectively, J_v and Y_v are the v-order Bessel function of the first kind, I_v and K_v are the v-order imaginary argument Bessel function of the first kind and the second kind, and H^1_v is the v-order Hankel functions of the first kind. n_a, n_d, in_m, n_o are the refractive index of the active layer, the isolation layer, the metal layer and the external layer, while the radius of the three interfaces are R_1, R_2 and R_3, respectively. For TE modes, the other two components E_r and E_φ are related to the z-directional magnetic field as:

$$E_r = \frac{-i}{kr\varepsilon_r}\frac{\partial H_z}{\partial \varphi}, \qquad E_\varphi = \frac{i}{k\varepsilon_r}\frac{\partial H_z}{\partial r}. \quad (3)$$

According to the boundary conditions, i.e., H_z and E_φ continuous at the boundaries $r = R_1$, R_2 and R_3, we can induce eigenvalue equations. The complex angular frequency ω can be solved as the eigenvalue with help of Newton iteration, and mode wavelength, mode Q factor and mode field pattern can be calculated based on the mode frequency. In addition, TM modes can be solved by the similar processing.

In the following simulation, the active layer is taken to be InGaAsP or AlGaInAs with a refractive index of 3.2, and the isolation layer of silicon dioxide with a refractive index of 1.45. The dispersive model of metallic materials are refers to [4] with mode wavelengths around 1.5μm.

978-1-4673-1602-6/12 $31.00 © 2012 IEEE

III. SIMULATION RESULTS

Firstly, we consider the microcircular directly confined by the metallic layer. Mode quality factors versus the metallic layer thickness are plotted in Fig. 2 for (a) $TM_{10,1}$ and (b) $TE_{9,1}$ modes in a circular microresonator with radius $R_1 = 1\mu m$, covered by Au, Al, Ag, Cu, and Ti, respectively. The mode wavelength is 1515 and 1524 nm for $TM_{10,1}$ and $TE_{9,1}$ in the circular microresonator in air without metal layer, respectively. For TM modes, mode Q factor firstly decreases with the increase of the metallic layer thickness due to the dissipation loss, but rapidly increases with the metallic layer thickness as the thickness is larger than 0.02~0.03 μm because of better confinement of the mode field pattern. The mode Q factor can be larger than 10^5 for the circular microresonator confined by Au, Al, Ag, or Cu layer with the thickness larger than 0.1μm. Al layer can even result in a higher mode Q factor. The metallic layers induce a larger dissipation loss for $TE_{9,1}$, and the largest mode Q factor is about 10^3 in Fig. 2(b). Furthermore, the results show that Ti layer, which is usually used for improving the adhesion of p-electrode, results lowest mode Q factor.

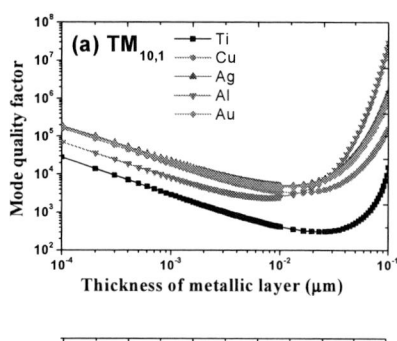

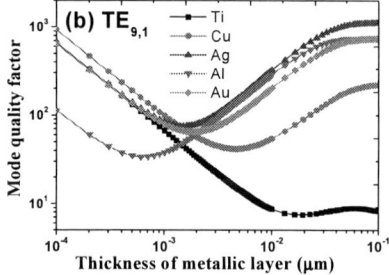

Fig. 2. Quality factors of $TM_{10,1}$ and $TE_{9,1}$ modes as the function of the thickness of different confinement metal

To reduce the dissipation loss for TE modes, an isolation layer is introduced between the active layer and the metallic layer. The Q factor of $TE_{9,1}$ mode is plotted in Fig. 3(a) as the function of the thickness of the isolation layer for the circular microresonator confined by the isolation layer and 100 nm metallic layer. The isolation layer has an optimal thickness of about 550nm for highest Q factor, and the mode Q factor takes the highest value for microresonator covered by the silver layer. In addition, Al layer is a little better than Au layer for realizing high Q factor. In Fig. 3(b), we plot the field patterns for $TE_{9,1}$

in the circular microresonator with the gold layer of 100 nm and the isolation layer of 0, 500 and 900 nm, respectively. At the isolation layer thickness of 500 nm, the mode field pattern is a mixture of $TE_{9,1}$ and $TE_{9,2}$ due to mode coupling, which results in the decrease of mode Q factor with the increase of the isolation layer thickness. In addition, the mode field pattern is mainly located in the isolation layer at the thickness of 900 nm.

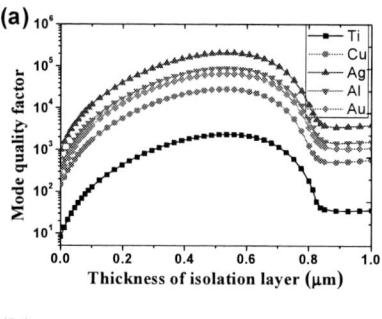

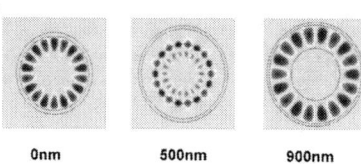

Fig. 3(a) Q factors of $TE_{9,1}$ mode versus the thickness of isolation layer with different metallic confinement and (b) the mode field patterns with the isolation of 0, 500 and 900nm

IV. CONCLUSSION

We compare mode quality factors for whispering-gallery modes in circular microresonator confined by different metallic materials based on analytical solutions. The results show that silver and aluminum layer can provide better optical confinement than gold layer. The titanium layer usually used in p-electrode will greatly reduce mode quality factor.

ACKNOWLEDGMENT

This work was supported by the National Nature Science Foundation of China under Grants 60838003, 61021003, 61106048, 61061160502 and 61006042, and the High Technology Project of China under Grant 2012AA012202.

REFERENCES

[1] M. T. Hill, Y. S. Oei, B. Smalbrugge, Y. Zhu, T. De Vries, P. J. Van Veldhoven, F. W. M. Van Otten, T. J. Eijkemans, J. P. Turkiewicz, H. De Waardt, E. J. Geluk, S. H. Kwon, Y. H. Lee, R. Notzel, and M. K. Smit, "Lasing in metallic-coated nanocavities," *Nature Photonics*, vol.1, pp.589-594, 2007.

[2] Y. D. Yang, Y. Z. Huang, and S. J. Wang, "Mode analysis for equilateral-triangle-resonator microlasers with metal confinement layers," *IEEE J. Quantum Electron.*, 45, pp. 1529-1536, Dec. 2009.

[3] K. J. Che, Y. D. Yang, and Y. Z. Huang, "mode characteristics for square resonators with a metal confinement Layer," *IEEE J. Quantum Electron.*, 46, pp. 414-420, Mar. 2010.

[4] E. D. Palik, handbook of optical constants of solid, 1st ed, Boston: academic press, 1985.

Surface Plasmon Resonance for Sensing

Zhihong Chen, [1,2*] and Chun Jiang[1]

[1] State Key Laboratory of Advanced Optical Communication System and Networks, Shanghai Jiao Tong University, Shanghai 200240, China

[2] College of Information, Mechanical & Electrical Engineering, Shanghai Normal University, Shanghai 201418, China

Abstract: We design a structure consisting of Ag strip pair arrays embedded in the background material to achieve localized surface plasmon resonance. Numerical simulation shows that one of the transmission dips of the structure is very sensitive to the background materials, which can be used to achieve high performance sensors.

When interacting with electromagnetic radiation the collective electronic oscillation in metal nanoparticles can generate localized surface plasmon resonance (LSPR) [1]. Such LSPR strongly depends on the size, shape, and surrounding dielectric environment of nanostructures. The latter dependence makes them especially attractive for refractive index sensing [2-3].

In this paper, we design a plasmonic structrue to achieve LSPR, where the structure consists of Ag strip pair arrays among the background material. Numerical simulation shows that one of the transmission dips for the structure is very sensitive to the background materials.

Fig.1 (a) shows the top view of the proposed plasmonic structure consisting of Ag strip pair arrays among the background material. Fig.1 (b) is the sectional view of the structure of x-z plane. The geometrical parameters are defined as in the figure: the Ag strip pair arrays with the length L_x and width L_y of the metal (Ag) block, in which the period of the metal block along x direction and y direction are d_x and d_y, respectively; The slits width and thickness of the Ag strip pair are W and H, respectively.

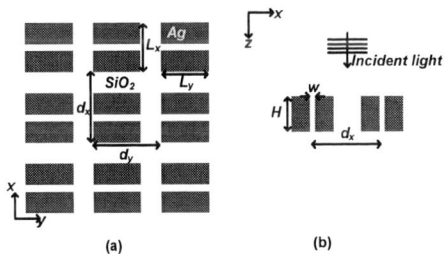

Figure1: (Color online). (a): top view of the Ag strip pair arrays structure. (b): the sectional view of the structure of x-z plane. Light propagates along z direction with the polarization along the x direction for TM mode.

We calculate the transmission spectra of Ag strip pair arrays structure among the background material of SiO_2 (refractive index n=1.5), using the finite-difference time-domain (FDTD) method [4] shown in Fig.2 (a). We also plot electric field intensity distributions of the structure (Fig.2 (b)). The computational domain consists of a single period of the structures with inhomogeneous mesh. We use the periodic boundary condition and perfect matched layer (PML) absorption boundary condition along the periodic direction and at the top and bottom boundaries, respectively. We describe the complex optical constants of metal Ag taking from experimental data [5]. Light propagates along z direction with the polarization along the x direction. For the structure, one Ag strip pair makes up a metal-dielectric-metal waveguide terminated by dielectric mirrors on both sides, and thus the array of Ag strip pairs form cavities that hold

978-1-4673-1602-6/12 $31.00 © 2012 IEEE

Fabry-Perot mode. Fig.2 (b) shows the localized surface plasmon mode between the Ag strip pair, and the energy distributes not only in the cavity, but also on the top and bottom surfaces of the Ag strips.

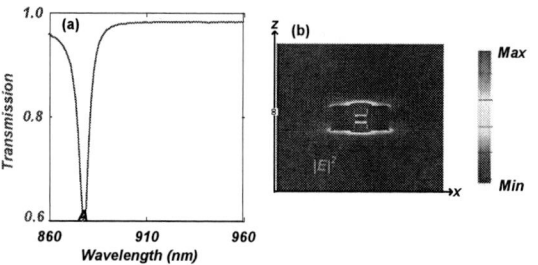

Figure2: (Color online). (a): Transmission spectra of the Ag strip pair arrays structure. (b): The electric field intensity distributions at the wavelength locations A marked in Fig.2 (a). (Lx= Ly=120nm, dx=dy =350nm, W= 30nm, H=120nm).

Meanwhile, one of the transmission dips is very sensitive to the background material; the structure may serve as a sensor. The parameters for the structure are same with Fig.2. Changes of the background material can be detected by measuring the shift of the sharp transmission dip. A clear shift of the transmission dip to longer wavelength is visible when the refractive index of the background materials increases from n=1.0 to n=2.0. The sensitivity (nm/RIU) of the structure is about 550nm/RIU.

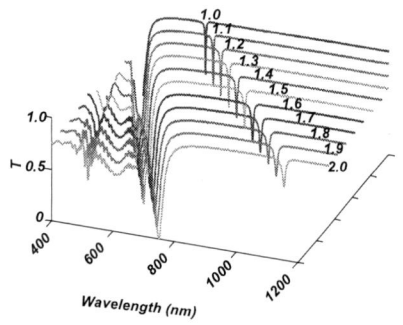

Fig.3 Electric Transmission spectra of the Ag strip pair arrays structure at different refractive index (Lx= L_y=120nm, d_x=d_y =350nm, W= 30nm, H=120nm).

In conclusion, we have explored a structure consisting of Ag strip pair arrays among the background material to achieve localized surface plasmon resonance. The simulated results indicated that one of the transmissions dips for the structure is very sensitive to the background material. The structure can be used to achieve high performance sensors. This will be a promising way for surface plasmon resonance applications toward sensing.

Reference

[1] H. Raether, Surface Plasmons on Smooth and Rough Surfaces and on Gratings, Springer, Berlin, 1988.

[2] Na Liu, Martin Mesch, Thomas Weiss, Mario Hentschel, and Harald Giessen, "Infrared Perfect Absorber and Its Application" Nano Lett., vol. 10, pp.2342–2348, Jun. 2010.

[3] N. Liu，T. Weiss，M. Mesch, L. Langguth, U. Eigenthaler, M. Hirscher, C. So¨nnichsen, and H. Giessen, "Planar Metamaterial Analogue of Electromagnetically Induced Transparency for Plasmonic Sensing," Nano Lett., vol.10, pp.1103-1107, Sept. 2010.

[4] Sullivan D M, Electromagnetic simulation using the FDTD method, 2000 IEEE Press, New York.

[5] Johnson P B, and Christy R W, Phys. Rev. B, vol. 6 , pp. 4370-4379, Dec. 1972.

Design of Terahertz Quantum-Well Photodetectors

X. G. Guo and J. C. Cao
Key Laboratory of Terahertz Solid-State Technology
Shanghai Institute of Microsystem and Information
Technology, CAS
Shanghai, China
xgguo@mail.sim.ac.cn

H. C. Liu
Key Laboratory of Artificial Structures and Quantum
Control, Department of Physics
Shanghai Jiao Tong University
Shanghai, China
h.c.liu@sjtu.edu.cn

Abstract—**Terahertz quantum-well photodetectors (QWPs) represent a new and emerging photon-type detector in terahertz region. We first discuss the many-particle effects on the accurate design of terahertz QWPs. Grating and metal-cavity light couplers for terahertz QWPs are introduced. At resonant coupling frequencies, the polarization of light field is effectively changed by the light couplers to fulfill the selection rule of intersubband transition. Meanwhile, the electric field intensities in the active multi-quantum-well region of terahertz QWPs are enhanced. The performance of terahertz QWPs with these light couplers is improved significantly.**

Keywords – terahertz; quantum-well photodetectors; grating; metal cavity

I. INTRODUCTION

Recently, terahertz QWPs have been realized [1]. In comparison with other terahertz detectors, intersubband-transition-based terahertz QWPs display some specific features [2]. Due to the intrinsic short lifetime of photon-excited electrons, terahertz QWPs can be operated with high response speed and therefore suited for high-frequency applications [2]. Terahertz QWPs are narrow band detectors because of the delta-function-like joint density of states of intersubband transitions. As a result, filters are not required in some laser-based concealed object imaging applications. The response peak frequency of a terahertz QWP is determined by the energy difference between the first and the second subbands of the quantum wells, which can be well designed and implemented with molecular beam epitaxy (MBE) growth technique. The mature semiconductor processing technique makes it possible to fabricate large-scale uniform, high resolution, and long-term stable focal plane array, which are important for realizing real-time terahertz imaging systems.

In this paper, we first discuss the design principle of GaAs/(Al,Ga)As terahertz QWPs. Two main many-particle interactions, the electron exchange-correlation potential and the depolarization effect are considered. Low intersubband absorption efficiency due to the low electron doping concentration in the quantum wells is a key factor in limiting the performance of terahertz QWPs. In order to improve the intersubband absorption efficiency, two light coupling schemes, metal diffractive grating couplers and metal-cavity couplers are investigated. High efficient light couplers are expected to effectively improve the responsivity and working temperature of terahertz QWPs.

II. BAND STRUCTURES AND PHOTOCURRENT OF TERAHERTZ QWPS

We calculate the band structures and the photocurrent spectra of two devices labeled as V266 and V267 reported in Ref. [1]. Numerical details are presented in Ref. [3].

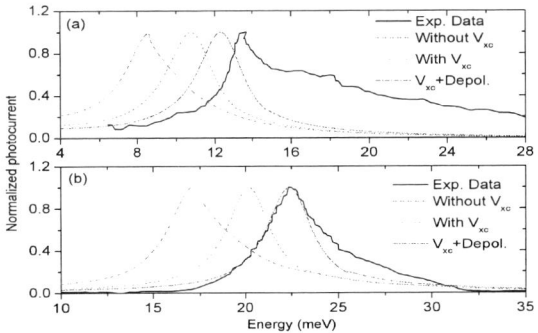

Fig. 1. Calculated and experimental photocurrent spectra of terahertz QWPs, (a) for V267 and (b) for V266. V_{xc} denotes the exchange-correlation potential.

The effects of the exchange-correlation potential and the depolarization interaction are explored [3]. For V266 and V267, when the electron Coulomb interaction is considered in Hartree approximation, only one localized subband exists in the quantum well, and the first excited subband is in alignment with the top of the barrier, which is coincidence with the design rule of bound-to-quasibound QWPs [2]. However, when the exchange-correlation potential is taken into account, the quantum well becomes deeper [4], which increases the energy difference between the ground subband and the first excited subband and pulls the first excited subband deep into the quantum well. The energy difference between the top of the barrier and the first excited subband ΔE is 3.2 meV for V266, and 3.0 meV for V267. The measured photocurrent spectra of the two devices indicate that the photon-excited electrons in the first excited subband can escape into the continuum states via scattering and electric-field-assisted tunneling mechanisms.

Theoretical and experimental photocurrent spectra for

V266 and V267 are shown in Fig. 1 [3]. The deviations of response peaks between theory and experiment for V266 and V267 are 5.6 meV (24.8%) and 4.8 meV (36.0%) without including any many-particle interaction. When the exchange-correlation potentials are taken into account, the deviations decrease to 2.4 meV (10.6%) and 2.6 meV (19.4%), respectively. The further improvements of theoretical response peak positions are achieved by considering the depolarization effects, and the discrepancies are about 0.2 meV (0.9%) and 1.1 meV (8.2%) for V266 and V267. The large remaining discrepancy between theory and experiment for V267 may originate from the fluctuation of Al fraction in barriers due to the small Al mole fraction (1.5%).

III. LIGHT COUPLERS FOR TERAHERTZ QWPS

Light coupling is another key factor for better performance of terahertz QWPs [2]. Since light absorption in terahertz QWPs is due to intersubband transitions, the selection rule dictates that terahertz QWPs cannot respond to normally incident light. We study two types of metal-grating-based light couplers for terahertz QWPs. The metal grating period is in X direction, the length of metal strips is infinite in Y direction, and the quantum well growth direction is along Z axis. We define a quantity γ, the normalized coupling efficiency of a light coupler to that of a 45-degree facet coupling scheme,

$$\gamma = \frac{2 \iiint_{MQWs} |E_Z|^2 dv}{\iiint_{MQWs} |E_0|^2 dv}, \quad (1)$$

where E_0 is the electric field intensity in the multi-quantum-well (MQW) region of a terahertz QWP with a 45-degree facet coupling scheme.

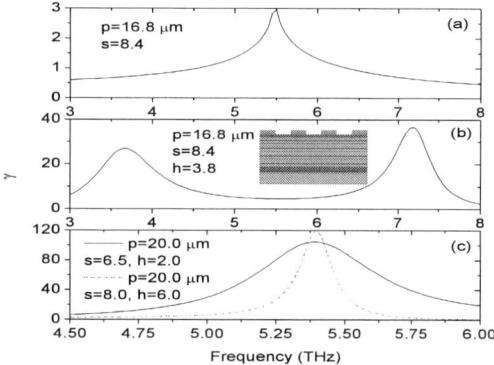

Fig. 2. Normalized coupling efficiencies of metal-grating couplers and metal-cavity couplers for terahertz QWPs, (a) metal-grating coupler, (b) metal-cavity coupler with the same grating parameters shown in (a), (c) optimal metal-cavity couplers for terahertz QWPs with the response peak frequency of 5.48 THz. The inset to (b) is a schematic of the cavity-coupled terahertz QWP.

The relative dielectric constant and the conductivity of gold, and the relative dielectric constant of GaAs are set to -1.80×10^4, 4.56×10^7 S/m, and 10.6, respectively in our calculations. The FEM [5] numerical resonant coupling peak of the metal grating is at 5.48 THz (Fig. 2(a)) [6]. However, in the metal-cavity-coupled terahertz QWPs, the original grating-determined resonant coupling peak at 5.48 THz disappears, and two other resonant coupling peaks with their maximum values at 3.65 THz and 7.20 THz emerge with the same grating parameters and the thickness of the cavity h=3.8 μm (Fig. 2(b)). In comparison with the case of metal-grating-coupled terahertz QWPs, the maximum coupling efficiency increases by about an order of magnitude [7]. The waveguide effects of the metal-cavity are responsible for the changes of the resonant coupling behaviors. A ray propagation method is successfully used to analyze the resonant behaviors in the metal-cavity-coupled terahertz QWPs qualitatively. The theoretical maximum value of normalized coupling efficiency γ is about 100 for the optimal metal cavity parameters (Fig. 2(c)).

IV. CONCLUSIONS

Due to the small barrier height and the energy difference between the ground subband and the first excited subband, many particle interactions play key roles in the band structure and the response peak frequency of a terahertz QWP. Because the exchange-correlation potential is minus, it increases the energy difference between the top of the barrier and the first excited subband ΔE. A blue shift of the response peak is introduced by both the exchange-correlation and depolarization interactions. Two types of light couplers are investigated. For the metal-cavity-based light couplers, at resonant coupling frequencies, the polarization of light field is effectively changed. Meanwhile, the electric field intensity in the active MQW regions of terahertz QWPs are substantially enhanced.

REFERENCES

[1] H. C. Liu, C. Y. Song, A. J. SpringThorpe and J. C. Cao, "Terahertz quantum-well photodetector," *Appl. Phys. Lett.*, vol. 84, pp. 4068-4070, May 2004.

[2] H. Schneider and H. C. Liu, "Quantum Well Infrared Photodetectors: Physics and Applications " (Springer, Berlin, 2007).

[3] X. G. Guo, Z. Y. Tan, J. C. Cao, and H. C. Liu, "Many-body effects on terahertz quantum well detectors," *Appl. Phys. Lett.*, vol. 94, pp. 201101-1−201101-3, Apr. 2009.

[4] X. G. Guo, R. Zhang, H. C. Liu, A. J. SpringThorpe, and J. C. Cao, "Photocurrent spectra of heavily doped terahertz quantum well photodetectors," *Appl. Phys. Lett.*, vol. 97, pp. 021114-1−021114-3, Jul. 2010.

[5] Multiphysics Modeling and Simulation Software-COMSOL: www.comsol.com.

[6] R. Zhang, X. G. Guo, C. Y. Song, M. Buchanan, Z. R. Wasilewski, J. C. Cao, and H. C. Liu, "Metal Grating Coupled Terahertz Quantum Well Photodetectors," *IEEE Electron Device Lett.*, vol. 32, pp. 659-661, May 2011.

[7] X. G. Guo, R. Zhang, J. C. Cao, and H. C. Liu, "Numerical Study on Metal Cavity Couplers for Terahertz Quantum Well Photodetectors," *IEEE J. Quantum Electron.* (2012, in press).

Crosstalk Suppressing Design of GaAs Microlenses Integrated on HgCdTe Infrared Focal Plane Arrays

Zhenhua Ye[a], Yang Li[a,b], Chun Lin[a], Xiaoning Hu[a], Ruijun Ding[a] and Li He[a]

[a] Key Laboratory of Infrared Imaging Materials and Detectors, Shanghai Institute of Technical Physics,
Chinese Academy of Sciences, Shanghai 200083, China
[b] Graduate School of the Chinese Academy of Sciences, Beijing 100039, China
Email: zhye@mail.sitp.ac.cn

Abstract— **Crosstalk suppressing design of dielectric GaAs microlenses integrated on traditional HgCdTe infrared focal plane arrays (IRFPAs) is presented in this paper, by exploiting the finite difference time domain (FDTD) technique. Responsive photocurrent and crosstalk between most adjacent IR detector pixels have been numerically simulated using Crosslight TCAD commercial software. An optimal curvature of GaAs microlenses has been achieved by maximizing its ability to focus the incoming infrared plane wave at a specific point near the interface of GaAs substrate and HgCdTe absorber layer.**

I. INTRODUCTION

The scale of infrared focal plane arrays (IRFPAs) is being enlarged by shrinking the pixel size. Ideally, the reduction in IRFPA pixel dimension can achieve improvements in image resolution without significant decrease of sensitivity. However, as the pixel size is continuously decreased, IRFPA detectors encounter a degraded optical efficiency, as well as an increase in the crosstalk between adjacent detector pixels. These issues can be resolved through suitable pixel design and placement of microlenses in front of each photodiode, to redirect and focus light into corresponding active detector regions.

Infrared detectors based on HgCdTe material have been widely used and largely investigated [1–5]. Many studies on design of microlenses have also been numerically carried out in terms of traditional geometrical optics for devices of different materials including HgCdTe [6–7]. When the pixel size is approaching to or less than the Airy spot of an IR system, it becomes necessary to investigate the optical energy distribution inside the device considering wave characteristics of incoming radiation, i.e., according to the Maxwell equations. Therefore, the finite difference time domain (FDTD) technique is used to simulate photo response and pixel crosstalk of microlensed HgCdTe IRFPAs, and then an optimized radius of dielectric GaAs microlenses is introduced in this paper.

II. MODEL AND METHOD

The cross-sectional structure of the microlensed HgCdTe planar array investigated in this work is schematically shown in Fig. 1. Three adjacent pixels are modeled in two-dimensional mode with a pitch of 20μm, and each of them consists of an ion-implanted vertical n^+-on-p Hg$_{1-x}$Cd$_x$Te homojunction ($x = 0.27$) and a spherical microlens formed on the back of GaAs substrate. Thickness of Hg vacancy doped p-type absorption layer is 9μm, and total thickness of GaAs substrate is 20μm

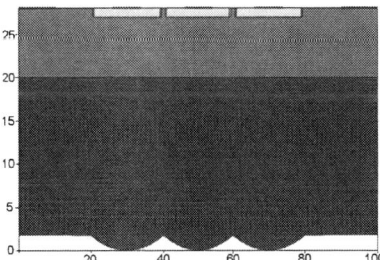

Figure 1. Cross-sectional structure of HgCdTe IRFPAs integrating GaAs microlens arrays against the corresponding ion-implanted *p-n* photodiodes.

including microlenses with radius of curvature adjustable. The width and depth of the *n*-type region are 18μm and 1μm, and the maximum donor and acceptor densities are 2×10^{17}cm^{-3} and 8×10^{15}cm^{-3}, respectively. The HgCdTe IRFPA detectors are modeled to operate at 77K, and the microlens corresponding to the central pixel is illuminated independently from backside by 5μm monochromatic infrared plane wave, to evaluate crosstalk of most adjacent photodiode as a ratio of photocurrents.

Firstly, optical energy distribution inside the entire device is obtained using the FDTD algorithm where boundary conditions of perfectly matched layer (PML) is utilized. Thereafter, the photoelectrical characteristics of IRFPA detectors are derived by solving the coupled system of continuity equations, Poisson equation and the drift-diffusion current equations. Structure setup and numerical calculations are conducted by applying computer aided design (TCAD) commercial software package of Crosslight Csuprem and Apsys. Then the optimal microlens radius can be determined in a reasonable way, even when the pixel size is shrunk near or even less than the Airy spot of infrared optical systems.

III. RESULTS AND DISCCUSION

A. Simulation Results

A series of simulations are carried out with the radius of microlens changed from 10μm to infinite. Three typical optical energy distributions of HgCdTe IRFPA detectors are depicted respectively in Fig. 2. In fact, focusing spots of the microlenses with radii of 11μm, 15μm and 30μm are located within GaAs substrate, at the interface of GaAs substrate and HgCdTe absorption layer and within HgCdTe layer, respectively.

Fig. 3 and 4 display the responsive photocurrent of central pixel and crosstalk between most adjacent infrared detectors

978-1-4673-1602-6/12 $31.00 © 2012 IEEE

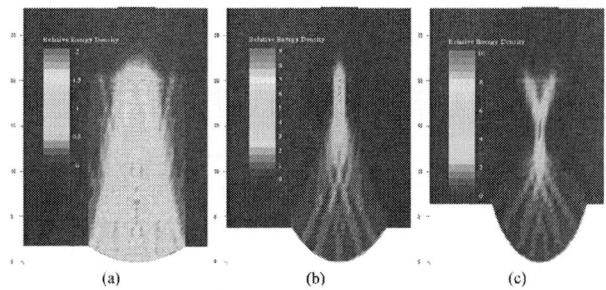

Figure 2. Optical energy distributions inside HgCdTe IRFPA intigrating GaAs microlens arrays with infrared plane wave back-side illuminated on single microlens with radius of (a) 30μm, (b) 15μm, (c) 11μm, respectively.

respectively with various electron lifetime values. As shown in Fig. 3, characteristics of responsive photocurrent versus radius of microlens are relatively complicated. As microlens radius increases, responsive photocurrent decreases a little firstly, and begins to increase to a maximum value when microlens radius equals 17μm, and then continues to decrease monotonically. As illustrated in Fig. 4, the minimum crosstalk value appears at microlens radius of 13μm, independent from the minority carrier lifetime. For minority carrier lifetime of 20ns, optimized crosstalk between most adjacent detector pixels is suppressed to only about 7.2%, less than half of that without integrating microlenses. Meanwhile, responsive photocurrent is enhanced by a factor of about 1.3. The optimal value of microlens radius can be adjusted from 13μm to 17μm, since the responsive photocurrent and crosstalk are insensitive to radius variation in this range.

B. Disccusion

As shown in Fig. 2, the maximum optical intensity does not

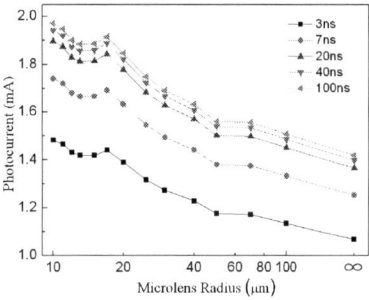

Figure 3. Responsive photocurrent of photodiode versus microlens radius with minority carrier lifetime of 3ns, 7ns, 20ns, 40ns and 100ns.

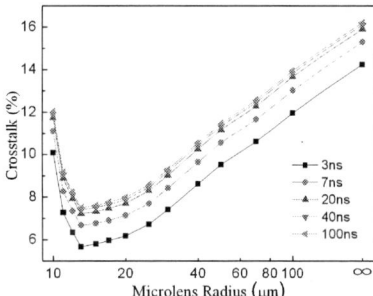

Figure 4. Crosstalk versus microlens radius with minority carrier lifetime of 3ns, 7ns, 20ns, 40ns and 100ns.

appear at microlens focusing spot that should be located deeply into HgCdTe absorption layer. This is thought to be caused by the high absorption coefficient of HgCdTe material, since most incoming infrared radiation will be absorbed within a thin layer near the interface. For this reason, the key to suppressing the crosstalk is to constrain carrier generation in a focusing spot as small as possible at the interface, which has been verified by our simulations.

In this paper, the photoresponse and crosstalk are simulated without taking account of recombination at the interface of HgCdTe absorption layer and GaAs substrate. Actually, the minority carrier recombination cannot be ignored because of large mismatch of crystal lattice between HgCdTe epitaxial layer and GaAs substrate. However, the effect of interface recombination is probably fairly similar to that of lifetime reduction of minority carrier, as they both mainly affect the number of electrons that can reach the *n*-region but make little difference in the distribution of light absorption.

I. CONCLUSION

The responsive photocurrent and crosstalk between most adjacent pixels of HgCdTe IRFPA integrating microlenses in front of their corresponding photodiodes have been simulated numerically in two dimensions by exploiting FDTD method of Crosslight TCAD software. An optimal radius of spherical GaAs microlens is achieved by maximizing its ability to focus the incoming infrared plane wave at a specific point near the interface of GaAs substrate and HgCdTe absorber layer. For the configuration in this paper, i.e., a pixel size of 20μm, radiation wavelength of 5μm and GaAs substrate thickness of 20μm, microlenses with radius around 15μm is very beneficial to increase responsive photocurrent and simultaneously reduce crosstalk.

ACKNOWLEDGEMENTS

We thank C. S. Xia, Y. Sheng and M. Yang from Crosslight Company Shanghai Office for technical assistance and helpful discussions. We also acknowledge support provided by the program (Grant cxjj-10-m29) and NSFC (Contract 6070612).

REFERENCES

[1] A. Rogalski, "HgCdTe infrared detector material: history, status and outlook," *Rep. Prog. Phys.*, vol. 68, pp. 2267–2336, 2005.

[2] J. Wang, X. S. Chen, W. D. Hu, L. Wang, W. Lu et al., "Amorphous HgCdTe infrared photoconductive detector with high detectivity above 200K," *Appl. Phys. Lett.*, vol. 99, p. 113508, 2011.

[3] F. Yin, W. D. Hu, B. Zhang, Z. F. Li, X. N. Hu et al., "Simulation of laser beam induced current for HgCdTe photodiodes with leakage current," *Opt. Quantum Electron.*, vol. 41, pp. 805–810, 2011.

[4] F. Yin, W. D. Hu, Z. J. Quan, B. Zhang, X. N. Hu et al., "Extraction of electron diffusion length in HgCdTe photodiodes using laser beam induced current," *Acta Phys. Sin.*, vol. 58, pp. 7884–7890, 2009.

[5] C. A. Musca, J. M. Dell, L. Faraone, J. Bajaj, T. Pepper et al, "Analysis of crosstalk in HgCdTe p-on-n heterojunction photovoltaic infrared sensing arrays," *J. Electron. Mater.*, vol. 28, pp. 617–623, 1999.

[6] N. Guo, W. D. Hu, X. S. Chen, C. Meng, Y. Q. Lv et al., "Optimization of microlenses for InSb infrared focal-plane arrays", *J. Electron. Mater.*, vol. 40, pp. 1647–1650, 2011.

[7] J. Huang, W. H. Zhou, W. T. Yin, "Fabrication and evaluation of the CdZnTe microlens arrays," *Proc. SPIE*, vol. 7658, p. 76584G, 2010.

978-1-4673-1602-6/12 $31.00 © 2012 IEEE

Impact of Duty Cycle and Nano-Grating Height on the Light Absorption of Plasmonics-Based MSM Photodetectors

Farzaneh Fadakar Masouleh[1], Narottam Das[2], *Senior Member, IEEE*, and Hamid Reza Mashayekhi[1]

[1]Physics Group, University of Guilan, Rasht, Iran,

[2]Department of Electrical and Computer Engineering, Curtin University, Perth, WA 6845, Australia

e-mails: fafadakar@msc.guilan.ac.ir, narottam.das@curtin.edu.au, mashhr@guilan.ac.ir

Abstract— We use finite difference time-domain (FDTD) method to calculate the light absorption enhancement of nano-grating assisted metal-semiconductor-metal photo-detectors (MSM-PDs). The simulated results show that the light absorption enhancement of nano-grating assisted MSM-PD is~9-times better than conventional MSM-PD.

Index terms— Subwavelength aperture, duty cycle, nano-grating, surface plasmon polariton, FDTD simulation, MSM-PDs.

I. INTRODUCTION

The application of periodic structures on the metal-semiconductor-metal photodetectors (MSM-PDs) leads to effective light absorption and transmission through the subwavelength apertures. They have significant appeal in optical fiber communication, high-speed sampling, and chip to chip interconnects. The MSM-PD is a symmetrical semiconductor device which is equivalent to two back-to-back connected Schottky diodes [1]. There are two distinct mechanisms to produce transmission in one dimensional metal grating with narrow slits, which are the excitation of horizontal and vertical surface resonances. The horizontal surface resonances are excited by the periodic structure of the nano-gratings. The vertical surface resonances correspond to Fabry–Perot-like resonances of the fundamental TM guided wave in the slits [2]. The metallic gratings can exhibit absorption anomalies. One of these particularly remarkable anomalies is observed for p-polarized light only, and is due to surface plasmon polaritons (SPPs) excitations [3]. The light incident on the metal nano-grating is converted into propagating SPPs that can absorb the light efficiently in extremely thin (10's~100's of nm's thick) layers. The extremely thin absorbing layers can act as a light concentrator which is essential for triggering the extra ordinary absorption (EOA) of light [4]. Subwavelength apertures have also been used to efficiently concentrate light into the deep subwavelength regions [5]. Finite-difference time-domain (FDTD) simulation results have demonstrated significant enhancement of light absorption for the design of ultrafast MSM-PDs [5-6].

II. DESIGN OF MSM-PD STRUCTURE

Figure 1 shows a simple plasmonics-based MSM structure with gold (Au) nano-gratings etched on top of a layer of the same metal. The structure design is shown with three separate parts, namely, the metal nano-gratings (top part), the subwavelength apertures (middle part) and the substrate (bottom part). The momentum of surface plasmons can be easily changed by adding thin layers of material on the metal surface or by changing the dielectric constant of the material deposited on it. Here, the gold (Au) metal nano-gratings were deposited on top of the layer containing subwavelength apertures and the layer is only on the semiconductor (GaAs) substrates.

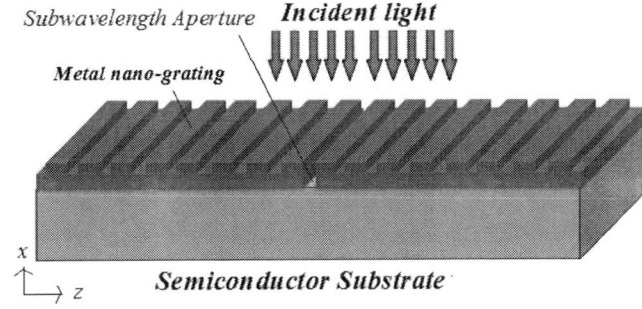

Fig. 1. Schematic diagram of the MSM-PD structure with rectangular shaped nano-gratings on top of the subwavelength apertures. The subwavelength apertures are just on top of the semiconductor (GaAs) substrates.

For a metal nano-grating period of Λ, the conservation of momentum in the direction parallel to the nano-gratings lead to the following relationship.

$$k_{x(out)} \cdot \sin\theta = k_{x(in)} \cdot \sin\theta \pm m k_g = \frac{\omega}{c}\sqrt{\frac{\varepsilon_m' \varepsilon_d}{\varepsilon_m' + \varepsilon_d}} \quad (1)$$

When the plasmonic excitations occur then the left side of equation (1) matches the wave vector of the excited SPP (K_{SPP}). Here, m is an integer corresponding to the order of the outgoing diffracted beam, $k_g = 2\pi/\Lambda$ is the grating wave vector, ω is the angular frequency of the incident light wave with θ as the angle of incidence and c is the speed of light.

III. RESULTS AND DISCUSSION

A. Impact of nano-grating height on LAEF

In this sub-section, we will discuss the influence of nano-grating height on the light absorption enhancement. Fig. 2 shows the light absorption enhancement factor (LAEF) spectra for different nano-grating heights with 60% duty cycle and the

978-1-4673-1602-6/12 $31.00 © 2012 IEEE

subwavelength aperture width is 100 nm. The assumed grating period is 810 nm. The TM mode of light was perpendicularly incident on the grooves with θ=0°. Different sets of results show that the amount of light transmitted into the active area of the MSM-PD changes with the variation of nano-gratings height. The peak wavelength is red shifted and it behaves like a sinusoidal manner.

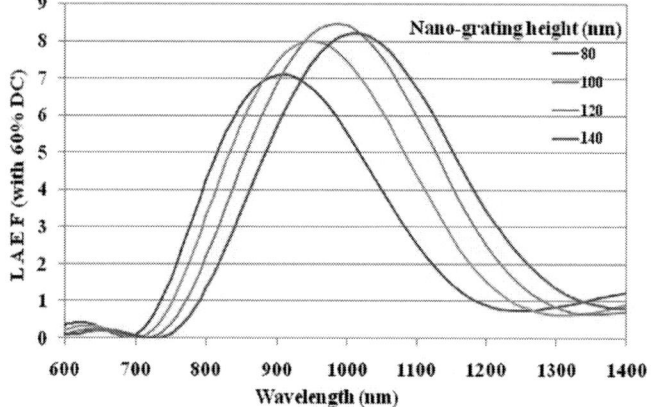

Fig.2. LAEF spectra for different nano-grating heights. Here, the duty cycle is 60% and subwavelength aperture width is 100 nm.

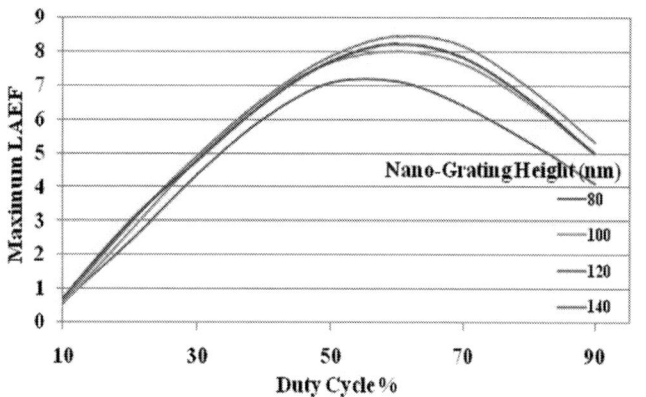

Fig. 3. Maximum (or peak) LAEF versus duty cycle characteristics with several nano-grating heights.

B. Influence of the duty cycles on LAEF with different nano-grating heights

In this sub-section, the LAEF spectra for several duty cycles which affect the amount of light flux transmitted into the active area is discussed. The duty cycle was varied from 10% to 90% while the subwavelength aperture was kept constant at 100 nm and the metal nano-grating heights were varied from 80 ~ 140 nm. The results illustrate that the amount of the LAEF grows gradually towards to 60% duty cycle and falls down moderately. The effects of the duty cycle on the transmitted power into the active region become more noticeable by plotting the value of the maximum LAEF as a function of the duty cycle, as shown in Fig. 3. It can be inferred that the maximum LAEF for each specific duty cycle increases from 80 nm to 120 nm of nano-grating height and decreases for 140 nm. It is clear that the duty cycle can affect the peak wavelength also the amount of light transmitted into the active area of the MSM-PDs.

C. Impact of nano-grating height and duty cycles on LAEF of MSM-PDs

It is noticed that the change of metal nano-grating height also affects the phase of the transmitted electric field, which is evident from the red shift exhibited when the metal nano-grating heights are increased. The LAEF for MSM-PD structure with different duty cycles is shown in Fig. 4. The maximum LAEF for each duty cycle can be obtained when the metal nano-grating height is around 120 nm that is the optimum nano-grating height for this typical device structure. As it was expected, the amount of power transmitted into the active region not only depends on the duty cycle but also on the nano-grating height which can be interpreted by changes in phase of the transmitted electric field.

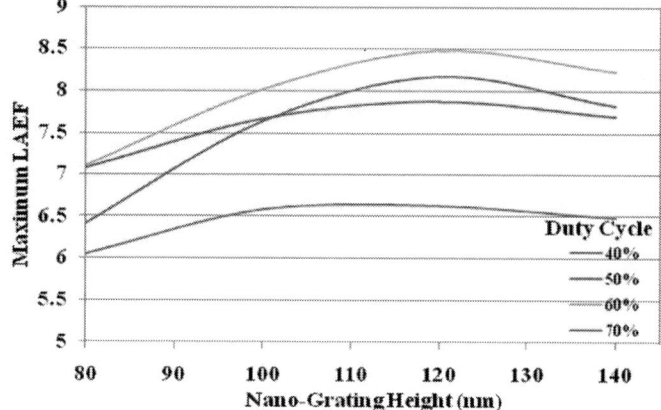

Fig. 4. Maximum (or peak) LAEF versus nano-grating height characteristics with several duty cycles.

IV. CONCLUSION

We have modeled the light absorption enhancement performance of a new MSM-PD structure with rectangular-shaped metal nano-gratings. The impact of metal nano-grating height and duty cycle on the LAEF of MSM-PD structures was analyzed using the FDTD technique to optimize these design parameters. Our simulation results show that the LAEF is ~9-times better than the conventional MSM-PDs with the subwavelength aperture width of 100 nm. These simulated results are useful for the design and development of high responsivity MSM-PDs.

REFERENCES

[1] A.D. Zebentout, Z. Bensaad, M. Zegaoui, A. Aissat, D. Decoster, "Effect of Dimensional Parameters on the Current of MSM Photodetector," *Microelectronics J*, Vol. 42, Issue 8, 1006-1009, 2011.

[2] S. Collin, F. Pardo, R. Teissier, J.L. Pelouard, "Horizontal and Vertical Surface Resonances in Transmission Metallic Gratings," *J.Opt. A: Pure Appl. Opt. vol. 4*, Issue 5, 2002

[3] J.S.White et al., "Extraordinary Optical Absorption Through Subwavelength Slits," *Optics Letters*, Vol. 34, Issue 5, 686-688, 2009.

[4] T. L'opez-Rios, D. Mendoza, F.J. Garc'ia-Vidal, J. S'anchez-Dehesa, B. Pannetier, "Surface Shape Resonances in Lamellar Metallic Gratings," *Phys. Rev. Lett*, Vol. 81, Issue 3, 665–668, 1998.

[5] N.Das, C.L. Tan, V.V. Lysak, K.Alameh, Y.T.Lee, "Light Absorption Enhancement in Metal-Semiconductor-Metal Photodetectors using Plasmonic Nanostructure Gratings," *Proc. HONET'09, Alexandria, Egypt, Dec. 28~30, 2009.*

[6] N. Das, A., M. Vasiliev, C. L. Tan, K. Alameh and Y. T. Lee, *Opt. Commun. 284, 1694, 2011.*

978-1-4673-1602-6/12 $31.00 © 2012 IEEE

High-efficiency Optical Coupling to Planar Photodiode using Metal Reflector loaded Waveguide Grating Coupler

G. Li, Y. Hashimoto, S. Ebuchi, T. Maruyama, and K. Iiyama

Natural Science and Technology, Kanazawa University,

Kakuma, Kanazawa, Ishikawa, 920-1192 Japan,

Tel: +81-76-234-4886, Fax: +81-76-234-4870, E-mail: maruyama@ec.t.kanazawa-u.ac.jp

Abstract- We propose a high-efficient vertical optical coupler using an amorphous Si waveguide grating coupler with top reflector. The coupling efficiency of 80% is calculated at the grating period of 380 nm and the duty ratio of 0.75 with top metal reflector.

1. Introduction

Recently, the operating speed of large-scale integrated (LSI) circuits is approaching a limit because global electrical inter-connection is becoming bottleneck. An optical inter-connection instead of the electrical interconnection on LSI is proposed to solve this problem [1]. Especially, a crystalline silicon (c-Si) optical waveguide have been studied intensively for optical interconnection at the 1.55 μm-wavelength range. The active device at this wavelength needs to introduce compound semiconductor such as GaInAs and AlInAs on c-Si substrate. These materials were difficult to grow epitaxially on Si substrate because of the lattice mismatch. A wafer bonding technique as one of the approaches to integrate on Si-LSI was reported [2].

On the other hand, c-Si can be used as the active device such as a photo-detector at 0.8 μm-wavelength range. We fabricated the c-Si avalanche photodiode by 0.18 μm-CMOS standard process [3]. The device realized a bandwidth of more than 1 GHz at 830 nm. However, c-Si is unsuitable as a waveguide at the 0.8 μm-wavelength range because of absorption at this wavelength. We propose the use of amorphous silicon (a-Si) as a waveguide material for the 0.8 μm-wavelength range because a band-gap energy of a-Si is about 1.4 eV-1.8 eV. We fabricated a-Si waveguides by photolithography and wet chemical etching, and obtained the propagation loss of 15 dB/cm at 830 nm[4]. It is promising for the realization of an all-Si optoelectronic integrated circuit (OEIC) using c-Si APDs and a-Si optical waveguides.

The optical coupling between APDs on Si-substrate and a-Si waveguides is a serious problem for realization of all-Si OEIC. To overcome this problem, grating couplers is the most qualified candidate [5]. In this paper, we describe a design of a high-efficient vertical optical coupler using the grating coupler between the APD on Si substrate and a-Si waveguide.

2. Calculation Results and Discussion

A simulation model of grating coupler is shown in Fig.1. The cross-sectional size of the a-Si single mode waveguide is 100 nm thickness and 300 nm width. The grating structure is fabricated into the a-Si waveguide on 2 μm-thick SiO_2 layer. The waveguide is covered with a BCB (1 μm-thickness) of dielectric material for top cladding layer. A reflecting mirror (Au) is located on top for higher efficient coupling.

The optical propagation in grating structure is analyzed by the two-dimensional finite element method. The incident wavelength and the refractive index of core is 850 nm and 3.39, respectively. The grating period (Λ= 340~380nm), the duty ratio (w/Λ= 0.25, 0.5, 0.75), and the etching depth (d= 5~100nm) dependence of the coupling efficiency, diffraction angular were analyzed.

Fig.2 shows the grating period dependence of coupling efficiency and diffraction angle. The coupling efficiency significantly reduced at the grating period of 360 nm. Because a strong Bragg reflection occurs at the Bragg period. On the other hand, the

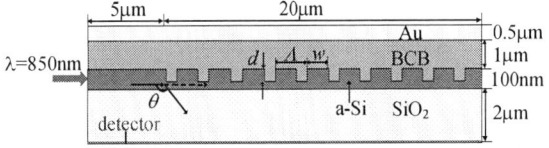

Fig.1: Analytical model of a-Si grating coupler

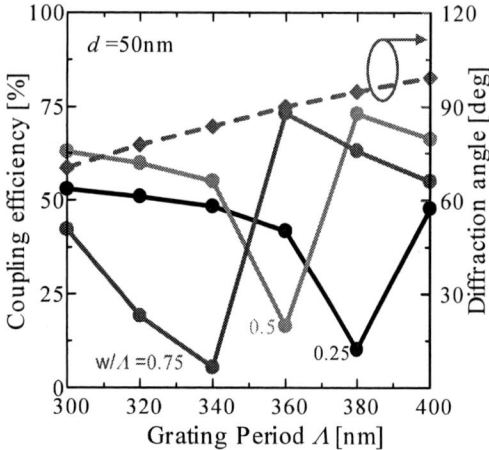

Fig.2: Grating period and duty ratio dependence of coupling efficiency and diffraction angle.

diffraction angle is independent of grating period.

The relationship between grating duty ratio and coupling efficiency is also shown in Fig.2. The grating period at dipping coupling efficiency becomes short at large duty ratio. The reason is that the equivalent refractive index of the waveguide becomes a high value at large duty ratio.

The etching depth dependence of the coupling efficiency is shown in Fig.3. A deeper etching depth, a stronger Bragg reflection and a longer Bragg period.

To improve the coupling efficiency, a metal reflector is loaded on top cladding layer. Fig.4 shows the etching depth dependence of coupling efficiency

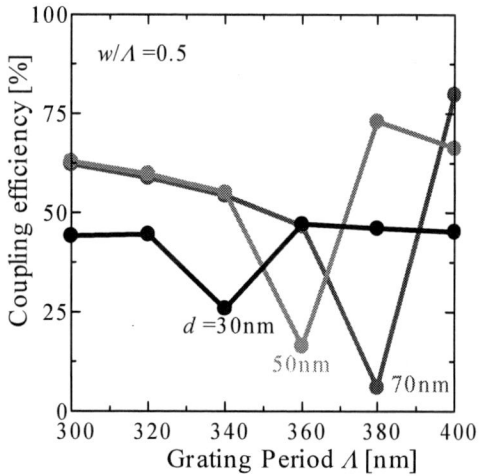

Fig.3: Etching depth dependence of coupling efficiency

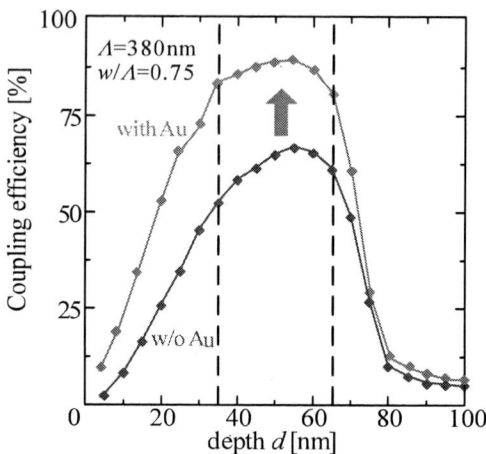

Fig.4: Etching depth dependence of coupling efficiency with and without top metal reflector.

with and without top metal reflector at grating period of 380nm and duty ratio of 0.75. It can be clearly observed that the coupling efficiency with the metal reflector achieved 1.4 times higher than that without the reflector.

3. Conclusions

We proposed and simulated the high-efficient vertical optical coupler using the grating coupler between the APD on Si substrate and a-Si waveguide with top metal reflector. The coupling efficiency of 80% was calculated at the metal loaded grating coupler. This coupling efficiency is 1.4 times higher than that without a reflector. This grating coupler can be expected to realize the high-efficiency optical coupler on all-Si OEIC.

References

[1] D. A. B. Miller, *Proc. IEEE* **97** (2009) 1166..

[2] D. Liang and J. E. Bowers, *Electron. Lett.* **45** (2009) 578.

[3] K. Iiyama, H. Takamatsu and T. Maruyama, *IEEE Photon. Technol. Lett.* **22** (2010) 932.

[4] T. Asukai, M. Inamoto, T. Maruyama, K. Iiyama, K. Ohdaira and H. Matsumura, *IEEE Group IV Photonics 2010,* **P2.16** (2010).

[5] R. Takei, K. Uchino and T. Mizumoto, Conference on Laser and Optics 2010, **CthP1** (2010)

New optical coupling structure of high light absorption quantum well Infrared photodetectors

Q. Li, N. Li, X. S. Chen, Z. F. Li[*], W. Lu[*]

State Key Lab of Infrared Physics, Shanghai Institute of Technical Physics, Chinese Academy of Sciences, 500 Yu Tian Road, Shanghai, China 200083

Abstract

A new optical coupling structure of high optical absorption quantum well infrared photodetectors is reported, in which 4 periods of $Al_{0.15}Ga_{0.85}As$/GaAs QWs was integrated with double gold films and a sandwiched structure of metal-QWs-metal gratings has been adopted. Normal incident light can be coupled and trapped in the dielectric layer in the form of transverse electromagnetic waves, when the structure is optimized. Therefore, the light absorption of quantum wells is greatly enhanced when the light travels back and forth in the dielectric layer. Numerical simulations are made via 2D finite-difference time-domain (FDTD) method, yielding consistent results with experiments, which shows the photocurrent response increase of 21 times to the 45 degree mesa photodetector. At the same time, we observe the Rabi splitting.

I. INTRODUCTION

For quantum well infrared photodetector (QWIPs), normal incident light can not be absorbed by quantum wells due to the inter-subband transition rule, thus proper optical coupling schemes should be employed for normal incident light [1,2]. Now we propose and demonstrate efficient photo-couplers for QWIP by exploiting SPP resonance occurring metal gratings in the metal-QWs-metal gratings structure as shown in Fig. 1. It shows that light can be coupled into left-going and right-going waves in the dielectric layer, thus standing wave resonances can be formed between neighboring grooves. A series of Fabry-Perot-like standing-wave modes associated with the impedance mismatch of the entrance and exit apertures is supported by an open-ended slit[3]. These resonant states can be used to enhance the optical absorption of photodetectors. And the high coupling efficiency of the proposed scheme suggests that such a structure may have great significance for quantum well infrared detectors.

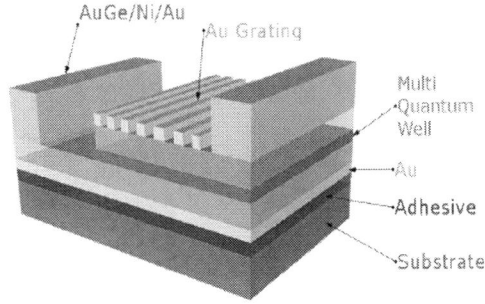

Fig. 1. (Color online) schematic diagram of the M-QWs-M structure of quantum well infrared photodetectors.

II. SIMULATION MODELS AND DEVICE STRUCTURE

The schematic view of unit structure is shown in figure 2. The contact/Quantum wells/contact (CQC) structure is located between a metal-reflection layer and a metal-grating layer. Grating period p, grating depth h and slit width d are marked in fig. 2. The thickness of between the top and bottom is h, and the thickness of both the Au-grating layer and the Au-reflection layer are h1. The contact /Quantum wells/ contact structure is regarded as a uniform medium with effective refractive index 3.34[4]. The frequency-dependent permittivity of Au is based on the Lorentz-Drude model [5].

In our simulation two-dimension finite-difference time-domain (FDTD) method is employed. Both periodic boundary conditions imposed on the left and right sides and finite structure are used. As many as 16-20 periods grating, for there is little difference between the two conditions, so periodic boundary conditions is used for convenience. And perfect matched layers are imposed at the front and back surfaces. To ensure the accuracy the grid size is carefully chosen.

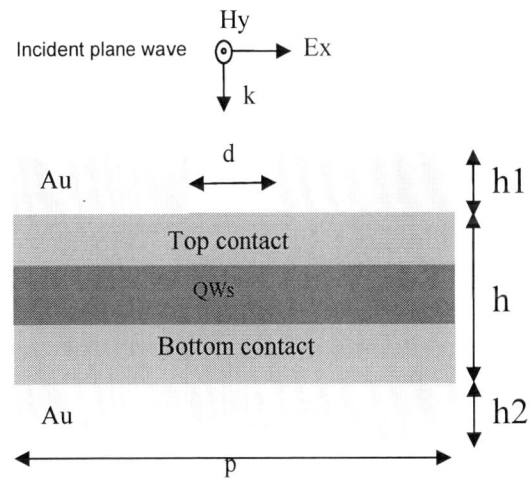

Fig. 2. (Color online) schematic diagram of the unit structure. The parameters are as shown.

III. RESULT AND DISCUSSION

The normal incident light has been redirected and trapped into GaAs/Quantum wells/GaAs layer, which makes such a structure is a perfect candidate as optical coupling scheme for QWIPs. Cavity modes are formed in the dielectric layer due to interference between the left-going and the right-going waves from neighboring slits: $E = 2E_0\cos(k*p)$, as period condition is imposed, and the resonant

---

[*] Corresponding author: zfli@mail.sitp.ac.cn, luwei@mail.sitp.ac.cn

condition can be given by k*p = 2n π, n = 1, 2, • • •, where n is the order of resonance and p is the grating period. The resonant frequency of the Fabry-Perot resonators can be exhibited as:

$$f_n = \frac{cn}{2n_r(p-d)},$$

Where c is the vacuum speed of light, n_r is the refractive index of dielectric of the CQC.

The black body photocurrent between the new M-CQC-M structure QWIP and standard 45° mesa detector with the same CQC material is compared. The black body response of the 45 degree mesa detector is 0.02(A/W) while that of the new structure can reach 0.15(A/W). The Fourier transform infrared spectroscopy (FTIR) method is used to characterize the photocurrent spectrum. Fig.3 displays the photocurrent spectrums of different structure. 45 degree mesa device (red), new M-CQC-M structure QWIP (black) with p=11.3um and d=2.9um and the yellow dots show the model transmission of the new structure. The resonance mode can be tell clearly when the new structure be used. The FWHM of the absorption of QWs is 7.5um while that of the resonance mode is about 1um. Between the spectrum of 13.8-14.6um, the response of the No.13 can be 22 time to the 45° mesa QWIP. And the numerical simulations show that the |Ez| field enhancements of at least 3.6 times.

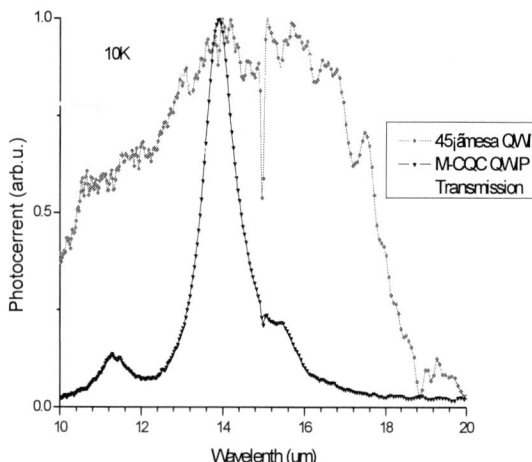

Fig. 3. (Color online) the photocurrent spectrums of different structure: 45 degree mesa device (red), new M-CQC-M structure QWIP (black) and the model transmission.

All the simulations above do not consider the absorption of the quantum wells. When cavity mode can not couple absorption, this influence can be ignored. When cavity mode and absorption are coupled, the Rabi splitting occurs [6], where the vacuum Rabi frequency can be expressed:

$$\Omega_R = \sqrt{\frac{e^2}{4\varepsilon_0 n_r^2 m^*} * \frac{f_{12} N_{2NEG} N}{h_2}} \quad [7],$$

in which e is the electron charge, m^* is the electron effective, n_r is the effective refractive index here is 3.34, f_{12} is the oscillator strength 0.96 be used[8], N is 4 as the quantum wells number, N_{2NEG} is $16.5*10^{10}$ cm^{-2} as he population difference between the two subbands and h_2 is 0.326 um as the thickness of the 4 periods of quantum wells. The absorption is Ω_R / π = 1.1 THz in this CQC material. In sample 3 P is 11.5um and d is 6um. When we simulate taken the effective

refractive index of CQC structure is 3.34, only crests at 10um, 13.4um and 14.7um can be seen. But there are two modes at 13.3um and 14um separated by 1.12 THz in the photocurrent spectrum, which is a good agreement with the Rabi frequency. To simulate this effect, the ISB absorption is add through a function of the electronic sheet density as a Drude-Lorentz dielectric tensor [9]. The simulation result is consistent to the experience.

IV. CONCLUSION

In conclusion, we experimentally demonstrate a new optical coupling structure of light absorption quantum well infrared photodetectors. The photocurrent response of such structure QWIP can increase 21 times to that of the 45 degree mesa QWIP. With an enhanced response, the new coupled QWIP can work at normal light incidence, at the same time, we observe the Rabi splitting. As far as we know, this is the first time to observe the Rabi splitting by the photocurrent spectrum, and the numerical simulations can consistent results with this effect. Introduction

ACKNOWLEDGEMENTS

This work was supported in part by the State Key Program for Basic Research of China (Grants No. 2007CB613206 and No. 2011CB922004).

REFERENCES

[1] J. Y. Adersson and G. Landgren, "Intersubband transition in single AlGaAs/GaAs quantum wells studied th Fourior transform infrared spectroscopy," J. Appl. Phys.61, 4123(1988).
[2] K. K. Choi, "The physics of quantum well infrared photodetectors", 1997.
[3] Alastair P. Hibbins and J. Roy Sambles, "Squeezing MillimeterWaves into Microns", PRL, 92, 14, 2004.
[4] Patrick Nickels, Shinpei Matsuda, Takeji Ueda, "Metal Hole Arrays as Resonant Photo-Coupler for Charge Sensitive Infrared Phototransistors", IEEE, 46, 3, (2010).
[5] A. D. Raki´c, A. B. Djuri si´c, J. M. Elazar, and M. L. Majewski, " Optical propertiesof metallic films for vertical-cavity optoelectronic devices" Appl. Opt. 44, 2332 (2005).
[6] Y. Todorov, A. M. Andrews, "Strong Light-Matter Coupling in Subwavelength Metal-Dielectric Microcavities at Terahertz Frequencies", PRL 102, 186402 (2009).
[7] Proceedings of the International School of Physics Enrico Fermi, Course CL, edited by B. Deveaud, A. Quattropani, and P. Schwendimann (IOS Press, Amsterdam, 2003)
[8] H. Schneider and H. C. Liu, Quantum Well Infrared Photodetectors: Physics and Applications. (2007)
[9] L. Wendler and T. Kraft, Phys. Rev. B 54, 11 436 (1996).

Modeling on Current-Voltage Characteristics of HgCdTe Photodiodes in Forward Bias Region

Yang Li [a,b], Zhenhua Ye [a,*], Chun Lin [a], Xiaoning Hu [a], Ruijun Ding [a] and Li He [a]

[a] Key Laboratory of Infrared Imaging Materials and Detectors, Shanghai Institute of Technical Physics,
Chinese Academy of Sciences, Shanghai 200083, China
[b] Graduate School of the Chinese Academy of Sciences, Beijing 100039, China

Abstract— Current-voltage (*I*–*V*) characteristics of HgCdTe photodiodes in the forward bias region have been modeled on account of mechanisms including diffusion and recombination currents, metal-semiconductor (*M-S*) contact and constant series resistance. Moreover, a data processing approach has been developed to obtain valuable physical parameters from measured *I*–*V* curves. This model and algorithm have also been verified to be available and promising by the fitting results on device parameters of HgCdTe photodiodes.

I. INTRODUCTION

Mercury cadmium telluride (HgCdTe) is the most widely used material system for infrared (IR) imaging applications [1–4]. Unfortunately, detectors based on this material suffer from very complex fabrication processes. Therefore, feasible models are required to simulate the device performance and acquire useful physical parameters. Many researches have been carried out on modeling the current-voltage (*I*–*V*) characteristics of HgCdTe photodiodes. However, most of them focused only on reverse bias or small bias region [2–3], and the information embedded in forward bias region have been dissipated.

In this work, forward-biased *I*–*V* characteristics of HgCdTe photodiodes have been modeled, considering diffusion current, recombination current, and the effects of metal-semiconductor (*M-S*) contact as well as a constant series resistance. Several device parameters were obtained from the measured resistance-voltage (*R*–*V*) data by a set of estimation and fitting techniques. The new method is applicable in evaluating detector quality, and the electrical properties of an *M-S* contact can be extracted directly from photodiodes instead of specially made samples.

II. METHOD

A. Physical Model

The "forward bias" here is a flexible concept, meaning the region where the current grows rapidly with increasing bias and the dynamical impedance accordingly decays to a low value. The diffusion current is given by the following formula:

$$I_{diff} = Aqn_i^2 \sqrt{\frac{kT}{q}} \left(\frac{\coth(W_p/L_n)}{N_a} \sqrt{\frac{\mu_n}{\tau_n}} + \frac{\coth(W_n/L_p)}{N_d} \sqrt{\frac{\mu_p}{\tau_p}} \right) \left[\exp\left(\frac{qV_1}{kT}\right) - 1 \right], \quad (1)$$

where V_1 is the effective bias applied on the *p-n* junction, L_n and L_p are the diffusion lengths of electrons and holes; W_p and W_n are the effective thicknesses of *p*- and *n*-region, respectively. The recombination current can be described as such a form that

$$I_{gr} = 2A \frac{n_i kT}{\tau_0} \sqrt{\frac{2\varepsilon_s \varepsilon_0 (N_a + N_d)}{qN_a N_d (V_{bi} - V_1)}} \sinh\left(\frac{qV_1}{2kT}\right) \cdot f(b), \quad (2)$$

and the meanings of all unexplained symbols in (1) and (2) can be found in [2]. Providing that Schottky barriers only form on *p*-type HgCdTe, *I*–*V* characteristics of the electrode contact can be expressed as

$$-I = SA^* T^2 \exp\left(\frac{-q\phi_{bp}}{kT}\right) \left[\exp\left(\frac{-qV_2}{nkT}\right) - \exp\left(\frac{qV_2}{n_2 kT}\right) \right], \quad (3)$$

where $1/n + 1/n_2 = 1$ and V_2 represents the actual voltage drop across the Schottky junction. All symbols' meanings in (3) can be identified by comparing (3) to equations in [4]. In addition, a constant series resistance R_s is incorporated, and the relations $I = I_{diff} + I_{gr}$ and $V = V_1 + V_2 + IR_s$ complete the entire model, in which I and V are the total current and total bias, respectively. This system of equations is to be solved by a two-fold Newton-Raphson method to calculate theoretical *I*–*V* or *R*–*V* curves.

A simplified alternative model is presented as follows to enable a preliminary estimation of contact parameters. In the high resistance region, the *p-n* junction bears most external bias and its behavior can be approximated as $I = I_1 \exp(qV_1/n_1 kT)$, where the saturation current I_1 and the ideality factor n_1 can be determined by the ln*I*-*V* plot. On the other hand, it is necessary to consider the bias divided by the contact in the low resistance region, and invoking $I = I_2 \exp(qV_2/n_2 kT)$ will easily lead to

$$V = IR_s + \frac{kT(n_1 + n_2)}{q} \ln I - \frac{kT(n_1 \ln I_1 + n_2 \ln I_2)}{q}, \quad (4)$$

$$R = \frac{kT(n_1 + n_2)}{q} \frac{1}{I} + R_s, \quad (5)$$

where I_2 can be recognized from (3). With I_1 and n_1 known, least-square analyses would give n_2 and R_s according (5), and thereafter give I_2 (also ϕ_{bp}) according to (4).

B. Fitting Algorithm

Total square relative error in *R*–*V* data acts as the objective function to be minimized. The fitting algorithm adopted here is a hybrid of the genetic algorithm and a "stirring-like" technique. At first, previously estimated contact parameters remain fixed while other fitting parameters change in a reasonable range in search of the minimum using a real-valued genetic algorithm. Then similar processes will be successively repeated, each time with a random portion of the fitting parameters allowed to vary, until the ultimate terminating condition has been reached. To be specific, the genetic algorithm is implemented via a variant tournament selection, in which the entire population is broken into small groups and several fittest individuals in each group survive to breed. Weighted arithmetic or geometric means of the parents are calculated to carry out crossover, and a gradient

---

* Correspondence email: zhye@mail.sitp.ac.cn

descent search technique fulfills mutation for those highest-ranking individuals to enhance local searching ability.

III. APPLICATION AND DISCUSSION

The photodiodes investigated were planar short-wavelength (SW) photovoltaic detectors fabricated by B^+ implantation on vacancy-doped p-type $Hg_{1-x}Cd_xTe$ ($x = 0.398$) layers grown liquid phase epitaxy (LPE). I–V characteristics were measured with zero field of view at 77K, and the fitting parameters were donor density (N_d), effective lifetime in the depletion region (τ_0), electron lifetime in the p-region (τ_n), and all three above-mentioned contact parameters, ϕ_{bp}, n and R_s. Nonadjustable parameters were assumed or measured as listed in Table I.

The measured R–V data and corresponding fitting results are shown in Fig. 1. A fairly good agreement can be observed between the calculated curve and the experimental one. As the bias increases, the total dynamical impedance is sequentially governed by the effects of recombination, diffusion and M-S contact. These results, especially the part magnified in the inset, could not be acquired using only a constant series resistance. Moreover, a well-fitted I–V curve was simultaneously achieved and is shown in Fig. 2. As we can see, the original current data would approach a stable negative value of -4.4×10^{-11}A under a small bias, which can be considered as the photocurrent caused by inevitable radiation. After correcting for this point, a more approximate curve to the predicted one was obtained.

In Table II list the extracted values of the fitting parameters, which are all in accordance with those typically reported for similar devices. The value of electron lifetime τ_n corresponds to a diffusion length larger than the p-region thickness, which indicates the necessity of using equation of such a form as (1).

TABLE I.   NONADJUSTALBE MODELING PARAMETERS

Parameter	Value	Parameter	Value	Parameter	Value
N_a (cm^{-3})	5.0×10^{15}	τ_n / τ_p (a.u.)	10.0	E_t/E_g (a.u.)	0.537
W_p (μm)	10.0	μ_p (m^{2}V^{-1}s^{-1})	0.04	A (μm^2)	896
W_n (μm)	1.0	μ_n (m^{2}V^{-3}s^{-3})	5.0	S (mm^2)	7.69

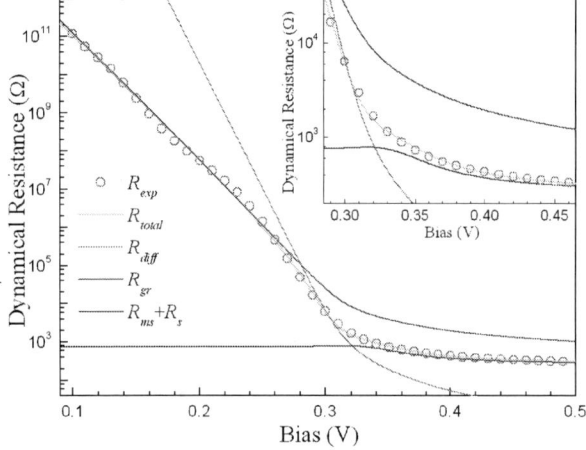

Figure 1.   Experimental R–V curve and corresponding fitting results.

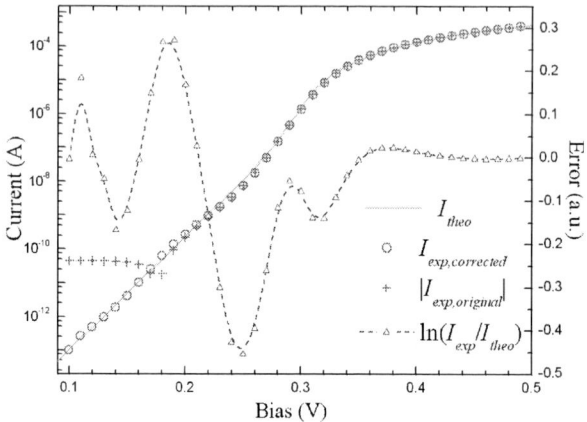

Figure 2.   Experimetal and predicted I–V curves, as well as corresponding relative errors. Both original and corrected data are depicted.

TABLE II.   VALUES OF EXTRACTED PARAMETERS

N_d (cm^{-3})	τ_n (ns)	τ_0 (ns)	ϕ_{bp} (V)	n (a.u.)	R_s (Ω)
6.46×10^{17}	24.0	0.461	0.144	1.75	254

The relatively large ideality factor n implies a notable deviation from pure thermionic emission theory, possibly resulting from tunneling, recombination and so on, which is beneficial to the formation of a low-resistance electrode contact.

IV. CONCLUSION

I–V characteristics of HgCdTe photodiodes in forward bias region have been successfully modeled, considering the effects of diffusion current, recombination current and especially the rectifying electrode contact. Several physical parameters, such as the donor concentration, the minority carrier lifetime, and the Schottky barrier height of electrode contact, were extracted from the experimental data of a typical SW planar device. Therefore, this novel model and method have been proven applicable in analyzing the properties and quality of HgCdTe photodiodes.

REFERENCES

[1] J. Wang, X. S. Chen, W. D. Hu, L. Wang, W. Lu et al., "Amorphous HgCdTe infrared photoconductive detector with high detectivity above 200 K", *Appl. Phys. Lett.*, vol. 99, p. 113508, 2011.

[2] F. Yin, W. D. Hu, B. Zhang, Z. F. Li, X. N. Hu et al., "Simulation of laser beam induced current for HgCdTe photodiodes with leakage current", *Optical and Quantum Electronics*, vol. 41, pp. 805–810, 2011.

[3] F. Yin, W. D. Hu, Z. J. Quan, B. Zhang, X. N. Hu et al., "Extraction of electron diffusion length in HgCdTe photodiodes using laser beam induced current", *Acta Physica Sinica*, vol. 58, pp. 7884–7890, 2009.

[4] W. D. Hu, X. S. Chen, F. Yin, Z. J. Quan, Z. H. Ye et al., "Analysis of temperature dependence of dark current mechanisms for long-wavelength HgCdTe photovoltaic infrared detectors," *J. Appl. Phys.*, vol. 105, p. 104502, 2009.

[5] V. Gopal, S. Gupta, R. K. Bhan, R. Pal, P. K. Chaudhary, V. Kumar, "Modeling of dark characteristcs of mercury cadmium telluride n^+–p junctions," *Infrared Phys. Technol.*, vol. 44, pp. 143–152, 2003.

[6] G. Bahir, R. Adar and R. Fastow, "The electrical properties of metal contact Au and Ti on p-type HgCdTe," *J. Vac. Sci. Technol. B*, vol. 9, pp. 266–272, 1991.

Simulation of InGaN/GaN light-emitting diodes with a non-local quantum well transport model

Chang Sheng Xia,[1,*] Z. M. Simon Li,[1] Yang Sheng,[1] Li Wen Cheng,[2] Wei Da Hu,[2] and Wei Lu[2]

[1]Crosslight Software Inc. China Branch, Suite 906, Building JieDi, 2790 Zhongshan Bei Road, Shanghai 200063, China

[2]National Lab for Infrared Physics, Shanghai Institute of Technical Physics, Chinese Academy of Sciences,

500 Yu Tian Road, Shanghai 200083, China

Abstract-Blue InGaN/GaN multiple quantum well (MQW) light-emitting diodes (LEDs) are simulated by the APSYS software with a non-local quantum well transport model. The simulation results are in good agreement with experiment and indicate the non-local quantum well transport model has a significant influence on the radiative recombination, the carrier transport and the current crowding of the InGaN/GaN MQW LEDs.

Ⅰ. INTRODUCTION

InGaN-based multiple quantum well (MQW) light-emitting diodes (LEDs) are receiving tremendous attention due to their potential applications in energy saving solid-state lighting [1]. Although significant improvements in material quality and device fabrication have been achieved during the past decade, there are still some problems such as electron leakage [2], poor hole injection efficiency [3], junction heating [4] and current crowding [5], which could largely degrade the optical and electrical performances of these devices. So it is imperative to resolve these issues by optimization of device structure.

Numerical simulation is an effective method to promote this process with low cost [6][7][8]. However, it is usually difficult to get good agreement between simulated results and experimental data for InGaN-based MQW LEDs by using traditional Drift-Diffusion model since this model is not adequate to describe the case where carriers directly fly over the quantum well (QW) without scattering. Recently, a non-local QW transport model has been implemented in our APSYS simulation software [9] to explain this effect. In this paper, we will carry out a 2D simulation for blue InGaN/GaN MQW LED with the APSYS software and show the importance of the non-local QW transport model to the LED simulation.

Ⅱ. THEORETICAL MODELS AND PARAMETERS

The APSYS simulation software is a finite-element based device simulator which self-consistently solves Poisson-Schrödinger equations, current continuity equations, heat transfer equations and hydrodynamic equations, including K·P models for MQW band structure, quantum tunneling model for heterojunction, heat flow model for self-heating,

*Corresponding author: xiachsh@crosslight.com.cn

ray-tracing model for photon extraction, spontaneous and piezoelectric polarization models for built-in electric field, as well as Shockley-Read-Hall (SRH) recombination and Auger recombination of carriers. The non-local QW transport model in APSYS is a modified version of Drift-Diffusion equation where QWs are treated as carrier traps with trapping rates determined by phonon scattering. The trapping rate is usually expressed as:

$$R_{qw} = (n - n_0)/\tau \tag{1}$$

where τ is capture time constant. n is carrier concentration and n_0 is that at equilibrium. High value of τ means that more carriers can directly flow over the QWs.

In this simulation, the SRH lifetime within QWs is estimated to be 100 ns. The built-in interface charges due to spontaneous and piezoelectric polarization are calculated by the methods developed by Fiorentini *et al.* [10], 50% of the theoretical value is used to account for the compensation by fixed defects and other interface charges. The AlGaN band offset ratio is assumed to be 50:50, according to the results of Piprek *et al.* [11] The Auger recombination coefficient is set to be 1.0×10^{-34} cm^{6}s^{-1} [12] which has a negligible influence on the emission efficiency of the LEDs.

Ⅲ. LED DEVICE STRUCTURE

The simulated structure in this paper is similar to that in [13], which consists of a 1-μm-thick n-type GaN:Si (5×10^{18} cm^{-3}), a 30 period GaN(4nm)/InGaN(4nm) superlattice:Si (5×10^{18} cm^{-3}), and six or nine periods of In$_{0.18}$Ga$_{0.82}$N(4nm)/GaN(20nm) MQW active regions, ending in a 16-nm-thick GaN barrier, followed by a 10-nm-thick undoped Al$_{0.15}$Ga$_{0.85}$N electron blocking layer and a 0.2-μm-thick p-type GaN:Mg (1×10^{18} cm^{-3}) cap layer. The device geometry is 526×315 μm^2.

Ⅳ. RESULTS AND DISCUSSIONS

Fig.1 shows the calculated and experimental [13] output power for the 6-QW and 9-QW LEDs. As shown in Fig. 1(a), there is small difference in output power for both of the 6-QW and 9-QW LEDs calculated by using the traditional Drift-Diffusion model. However, when the non-local QW transport model is employed, the output power for both of them is decreased, but the difference between them is increased gradually with the increase of τ as indicated in Fig. 1(b)-1(d). Moreover, good agreement between experimental data and simulated results is demonstrated when $\tau = 1.0 \times 10^{-6}$ s.

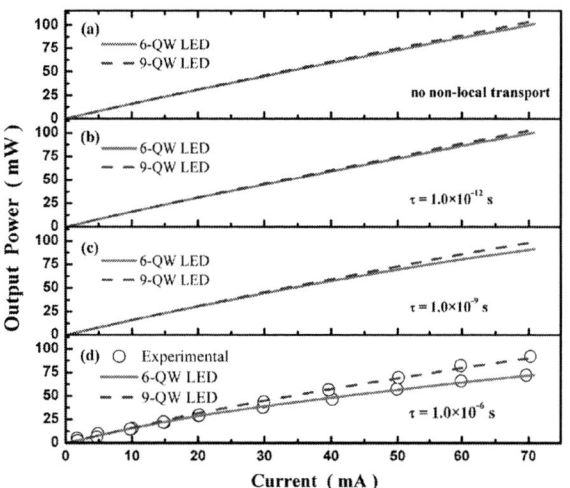

Fig. 1 Experimental and simulated output power for the 6-QW and 9-QW LEDs with and without the non-local QW transport model

The simulated current-voltage (I-V) curves for the 6-QW and 9-QW LEDs are plotted in Fig. 2. It is apparent that the I-V characteristics of these two LED devices are both improved markedly by using the non-local QW model with a higher value of τ compared to those based on the traditional Drift-Diffusion model. As shown in Fig. 2(d), the forward voltages of the 6-QW and 9-QW LED at 20 mA are similar to those that have been reported in experiment [13]. These results indicate that there are more carriers that can directly fly over the QWs to take part in the transport process other than being captured by the QWs in the realistic LED devices.

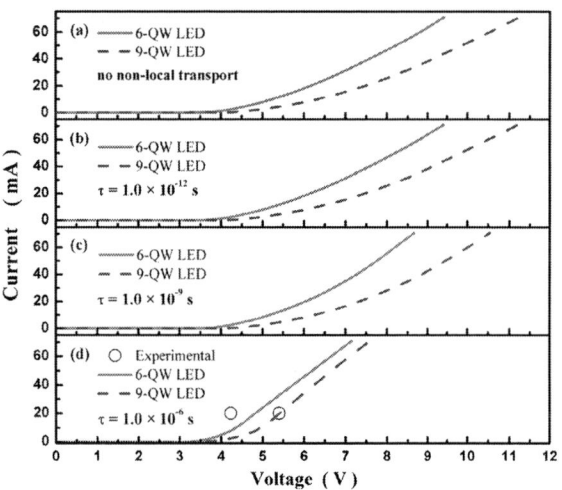

Fig. 2 Simulated I-V curves for the 6-QW and 9-QW LEDs with and without the non-local QW transport model.

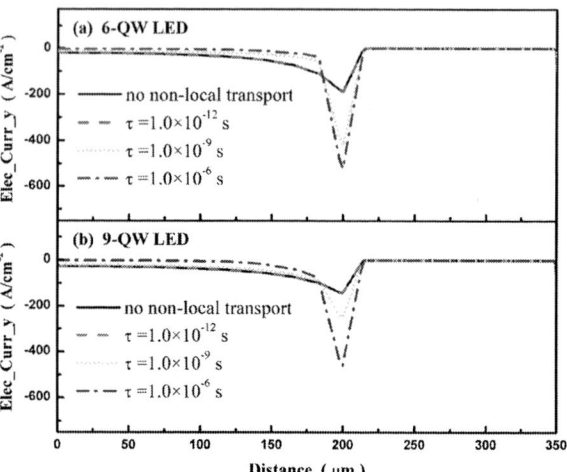

Fig. 3 Local electron current density in the horizontal direction at the middle of n-GaN layer for the 6-QW and 9-QW LEDs with and without the non-local QW transport model.

The horizontal profiles of the local electron current density at the middle of n-GaN layer for both of the LEDs are shown in Fig. 3 and they reveal that the non-local QW transport model with a higher value of τ leads to stronger current crowding due to the increasing number of transportable carriers. Furthermore, the stronger current crowding in the 6-QW LED as compared to that in the 9-QW LED may result in the lower output power for the 6-QW LED as indicated in Fig. 1.

V. CONCLUSIONS

We have simulated the electrical and optical characteristics of blue InGaN/GaN MQW LEDs with a non-local QW transport model. The simulation results are in good agreement with experimental data. We found that the non-local transport model can explain the experimental phenomena well and has a significant influence on the radiative recombination, the carrier transport and the current crowding of the InGaN/GaN MQW LEDs.

REFERENCES

[1] S. Pimputkar, J. Speck, et al., Nat. Photonics **3**, 180 (2009).
[2] M. H. Kim, M. F. Schubert, et al., Appl. Phys. Lett. **91**, 183507 (2007).
[3] I. V. Rozhansky and D. A. Zakheim, Semiconductors **40**, 839 (2006).
[4] A. A. Efremov, N. I. Bochkareva, et al., Semiconductors **40**, 605 (2006).
[5] Y. K. Su, S. J. Chang, et al., IEEE Electron Device Lett. **26**, 891(2005).
[6] Y. K. Kuo, J. Y. Chang, et al., Proc. of SPIE **7933**,793317 (2011).
[7] C. H. Wang, C. C. Ke, et al., Appl. Phys. Lett. **97**, 261103 (2010).
[8] C. S. Xia, Z. M. S. Li, et al., Appl. Phys. Lett. 99, 233501 (2011).
[9] Crosslight Software Inc., Burnaby, Canada (http:// www. crosslight.com).
[10] V. Fiorentini, F. Bernardini, et al., Appl. Phys. Lett. 80, 1204 (2002).
[11] J. Piprek and S. Li, Opt. Quant. Electron. 42, 89 (2010).
[12] J. Hader, J. V. Moloney, et al., Appl. Phys. Lett. 92, 261103 (2008).
[13] S. Tanaka, Y. Zhao, et al., Electron. Lett. 47, 335 (2011).

Simulation of InGaN/GaN light-emitting diodes with Patterned Sapphire Substrate

Yang Sheng,[1,*] Chang Sheng Xia,[1] Z. M. Simon Li,[1] and Li Wen Cheng[2]

[1]Crosslight Software Inc. China Branch, Suite 906, Building JieDi, 2790 Zhongshan Bei Road, Shanghai 200063, China

[2]National Lab for Infrared Physics, Shanghai Institute of Technical Physics, Chinese Academy of Sciences,

500 Yu Tian Road, Shanghai 200083, China

Abstract-**Blue InGaN/GaN multiple quantum well (MQW) light-emitting diodes (LEDs) with patterned sapphire substrate (PSS) are simulated by the APSYS software. Approach of combining finite-difference time-domain (FDTD) method and ray tracing technique is applied to perform light extraction. The simulation results show that PSS dramatically increases extraction efficiency of light power, in agreement with experiment. It is found that extraction efficiency can be maximized by changing the shape of PSS. This work presents a new approach to combine electrical simulation with FDTD and raytracing for both accuracy and efficiency in 3D TCAD simulation of GaN-LED.**

I. INTRODUCTION

Gallium-nitride-based (GaN-based) multiple quantum well (MQW) light-emitting diodes (LEDs) have been applied to many energy-saving fields such as out-door full-color displays, high performance back-lighting for liquid crystal displays [1][2]. However so far, the external quantum efficiency (EQE) of MQW LEDs is still not large enough compared with what is expected [3]. The EQE is a product of carrier injection efficiency (CIE), internal quantum efficiency (IQE) and light extraction efficiency (LEE), and each term in the product is the focus of researches. CIE and IQE can be improved by specific designs such as electron blocking layer (EBL) [4], ITO as current spreading layer [5] and the last quantum barrier $In_{0.03}Ga_{0.97}N$ for improving IQE [6]. To improve LEE, the techniques of surface roughening [7], nanoprinting [8] have been used but not effectively [3]. In the past few years, GaN based-LEDs grown on patterned sapphire substrate (PSS) have attracted more and more interest from researchers [9][10]. The PSS reduces threading dislocation density in GaN epi-layers as well as enhances LEE of LEDs due to increased light escape probability [11]. Several types of PSS have been successfully applied to fabrication of high-efficiency LEDs [12] [13]. Also, comparison of the LED EQE between un-patterned sapphire substrate (u-PSS) and PSS has been carried out experimentally [14][15]. However, theoretical analysis of LEE on PSS LEDs is a major challenge. Ray tracing (RT) technique has been applied to LEE analysis for u-PSS LEDs previously. For PSS LEDs, some simplified RT methods were proposed in recent years but all were based on geometric optics [16][3]. When PSS size

becomes comparable with wavelength of light emitted by LEDs, wave properties of light must be considered. In this work, we carried out 2D/3D simulation of InGaN/GaN MQW LED with PSS by the APSYS software. Light extraction processing is implemented by a new RT method which combines both geometric and wave behavior of light emitted from MQW. By properly changing shape of PSS, optimized LEE has been achieved.

II. THEORETICAL MODELS

The APSYS software [17] self-consistently combines carrier transport and self-heating. Various mechanisms which are critical for LED analysis, such as self-consistent quantum well based spontaneous emission, non radiative recombination and injection current overflow have been included in the software. The transport model includes drift and diffusion of electrons and holes, Fermi statistics, built-in polarization and thermionic emission at heterointerfaces. An efficient RT module in APSYS proves to be especially useful for the study of light extraction. MEEP, a finite-difference time-domain (FDTD) simulation software package developed at MIT, was also included in APSYS using special data interface.

FDTD simulation calculates angular distribution of reflection and transmission coefficients with different incident angles to PSS interface. Fig. 1(a) shows the schematic of FDTD simulation. Light emitted by LED is treated as plane wave on PSS with an incident angle. Periodical boundary condition (P.B.C.) is considered.

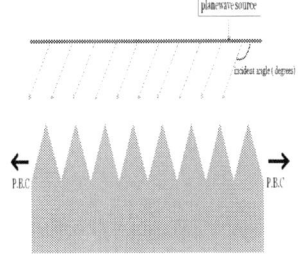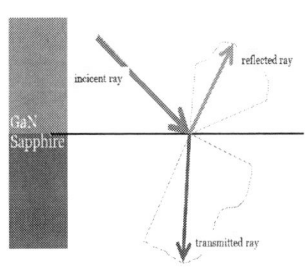

Then in RT simulation, when an incident ray hits PSS interface, *Fresnel* conditions will not be applied. Effective reflection and transmission coefficients are imported from FDTD results. In other words, PSS interface is considered as a

---

*corresponding author: shengyang@crosslight.com.cn

special interface with wave behavior. Fig. 1(b) presents the schematic of applying FDTD results on PSS in RT simulation. The reflected ray points in a direction with probability based on the angular profile of reflected radiation in the FDTD calculation. The refracted ray is handled in the same way.

III. LED DEVICE AND PSS STRUCTURE

The LED structure in this paper stands on a 200μm sapphire. The layer structure consists of 2μm un-doped GaN, 1μm n-doped (5×10^{18} cm^{-3}) GaN, four periods of un-doped In$_{0.11}$Ga$_{0.89}$N(4nm)/GaN(20nm) MQW active region, topped with 0.2μm p-type (8×10^{18} cm^{-3}) GaN cap layer. Fig. 3(a) shows the schematic structure of the LED with setting of contact positions.

Unit structure of PSS is formed by a half circle (sphere in 3D view) with radius of r and distance of s to next unit, which is shown in Fig. 3(b). The unit structure will be periodically extended according to the P.B.C. in the FDTD simulation.

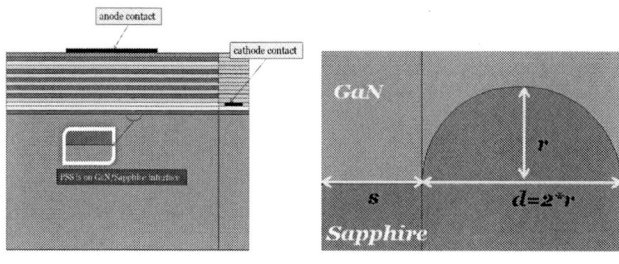

IV. RESULTS AND DISCUSSIONS

To investigate how PSS shape affects light extraction, we settled on four shapes of PSS unit with sizes ranging from much smaller to much larger than peak wavelength of the LED: case A, B, C and D with r =0.05μm, 0.2μm, 0.5μm and 1μm while s is fixed as 0.5μm, respectively. Fig. 3 illustrates a typical angular distribution of reflection and transmission coefficients for case C with incident angle of 15°.

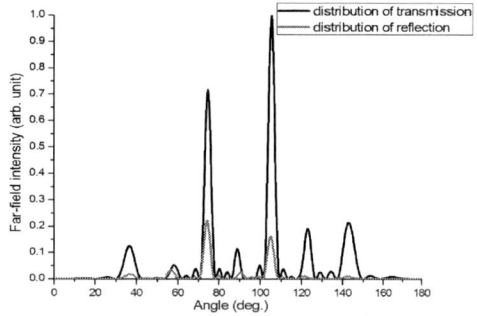

Integration of each distribution curve gives total reflected power (reflectivity R) and transmitted power (transmissivity T). Fig. 4 shows the changing trends of R and T for the four cases.

RT simulation is finally carried out for the LED device with the four cases of PSS. Light sources of RT simulation are automatically located by getting results from opto-electrical simulation of APSYS software. RT simulation for u-PSS LED is also performed. Fig. 5 shows the percent of transmitted light

power out of LED device. All PSS patterns used in this study

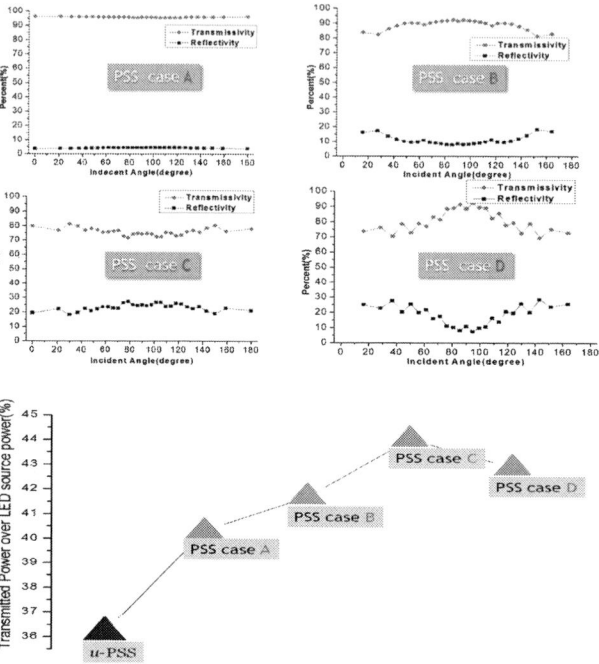

yield a higher LED light power compared with u-PSS LED. Moreover, for PSS cases, as r increases, transmitted light power increases firstly and then drops down after a peak at r=0.5μm. This indicates that micro-structure width (2r) has a strong effect on LED light extraction.

V. CONCLUSIONS

An LED with PSS has been accurately simulated by combining FDTD and ray tracing approaches and different PSS configurations has been compared. An important finding of this study is that for a specific PSS geometry, the peak of light extraction efficiency can be found at a PSS size. This work can be widely extended to investigations on any shape of PSS for the optimization of LED performance.

REFERENCES

[1] S . Nakamura, N. Senoh, et al., Jpn. J. Appl. Phys. **34**, L797(1995).
[2] L. Q. Zhang, S. M. Zhang, et al., Chin. Phys. B **18** 5350(2009).
[3] X. H. Huang, et al., IEEE Photonics Technology Lett. **23**, 14(2011).
[4] Y. K. Kuo, J. Y. Chang, et al., Optics Lett. **35**, 19(2010).
[5] F D. V. Morgan, I. M. Al-Ofi, et al., Semicond. Sci. Technol. **15**, 67(2000).
[6] G E. H. Park, J. Jiang, et al., Appl. Phys. Lett. **93**, 101112 (2008).
[7] H W. C. Peng and Y. S. Wu, Appl. Phys. Lett. **88**, 18117(2006).
[8] H. Huang, C. H. Lin, et al., Nanotechnology **19**, 185301(2008).
[9] T. Y. Tang, W. Y. Shiao, et al., J. Appl. Phys. **105**, 23501(2009).
[10] K. T. Lee, Y. C. Lee, et al., J. Electrochem. Soc. **155**, H638(2008).
[11] H. Y.Gao, F. W.Yan, et al., Solid State Electron. **52**, 962(2008).
[12] H. C. Lin, H. H. Liu, et al., J. Electrochem. Soc. **157**, H304(2010).
[13] H. C. Lin, R. S. Lin, et al., IEEE Photon. Technol. Lett. **20**, 1621(2008).
[14] J. C. Song, S. H. Lee, et al., J. Cryst. Growth **308**, 321(2007).
[15] H. Lin, S. Liu, et al., Acta Phys. Sin. (China) **58**, 959(2009).
[16] C. L. Xu, T. J. Yu, et al., Phys. Status Solidi C **9**, 757(2012).
[17] Crosslight Software Inc., Burnaby, Canada (http:// www. crosslight.com).

Two-dimensional Simulation of Mid-infrared Quantum Cascade Lasers: Temperature and Field Dependent Analysis

Ying-Ying Li[1], Z.-M. Simon Li[2], Guo-Ping Ru[1*]

[1]State Key Laboratory of ASIC and System, Department of Microelectronics, Fudan University, Shanghai 200433,China;
[2]Crosslight Software Inc., 121-3989 Henning Dr. Burnaby, BC, V5C 6P8, Canada
* Email: gpru@fudan.edu.cn

Abstract-We report on a 2D simulation study of a couple of mid-infrared quantum cascade lasers based on the integration of a number of optoelectronic models. Quantum mechanical computation was performed to find the quantization states and a rate equation approach was used to compute the optical gain. Temperature and field dependence effects are taken into account in optical gain model to make a realistic simulation of QCLs. The simulation study compared the integrated models with experimental data with different structures and at different temperatures. Reasonable agreements between experiment and simulation have been obtained.

I. INTRODUCTION

Quantum cascade lasers (QCLs) can be modeled in various ways, from rate equation model, Monte-Carlo simulations, non-equilibrium Green's function, to density matrix model. Most of these models focus on the carrier dynamics and optical gain in a short range, usually one and a half periods, without a comprehensive simulation of the global devices.We have reported a comprehensive simulation model of QCLs based on integration of a number of optoelectronic models on both microscopic and macroscopic scales[1].On the microscopic scale, quantum mechanical computation was performed to find the quantization states and a rate equation approach was used to compute the optical gain. On the macroscopic scale, we solved transport equations based on nonlocal transport model to account for long-range carrier transport in the device. This model has been implemented in commercial simulation software LASTIP and PICS3D[2].

In our previous articles[3,4], simulation studies of carrier transport properties of mid-infrared QCLs based on non-local transport model have been discussed. In this paper we will focus on a temperature and field dependent analysis of gain model based on simulation of a couple of mid-infrared QCLs with different structures at different temperatures.

II. OPTICAL GAIN MODEL

A critical quantity in QCL modeling is the optical gain which can be expressed as follows for sub-band transition from level j to level i [5]:

$$g = g_0 (n_j - n_i) \qquad (1)$$

Here g_0 is the gain coefficient:

$$g_0 = \frac{\pi E \, |M_{ij}|^2 \, f_b(E - E_{ij})}{h n_r c_0 \varepsilon_0 t_p} \qquad (2)$$

where E is the photon energy, n_r is the real part of the refractive index and t_p is the QCL period thickness. The dipole moment can be written as

$$|M_{ij}|^2 = q^2 |<i|z|j>|^2 \qquad (3)$$

and the normalized gain spectrum broadening function is given by

$$f_b(E - E_{ij}) = \frac{\tau_g}{h\sqrt{2\pi}} \exp\{-(1/2)[(E - E_{ij})\tau_g / h]^2\} \qquad (4)$$

where τ_g is the gain broadening lifetime.

The optical gain described above is sufficient for low temperature and unity injection from injector to active region. Usually there are four major factors which will reduce gain when temperature is high or resonant condition is not satisfied[6]. Firstly,the width of gain spectrum is broadened at higher temperatures. Secondly, electron scattering times for LO phonon emission are reduced by the Bose factor through the larger phonon population at higher temperatures. In addition, extrinsic electrons from the injector region can be thermally excited back into the bottom laser level, where they reduce population inversion. Finally, injection into upper laser level is most efficient at a resonant field and it will reduce when resonant between ground level of injector and upper laser level is not satisfied. To make a realistic simulation of QCLs,temperature and field dependent effects described above should be taken into account, which we will give a detailed discussion in the full paper.

III. RESULTS AND DISCUSSION

The structure studied here is a three-well active region QCL with 75 periods and lasing at 7.8 um based on $In_{0.53}Ga_{0.47}As/Al_{0.48}In_{0.52}As$ multiple quantum wells lattice matched to InP. Specific details and experimental results can be found in [7].

By solving the Schrodinger's and Poisson's equations self-consistently in the microscopic project, one can obtain sub-band diagram as well as the electron wave functions for a QCL. A two-period band diagram and the squared amplitude of the wave functions of the modeled device under resonant field

978-1-4673-1602-6/12 $31.00 © 2012 IEEE

of 45 kV/cm are shown in Fig. 1. The levels involved in lasing transition are plotted with red lines. With the sub-band electron distribution obtained from rate equation, all the possible inter-subband transition gains are calculated, as shown in Fig. 2. With the current increasing from 10 to 1000 mA, a strong gain peak at $\lambda = 8$ μm can be seen, corresponding to energy difference $E_{32} = 159$ meV as shown in Fig. 1.

Multiple lateral optical modes were computed by solving a scalar wave equation as an eigenvalue problem. 2D optical field distribution is shown in Fig.3. Finally, multiple lateral mode laser cavity photon rate equations were solved with the transport equations in a self-consistent manner to predict the lasing characteristics of a quantum cascade laser. Simulated I-L characteristics at different temperatures show reasonable agreement with experimentalresults, as illustrated in Fig.4.

IV. CONCLUSION

We have presented a 2D simulation of a conventional InGaAs/InAlAs QCL, starting from microscopic rate equation to 2D electrical and optical simulations. To verify the efficiency of this model, several other InGaAs/ InAlAs QCLs over different periods at different operating temperatures are also simulated, which will be shown in a full paper in the future.

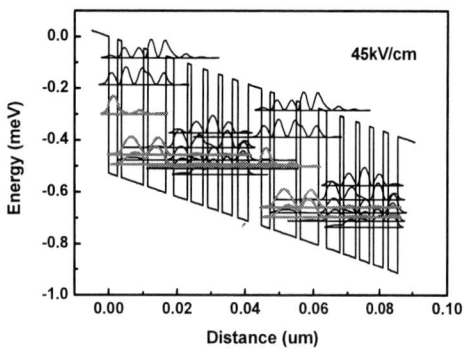

Fig.1 Conduction band diagram and the squared amplitude of the wave functions in two periods of active/injector region at 45 kV/cm; levels involved in lasing transition are plotted with red lines.

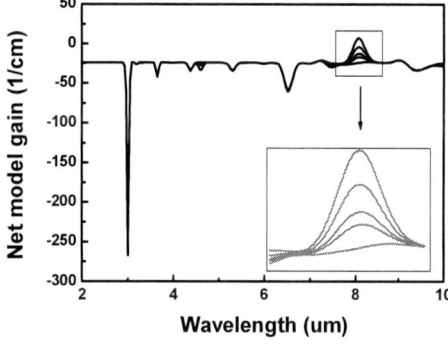

Fig.2 Net modal gain spectra at current ranging from 10 to 1000 mA with an increment of 200 mA. Inset: partial enlarged view of gain peak

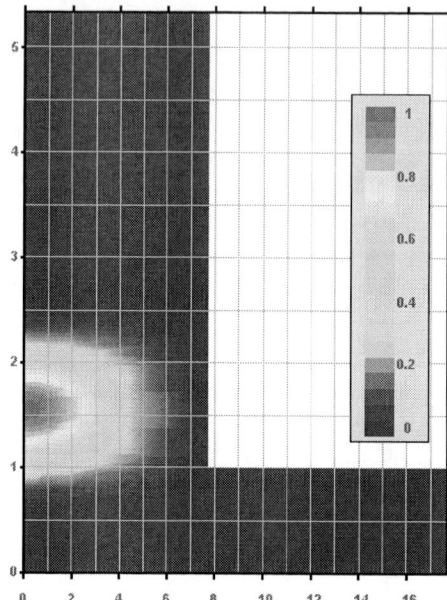

Fig.3 2D optical field distribution of a deep etched ridge waveguide QCL

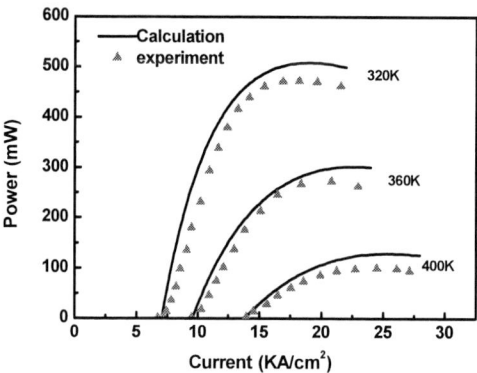

Fig.4 Simulated and experimental J-L characteristics of a 75-period QCLat 320, 360, and 400 K.

REFERENCES

[1] Z. M. Simon Li, Y. Y. Li, and G. P. Ru, "Simulation of quantum cascade lasers," J. Appl. Phys.,110: 093109 , 2011.
[2] See http://crosslight.com for information about LASTIP and PICS3D simulation software packages.
[3] Y. Y. Li, G. P. Ru, and Z. M. Simon Li, "Simulation of carrier transport in quantum cascade lasers," in IEEE 9th International Conference on ASIC, 866-869, Xiamen, China, Oct. 25-28, 2011.
[4] Y. Y. Li, G. P. Ru, and Z. M. simon Li, "Simulation of transport properties in mid-infrared quantum cascade lasers," J. Infrared Millim. Waves, 2012 (in press).
[5] J. Kim, M. Lerttamrab, S.L. Chuang, et al."Theoretical and experimental study of optical gain and linewidth enhancement factor of type-I quantumcascade lasers" IEEE J. Quantum Electron. 40, 12 ,2004.
[6] Gmachl, C., F. Capasso, et al. (2001). "Recent progress in quantum cascade lasers and applications." Reports on Progress in Physics 64(11): 1533-1601.
[7] C. Gmachl, A. Tredicucci, F. Capasso, et al. "High-power gimel approximate to 8um quantum cascade lasers with near optimum performance," Appl. Phys. Lett., 72(24): 3130-313, 1998.

All-Optical Gate Switches Employing the Quasi-Phase Matched Cascaded Second-Order Nonlinear Effect: Effect of Fabrication Errors

Ryohei Oshige, Yusuke Osawa, and Yutaka Fukuchi

Department of Electrical Engineering, Faculty of Engineering, Tokyo University of Science
1-3 Kagurazaka, Shinjuku-Ku, Tokyo 162-8601, Japan
Tel: +81-3-3260-4271, Fax: +81-3-5213-0976, E-mail: j4311612@ed.tus.ac.jp

Abstract-**We numerically calculate characteristics of all-optical gate switches using the cascaded second-order nonlinear effect in quasi-phase matched lithium niobate waveguides. Small amount of the domain length error causes significant decrease of the switching efficiency.**

I. INTRODUCTION

The cascaded $\chi^{(2)}$ effect in quasi-phase matched (QPM) lithium niobate (LN) waveguide devices can be applied to all-optical ultra-fast gate switches [1-12]. In these devices, the QPM wavelength can be arbitrarily controlled by the period of the $\chi^{(2)}$ grating. When the wavelength of the gate pulse is set to the QPM wavelength, the gate pulse can switch the signal pulse through the cascaded $\chi^{(2)}$ process. For example, in all-optical demultiplexing switches used in optical time-division multiplexed (OTDM) systems, the clock pulse restored from the received OTDM signal acts as the gate pulse for the switch [6,9].

In our previous papers, we have numerically shown the possibility of efficient ultra-fast operation of such switches [4,5,8,10]. However, in an actual case, the switching efficiency is much smaller than the value estimated from the numerical analyses [6,7,9]. This might be attributed to device fabrication errors of the QPM-LN waveguides [12].

In this paper, taking the fabrication errors into account, we numerically calculate the switching efficiency of the all-optical gate switches using the cascaded $\chi^{(2)}$ effect in QPM-LN waveguide devices. We find that the domain length error of the device decreases the switching efficiency significantly.

II. STRUCTURE AND PRINCIPLE

Figure 1 shows the structure of the all-optical gate switch using the cascade of second harmonic generation (SHG) and difference frequency mixing (DFM) in the QPM-LN waveguide device. The signal pulses are launched on the QPM-LN waveguide device together with the gate pulses. When the center wavelength of the input gate pulses is set to the QPM wavelength determined by the domain length d_{QPM}, the second harmonic (SH) of the input gate pulses is first generated. Hereafter, we refer to such frequency-doubled gate pulses as the SH gate pulses. Then, the DFM between the SH gate pulses and the signal pulses creates the switched pulses. Since the center

wavelength of the switched pulses is different from those of the signal and the fundamental and SH gates, the switched pulses can be filtered out by an optical bandpass filter (OBPF) with an appropriate bandwidth.

The maximum bit rate R_{max} processed by the QPM-SHG/DFM-LN waveguide switch is limited by crosstalk [4-12]. In the 1550-nm-band, the group-velocity mismatch (GVM) between the fundamental and SH gate pulses is as large as 350 ps/m. Due to such a large GVM, the walk-off delay is induced between these two pulses. The delayed SH gate pulse then overlaps with the bit of the signal succeeding the switched bit, generating crosstalk. For a given device length, the amount of crosstalk increases as the bit interval becomes shorter; hence, the crosstalk limits R_{max}. In our previous papers, we have both experimentally and numerically shown that the product of the device length and R_{max} is about 4 Gbps·m [4-12]. For example, R_{max} of the 2-cm-long device is 200 Gbps.

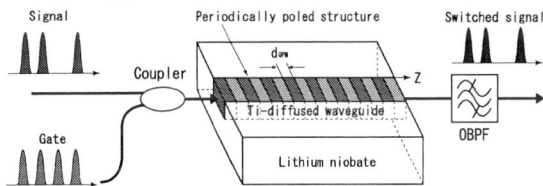

Fig. 1. Structure of the all-optical gate switch using the QPM-LN waveguide device.

III. TYPES OF DEVICE FABRICATION ERROR

In the QPM-SHG/DFM-LN waveguide device, we can consider three types of the fabrication error: random boundary position error, random domain length error, and random cross-section error.

As shown in figure 2 (a), the random boundary position error is the stochastic variation of boundary positions around ideal positions. This type of error can be found in the lithographic process with an accurate lithography mask, and the lengths of adjacent domains have a negative correlation.

The random domain length error shown in figure 2 (b) is the stochastic variation of the domain lengths. This type of error is attributed to the fabrication error of the lithography mask, and the lengths of adjacent domains are uncorrelated.

On the other hand, the random cross-section error shown in figure 2 (c) is the variation of the cross-section area of the waveguide along the device length. This type of error can be found in the fabrication process of the waveguide by titanium-diffusion. As shown in reference [12], this type of error is apparently equivalent to the random domain length error. Therefore, in the following analyses, we investigate effects of the random boundary position error and the random domain length error.

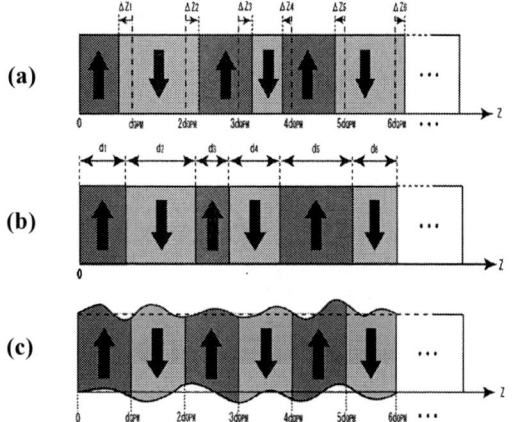

Fig. 2. Device fabrication errors. (a): Random boundary position error. (b): Random domain length error. (c): Random cross-section error.

IV. NUMERICAL RESULTS AND DISCUSSIONS

We consider a 2-cm-long QPM-LN waveguide device with an average domain length $<d_{QPM}>$ of 8.1 μm. This average domain length is required for SHG using d_{33} (= 25.9 pm/V) when the center wavelength of the input gate pulse is 1550 nm. The GVM between the fundamental and SH pulses is assumed to be 350 ps/m. The effective cross-section area of the waveguide is 12 μm². The center wavelength of the input signal pulse is set to 1520 nm. The input signal pulse completely overlaps with the input gate pulse. These input pulses are assumed to be Gaussian having the same pulse width parameter T_0 of 1 ps. The peak powers of the input gate and signal pulses are fixed at 1 W and 100 mW, respectively. In the analyses, we numerically calculate evolution of waveforms of the fundamental gate, SH gate, signal, and switched pulses along the device length by using the nonlinear coupled-mode equations given in reference [3].

First, we investigate effect of the random boundary position error shown in figure 2 (a). In the analyses, the boundary positions are assumed to have Gaussian distribution around ideal positions whose standard deviation is $<d_{QPM}>\times\Delta z$. Figure 3 (a) shows the switching efficiency η calculated as a function of Δz, where η is defined as the ratio of the peak power of the switched pulse to that of the input signal pulse. We find that η is almost independent of Δz. The reason is as follows: In the random boundary position error, the lengths of adjacent domains have a negative correlation as shown in figure 2 (a). In such a case, since the domain inversion period is preserved, the effect of this error is not accumulated.

Next, we investigate effect of the random domain length error shown in figure 2 (b). In the analyses, the domain lengths are assumed to have Gaussian distribution around ideal value $<d_{QPM}>$, whose standard deviation is $<d_{QPM}>\times\Delta d$. Figure 3 (b) shows η calculated as a function of Δd. We find that small amount of Δd causes significant decrease of η. For example, η at $\Delta d = 2$ % is about 3 dB smaller than that at $\Delta d = 0$ %. The reason is as follows: In the random domain length error, the lengths of adjacent domains are uncorrelated as shown in figure 2 (b). In such a case, since the domain inversion period can no longer be preserved, the effect of this error is accumulated.

Thus, the domain length error of the QPM-LN waveguide device must be reduced significantly for high efficient operation.

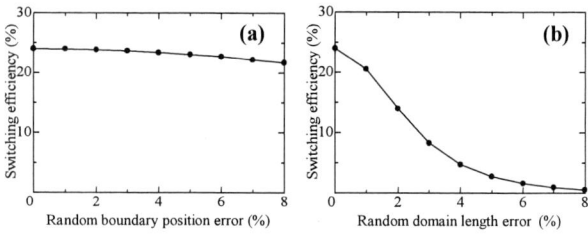

Fig. 3. (a): Switching efficiency η as a function of the random boundary position error Δz. (b): Switching efficiency η as a function of the random domain length error Δd.

V. CONCLUSIONS

We have numerically calculated the switching efficiency of the all-optical gate switch using the cascade of SHG and DFM in the QPM-LN waveguide device taking the fabrication errors into account. In the analyses, we have investigated the effects of the boundary position error and the domain length error of the device. We have found that the switching efficiency is almost independent of the boundary position error. On the other hand, we have found that small amount of the domain length error causes significant decrease of the switching efficiency.

REFERENCES

[1] S. Kawanishi, M. H. Chou, K. Fujiura, M. M. Fejer, and T. Morioka, *Electron. Lett.*, vol. 36, pp. 1568–1569, August 2000.

[2] S. Kawanishi, T. Yamamoto, M. Nakazawa, and M. M. Fejer, *Electron. Lett.*, vol. 37, pp. 842–844, June 2001.

[3] H. Ishizuki, T. Suhara, M. Fujimura, and H. Nishihara, *Opt. Quantum Electron.*, vol. 33, pp. 953–961, July 2001.

[4] Y. Fukuchi and K. Kikuchi, in *Proc. Tech. Dig. Conf. Lasers Electro-Optics (CLEO'2001)*, Baltimore, Maryland, USA, May 6–11, 2001, Paper CWB6, pp. 317–318.

[5] Y. Fukuchi and K. Kikuchi, in *Proc. Tech. Dig. Eur. Conf. Optical Communication (ECOC'2001)*, Amsterdam, The Netherlands, September 30–October 4, 2001, Paper We.P.9, pp. 396–397.

[6] K. Kikuchi, Y. Fukuchi, A. Suzuki, D. Kunimatsu, and H. Ito, in *Proc. Tech. Dig. Eur. Conf. Optical Communication (ECOC 2002)*, Copenhagen, Denmark, September 8–12, 2002, Paper 8.4.2.

[7] Y. Fukuchi, T. Sakamoto, K. Taira, K. Kikuchi, D. Kunimatsu, A. Suzuki, and H. Ito, *IEEE Photon. Technol. Lett.*, vol. 14, pp. 1267–1269, September 2002.

[8] Y. Fukuchi and K. Kikuchi, *IEEE Photon. Technol. Lett.*, vol. 14, pp. 1409–1411, October 2002.

[9] Y. Fukuchi, T. Sakamoto, K. Taira, and K. Kikuchi, *Electron. Lett.*, vol. 39, pp. 789–790, May 2003.

[10] Y. Fukuchi, M. Akaike, and J. Maeda, in *Proc. Tech. Dig. Nonlinear Optics Conf. (NLO'2004)*, Waikoloa, Hawaii, USA, August 2–6, 2004, Paper WD9.

[11] Y. Fukuchi, M. Akaike, and J. Maeda, in *Proc. Tech. Dig. IEEE Lasers and Electro-Optics Society Conf. (LEOS'2004)*, Rio Grande, Puerto Rico, November 7–11, 2004, Paper ThBB3.

[12] Y. Hata, Y. Fukuchi, and M. Akaike, *The 5-th Japan-Korea Joint Workshop on Microwave and Millimeter-wave Photonics*, Otsu, Shiga, Japan, January 29–30, 2004, Paper T4-12.

The Hybridization of Plasmons in GaN-based Two-Dimensional Channels

Lin Wang, Weida Hu[*], Xiaoshuang Chen[*], and Wei Lu

National Lab for Infrared Physics, Shanghai Institute of Technical Physics, Chinese Academy of Sciences, 500 Yu Tian Road, Shanghai, China 200083

Abstract

This paper displays the plasmon resonance phenomenon in single channel and double channel (DC) devices with varying dimensions in grating-gate period, slit and spacing between two channels in DC structures at terahertz domain. The results indicate that higher order plasmon can be excited in devices with longer period and narrow slit grating due to the enhanced coupling between plasmon and terahertz radiation. Splitting of plasmon resonance takes places in double channel device due to the hybridization between plasmons, which will improve the tunability of terahertz plasmonic device.

I. INTRODUCTION

Since the last decades, terahertz (THz) technology such as sensing of drugs and explosive materials has attracted great interest due to its inherent advantage in biomedical imaging and security imaging [1] [2]. To sense the THz radiation, detectors available now include bolometers, Schottky diodes [3], and photoconductive detectors. However, these detectors are not frequency-agile and require mechanical motion of external optics to generate spectral information [4, 5]. Recently, a new detection mechanism utilizing hydrodynamic nonlinearity of plasma wave (plasmon) in the channel of field effect transistors (FETs) has been proposed [6]. Plasma waves in FETs have a linear dispersion [7], $\omega = sk$, where s is the wave velocity. The plasma wave velocity $s = (e^2 n / m^* C)^{0.5}$ depends on the carrier density as controlled by the gate voltage and the gate-to-channel capacitance per unit area $C = \varepsilon / 4\pi d$. For the device with gate length L, the channel can serves as resonant "cavity", and the frequencies of plasma waves are discretized as $\omega_N = sk$ ($k = (2n-1)\pi/L$, and $n = 1, 2, 3, \ldots$). The plasmon is the spatio-temporal collective vibration of carrier densities under the excitation of external radiation, which will lead to the rectification of ac component gate voltage as induced. In the III-V compound semiconductor heterostructures with $L \sim 1\mu m$, the typical frequencies of plasma waves are located at THz band. Further, the plasmon is result of classical excitation and does not saturate with temperature, which will eliminate expensive cooling system [8]. Thus, a low-cost and frequency-tunable THz spectrometer/detection system will compact will be realized based on the state-of-art FETs.

Most of previous work focus on the the GaAs or InP material system with deep-submicron meter gate-length operating in the THz/sub-THz regime [9][10]. While Murovjov et al. illustrate that wider frequency tunability of plasmon resonance in grating-gate GaN HEMTs can be reached with finger-length around 1 μm [11], which is benefit for the designing of coupling elements with area-matched to the collimated incident wave. This paper aims to present the resonance properties of plasma wave in single channel and their interaction in double-channel GaN HEMTs, the appearence of new mode may be utilized for the resonant detection.

II. DEVICE DESCRIPTION AND DISCUSSION

Figs. 1 (a) and (b) show the structure of grating-gated single-channel (SC) and double–channel (DC) heterojunction FETs (HFET). The grating-gate can be used both as the electrodes controlling the sheet electron density and polarizer for the incident waves. Both of these devices consist of 2μm $Al_yGa_{1-y}N$ buffer layer (note that the mole fraction y should not exceed 0.15, which will cause the depletion of electron density in the channel) and 30nm channel layer or upper and lower channel

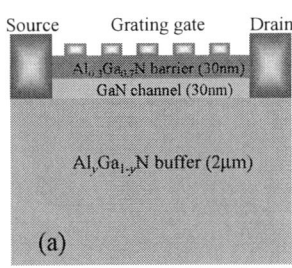 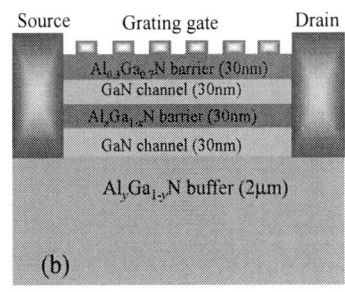

Fig.1. Schematic of device structures: (a) grating-coupled single-channel HEMT; (b) double-channel HEMT with $Al_xGa_{1-x}N$ inter-layer separating the upper and lower channels.

layers. The fabrication of these devices can be completed after the deposition and patterning of gate electrodes with period L and slit S. Room temperature Hall measurement indicate that the sheet electron densities in these devices can reached at about $2 \times 10^{13} cm^{-2}$ or even higher and mobility is around $1200 cm^2/Vs$~relaxation time is around 0.18ps. It has been known that in a single channel device contains two elementary plasmons with symmetrical (Ω_+) and asymmetrical charge (Ω_-) distribution across the channel can be excited by the THz radiation (and plasmon energy $\Omega_+ > \Omega_-$). While in double channel device the case is even more complicated due to the

---

[*] Corresponding author: wdhu@mail.sitp.ac.cn, xschen@mail.sitp.ac.cn.

interaction between these two kinds of plasmons in upper and lower channels, which will lead to the splitting of plasmon resonance further. The interactions between these plasmons can be separated into three categories according to the permutations and combinations: (a) symmetrical plasmon vs. symmetrical plasmon; (b) symmetrical plasmon vs. asymmetrical plasmon; (c) asymmetrical plasmon vs. asymmetrical plasmon. However, the splitting phenomenon does not always happen, which depends on the dipole distribution and resonant frequencies of these plasmons. As examples, the splitting of plasmons takes places when the frequencies of asymmetrical/symmetrical plasmons in the upper and lower channels approach with each other leading to the formation of (Ω_{+-} and Ω_{++})/(Ω_{-+} and Ω_{--}) new dispersion branches, however it does not happen when the symmetrical plasmon and asymmetrical plasmon are near resonant with each other. Further, in a single channel device, the change in strength and frequencies (the regime Ω_{-} and Ω_{+}) of plasmon resonance can be obtained through the change of grating period. Higher order plasmons (Ω_{-}) can be excited in the channel with longer period and narrow slit gratings. This is because the coupling strength between THz wave and plasmon is enhanced due to larger net dipole moment and stronger near-field in long period and narrow slit samples, respectively.

III. CONCLUSIONS

A finite-difference method is employed to describe the local response of plasmonic oscillation in single and double channel devices. Our results indicate that in long period/narrow slit samples, higher order plasmon resonances are being activated due to the improvement of field coupling. In addition, the interaction between plasmons in different channels leads to the splitting and enhancement of plasmon resonance, which is benefit for the wider tunability of THz device.

ACKNOWLEDGEMENTS

The authors acknowledge the support provided by the National Natural Science Foundation of China (61006090), and the Aviation Science Fund (20110190001).

REFERENCES

[1] N. Pala and M. S. Shur, "Plasma wave terahertz electronics", *Electronics Letters*, vol. 44, pp. 1391, Nov. 2008.

[2] T. A. Elkhatib, V. Y. Kachorovskii, W. J. Stillman, D. B. Veksler, K. N. Salama, X. C. Zhang, "Enhanced plasma wave detection of terahertz radiation using multiple high electron mobility transistors connected in series." *IEEE Trans. Micro. Theo. and Tech.*, vol. 58, pp. 331-337, Feb. 2010.

[3] L. Wang, X. S. Chen, W. D. Hu, J. Wang, J. Wang, X. D. Wang, and W. Lu, "The plasmonic resonant absorption in GaN double-channel high electron mobility transistors," *Appl. Phys. Lett.*, vol. 99, pp. 063502, Aug. 2011.

[4] L. Wang, W. D. Hu, J. Wang, J. Wang, X. D. Wang, S. W. Wang, X. S. Chen, and W. Lu, "Plasmon resonant excitation in grating-gated AlN barrier transistors at terahertz frequency," *Appl. Phys. Lett.*, vol. 100, pp. 123501, Mar. 2012.

[5] L. Wang, X. S. Chen, W. D. Hu, W. Lu, "Spectrum analysis of 2D plasmon in GaN based high electron mobility transistors," DOI: 10.1109/JSTQE.2012.2188381 , 2012.

[6] M. Dyakonov and M. S. Shur, "Plasma wave electronics: Novel terahertz devices using two dimensional electron fluid." *IEEE Trans. Electron Devices*, vol. 43, pp. 1640-1645, Oct. 1996.

[7] M. Dyakonov and M. S. Shur, "Shallow water analogy for ballistic field effect transistor: New mechanism of plasma wave generation by dc current." Phys.Rev. Lett., vol. 71, pp. 2465-2468, Oct. 1993.

[8] S. Kim, J. D. Zimmerman, P. Focardi, A. C. Gossard, D. H. Wu, and M. S. Sherwin, "Room temperature terahertz detection based on bulk plasmons in antenna-coupled GaAs field effect transistors." *Appl. Phys. Lett.*, vol. 92, pp. 253608, May. 2008.

[9] T. A. Elkhatib, V. Yu. Kachorovskii, W. J. Stillman, S. Rumyantsev, X. –C. Zhang, and M. S. Shur, "Terahertz response of field-effect transistors in saturation regime." Appl. Phys. Lett., vol. 98, pp. 243505, June. 2011.

[10] F. Teppe, M. Orlov, A. EI Fatimy, W. Knap, J. Torres, V. Gavrilenko, A. Shchepetov, Y. Roelens, and S. Bollaert, "Room temperature tunable detection of subterahertz radiation by plasma waves in nanometer InGaAs transistors." *Appl. Phys. Lett.*, vol. 89, pp. 222109, Nov. 2006.

[11] A. V. Muravjov, D. B. Veksler, V. V. Popov, O. V. Polischuk, N. Pala, X. Hu, R. Gaska, H. Saxena, R. E. Peale, and M. S. Shur, "Temperature dependence of plasmonic terahertz absorption in grating-gate gallium-nitride transistor structures." *Appl. Phys. Lett.*, vol. 96, pp. 042105, Jan. 2010.

Light Extraction Enhancement Analysis of GaN-Based LED with Surface Spherical Crown Array

Xiaomin Wang,[1,2] Kang Li,[1,*] Fanmin Kong,[1] Zhenming Zhang,[1]

[1]Shandong University, Jinan, Shandong 250100,China
[2] Biomedical Engineering Department, Shandong Provincial Hospital affiliated to Shandong University,
Jinan, Shandong 250014,China, antwxm@163.com
kangli@sdu.edu.cn

Abstract-For promoting the light extraction efficiency (LEE) of GaN-LED with nano-spherical hexagonal arrays, finite-difference time-domain (FDTD) method was used for optimizing the structure parameters such as spherical radius and height. The LEE of GaN spherical crown hexagonal array with 473nm radius and 250nm height over the LED surface exhibited 5.7 times enhancement than that of the planar LED, better than the LEE of whole-sphere array and pure-hemisphere array both.

I. INTRODUCTION

With the development of recent III-V semiconductor industry, great energy-saving impulse is now pushing GaN-based light-emitting diode (LED) research ahead. With the well known attributes of long life, high reliability and energy saving, the GaN-based LED plays an important role in daily life such as medical application, traffic light, back light, advertise outdoor and so on.

However, in a real LED, the vertical light extraction efficiency (LEE) of the LED, due to total internal reflection (TIR) at the interface of the semiconductor and the outer medium (air in common), not all the power emitted from the active region is emitted into free space. Due to Fresnel reflection at the boundary, even with 100% internal quantum efficiency, LEE is limited to about 4%.

During the past 20 years, many approaches have been applied for improving the extraction efficiency of LEDs, such as surface roughening[1], patterned surface[2], patterned sapphire[3], photonic crystal[4], nano-rode[5], surface plasmon[6], and so on. The patterned sapphire and photonic crystals process use either e-beam lithography or holography lithography. These methods need high cost and can hardly be invested for the large-scale production. There are reports about the fabrication method of self-assembled 2-D SiO2/polytstyrene (PS) microlens arrays with various PS thickness on top-emitting InGaN quantum well LEDs, the hemisphere of GaN and SiO2 array also brought the LEE well increase [7, 8]. The approaches based on inexpensive methods are highly desirable for commercialization.

Although there has been some experiments on the LED with surface nano-sphere structures, research for systematic theoretical optimization is rarely reported in the literatures. Meanwhile, the Monte Carlo ray tracing methods are usually

applied for surface microlens structure, but compared with the wavelength of light, the size of microlens is sometimes too small to be simulated. For the sake of optimizing the parameters of LED with surface microlens structure and providing a theoretical reference, the comprehensive research is desired. In this report, the FDTD method has been applied for the SiO2 and GaN spherical crown array with different radius, followed by the comparison of LEE of LED with height varied spherical crown above LED surface. With the geometrical parameters optimized, more than 5 times LEE enhancement has been achieved than that of the ordinary LED.

II. NUMERICAL METHODS IN SIMULATION

In this study, as in Fig.1, a three dimension model and vertical cross-section of the simulated structure are schematically depicted. The LED layers are grown on a sapphire substrate. The refractive index of GaN in the visible spectrum is 2.5,while the refractive index of sapphire substrate and SiO2 are 1.7 and 1.46.

A 4μm×4μm×3μm three-dimension field distribution was calculated using the FDTD method, also a space and time discretion of Maxwell curl equation. A single dipole source was chosen in the simulation. The source in this study is short Gaussian pause. A perfectly matched layer(PML) enclosing the computing domain was used to absorb outgoing waves and avoid non-physical reflections.

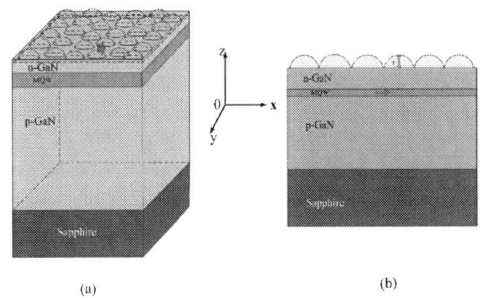

Fig.1. Schematically diagrammed model of LED simulation. Structure used to calculate the extraction efficiency and far field radiation pattern of LED with (a)3D and (b)2D cross section of the spherical crown array.

The extraction efficiency was calculated from the power flux extracted from the structure with respect to the overall emitted power from the source:

This work was supported by the National Natural Science Foundation of China (61077043), the National Basic Program of China (973 program) through Grand No. 2009CB930503, 2009CB930501, and 2007CB613203, the Reward Fund of Outstanding Youth and Middle Age Scientist of Shandong Province under Contract No. BS2009NJ002.

$$\eta_{extr} = \frac{p^+_{z,out}}{p^+_{x,in} + p^-_{x,in} + p^+_{y,in} + p^-_{y,in} + p^+_{z,in} + p^-_{z,in}} \qquad (1)$$

the $p^+_{z,out}$ means the power flow integrated over a plane just above the LED structure and $p^+_{x,in}$, $p^-_{x,in}$ are the integrated power flux through the plane normal to x, as shown in Fig.1. The + and - indicate power flow parallel or anti-parallel to the axis. The vertical LEE enhancement factor F is designed as follow:

$$F = \frac{\eta_{extr}}{\eta_0} \qquad (2)$$

The η_{extr} is the vertical LEE of nano-sphere LED while η_0 is the vertical LEE of the ordinary LED.

III. SIMULATION RESULT

In simulation, the position of dipole was fixed at 200nm below the top GaN surface. As is depicted in Fig.2, with the variation of the angle θ, both of the radiuses of sphere (R) and bottom circuit (r) change accordingly. The calculated LEE enhancement of LED models with GaN/SiO$_2$ spherical crown hexagonal array on the top surface is shown in Fig.2. From the results in Fig.2, It is showed that the LEE of LED with GaN and SiO$_2$ spherical crown hexagonal array both rise with θ increasing. The key point is the line for LED surface structure made of GaN rises more obviously than that of SiO$_2$. The LEE reached climax when $\theta=65°$, with the corresponding spherical radius of 473nm.

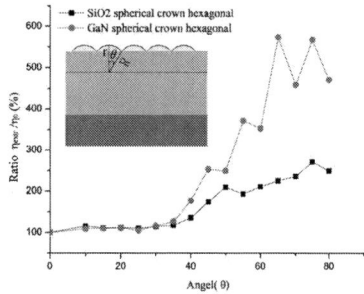

Fig.2. The LEE enhancement factor changed as the function of the incident angel θ. Fixed the dipole at the spherical center, the black curve is corresponding to the SiO$_2$ spherical crown, while the red one is corresponding to the GaN. The inset indicates the schematic diagram of simulation structure of spherical crown arrays.

For the sake of further optimization, the radius of sphere was set to 473nm and change the vertical position of the sphere so that the heights of spherical crowns vary as was shown in Fig.3.

The LEE enhancement factors of LED model with spherical crown of various heights were also shown in Fig.3. From the calculated results, it could be noted that the line for LEE of the LED with SiO$_2$ hemisphere structure does not change very extensively with respect to the height. However, the LEE of the LED with GaN spherical structure changed remarkably. The LEE enhancement factor reached 570% when the height is about 250nm, which is a very promising value of LEE enhancement.

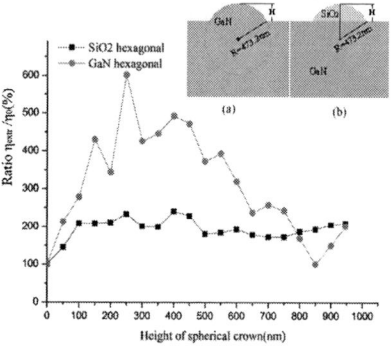

Fig.3. The LEE changed as function of the height of spherical crown above the LED surface. Fixed with 473nm radius, the black one is corresponding to the SiO$_2$, while the red one is corresponding to the GaN spherical crown. The insets above indicate the single unit among the simulate structure (left one is GaN spherical crown, right one is SiO$_2$).

IV. CONCLUSION

Comprehensive research to optimize the LEE of GaN-based LED with SiO$_2$ and GaN spherical crown arrays in different parameters has been carried out. The results showed that GaN is more suitable than SiO$_2$ to compose the surface hemisphere microlens structures. When the radius of sphere is 473nm, the height above the top surface of LED is 250nm, the LEE could obtain 5.7 times enhancement than LED model without surface microlens structures. The optimized results provide important reference for further experiments and commercial application.

REFERENCES:

[1] T. Fujii, Y. Gao, R. Sharma, E. L. Hu, S. P. DenBaars, and S. Nakamura, "Increase in the extraction efficiency of GaN-based light-emitting diodes via surface roughening," *Applied physics letters*, vol. 84, p. 855, 2004.

[2] Y. H. Cheng, J. L. Wu, C. H. Cheng, K. C. Syao, and M. C. M. Lee, "Enhanced light outcoupling in a thin film by texturing meshed surfaces," *Applied Physics Letters*, vol. 90, p. 091102, 2007.

[3] J. H. Lee, D. Y. Lee, B. W. Oh, and J. H. Lee, "Comparison of InGaN-based LEDs grown on conventional sapphire and cone-shape-patterned sapphire substrate," *Electron Devices, IEEE Transactions on*, vol. 57, pp. 157-163, 2010.

[4] Y. C. Shin, D. H. Kim, D. J. Chae, J. W. Yang, J. I. Shim, J. M. Park, K. M. Ho, K. Constant, H. Y. Ryu, and T. G. Kim, "Effects of Nanometer-Scale Photonic Crystal Structures on the Light Extraction From GaN Light-Emitting Diodes," *Quantum Electronics, IEEE Journal of*, vol. 46, pp. 1375-1380, 2010.

[5] H. M. Kim, Y. H. Cho, H. Lee, S. I. Kim, S. R. Ryu, D. Y. Kim, T. W. Kang, and K. S. Chung, "High-brightness light emitting diodes using dislocation-free indium gallium nitride/gallium nitride multiquantum-well nanorod arrays," *Nano letters*, vol. 4, pp. 1059-1062, 2004.

[6] K. C. Shen, C. Y. Chen, H. L. Chen, C. F. Huang, Y. W. Kiang, C. C. Yang, and Y. J. Yang, "Enhanced and partially polarized output of a light-emitting diode with its InGaN/GaN quantum well coupled with surface plasmons on a metal grating," *Applied Physics Letters*, vol. 93, p. 231111, 2008.

[7] X. H. Li, R. Song, Y. K. Ee, P. Kumnorkaew, J. F. Gilchrist, and N. Tansu, "Light extraction efficiency and radiation patterns of III-nitride light-emitting diodes with colloidal microlens arrays with various aspect ratios," *IEEE Photon. J*, vol. 3, pp. 489-499, 2011.

[8] Y. K. Ee, P. Kumnorkaew, R. A. Arif, H. Tong, H. Zhao, J. F. Gilchrist, and N. Tansu, "Optimization of light extraction efficiency of III-Nitride LEDs with self-assembled colloidal-based microlenses," *Selected Topics in Quantum Electronics, IEEE Journal of*, vol. 15, pp. 1218-1225, 2009.

Electro-optical characteristics for AlGaN solar-blind *p-i-n* photodiode: Experiment and simulation

X. D. Wang[a], W. D. Hu[a,*], X. S. Chen[a,*], J. T. Xu[b], L. Wang[b], X. Y. Li[b], and W. Lu[a,*]

[a] National Lab for Infrared Physics, Shanghai Institute of Technical Physics, Chinese Academy of Sciences, 500 Yu Tian Road, Shanghai, China 200083

[b] State Key Laboratories of Transducer Technology, Shanghai Institute of Technical Physics, Chinese Academy of Sciences, Shanghai, China 200083

Abstract

The fabrication and modeling for solar-blind AlGaN-based *p-i-n* photodiode have been presented. The simulated dark current characteristics are in good agreement with the experiments. It is found that the peak responsivity of 0.005A/W can be achieved at 265nm corresponding to the cutoff wavelength of the $Al_{0.45}Ga_{0.55}N$ absorption layer. The transmission spectra drop to nearly zero due to the intense light absorption of n-type $Al_{0.65}Ga_{0.35}N$ layer.

I. INTRODUCTION

The atmospheric ozone layer filters out the ultraviolet (UV) solar radiation with wavelengths shorter than 290nm, preventing it from reaching the Earth's surface. UV photodetectors with cutoff wavelengths below 290nm, the so-called solar-blind detectors (SBDs), are thus capable of detecting feeble UV signals under solar background radiation with high signal to background ratio. Owing to this advantage, SBDs are in high demand for a number of applications such as flame detection, ozone layer monitoring, UV astronomy, water purification, submarine communication, and medical researches [1-5]. $Al_xGa_{1-x}N$-based photodetectors potentially offer significant advantages over the current photomultiplier tube and silicon-based solar-blind detector technology in terms of size, complexity, cost, robustness, stability, power demands, and bandwidth [6]. Moreover, its intrinsic solar blindness (for $x>0.4$) and the ability of operation under harsh conditions (high-temperature and high power levels) resulting from its wide band gap makes $Al_xGa_{1-x}N$-based photodetectors attractive for high-performance solar-blind detection applications.

A major obstacle in developing high-performance AlGaN SBDs is the poor AlGaN crystalline quality owing to the lack of lattice and thermal match substrates. The high Al-composition $Al_xGa_{1-x}N$ epilayers grown on popular substrates such as sapphire are usually with high densities of threading dislocations (TDs) or, even more serious, macroscopic cracks, caused by the large lattice and thermal mismatches between the substrate and the subsequent epilayers. It has been well pointed out that high-density TDs are the primary reason for the leakage current and reduced spectral rejection ratio in AlGaN based photodiodes [7, 8]. Accordingly, the substrate becomes a key factor concerning the epilayer quality and its influence on detector performance.

In order to reduce dislocation densities and improve AlGaN crystalline quality, we present, in this paper, the growth, processing, and modeling of $Al_{0.45}Ga_{0.55}N$ solar-blind *p-i-n* photodiodes on AlN/sapphire template. The results show that high optical responsivity and low dark current have been achieved as a result of the usage of AlN/sapphire template.

II. SIMULATION MODELS AND DEVICE STRUCTURE

The steady-state two-dimensional numerical calculations were performed using Sentaurus Device, a commercial package by Synopsys [9]. For plain drift-diffusion simulation the well known Poisson equation and continuity equations are used. The carrier generation-recombination process consists of Shockley-Read-Hall, Radiative, Auger, and optical generation-recombination terms. Additionally, the trap-assisted tunneling is included in the continuity equations. Moreover, we assume a same single acceptor type electron bulk trap level in all the epilayers. The values of the trap levels are extracted from the experimental data for the $Al_xGa_{1-x}N$. The trap density of AlGaN is $N_{AlGaN}=5\times10^{16}cm^{-3}$ with a capture cross section of $\sigma_{AlGaN}=1.0\times10^{-15}cm^{-2}$, locating approximately 2.2eV below the conduction band [10].

Samples were grown by metal-organic chemical-vapor deposition (MOCVD) on transparent AlN templates on double side polished *c*-plane sapphire substrates. First, a 0.15-μm-thick undoped $Al_{0.65}Ga_{0.35}N$ layer was grown on the top of the AlN/sapphire template to improve the material quality for the subsequent device layers by reducing the defect density. After this, there were the layers of *p-i-n* structure, which are 0.5-μm-thick Si doped *n*-type $Al_{0.65}Ga_{0.35}N$ layer, and 0.15-μm-thick unintentionally doped $Al_{0.45}Ga_{0.55}N$ absorption layer, and 0.15-μm-thick *p*-type $Al_{0.45}Ga_{0.55}N$ layer. To reduce the metal *p*-contact resistance and facilitate carrier collection, a 35-nm-thick *p*-type GaN layer were grown on top of the *p-i-n* structure. Device fabrication was completed by a series of processing, including photolithography, inductively coupled plasma (ICP) dry etching, metal evaporation, and SiO$_2$ passivation. The detailed structural information of AlGaN solar-blind *p-i-n* photodiode is shown in the inset of Fig. 1.

III. RESULT AND DISCUSSION

---

* Corresponding author: wdhu@mail.sitp.ac.cn, xschen@mail.sitp.ac.cn, luwei@mail.sitp.ac.cn.

Current-voltage (*I-V*) characteristics were measured with a keithley 4200-SCS semiconductor characterization system. The simulated and measured dark current characteristics are both shown in Fig. 1. The simulated *I-V* curve is in good agreement with the experiment, confirming the validity of $Al_{0.45}Ga_{0.55}N$ solar-blind *p-i-n* numerical model. The forward current tends to saturate when the bias exceeds 1V due to the large series resistance effect. The reverse current almost exponentially increases with bias because of the large dislocation-induced trap-assisted tunneling effect.

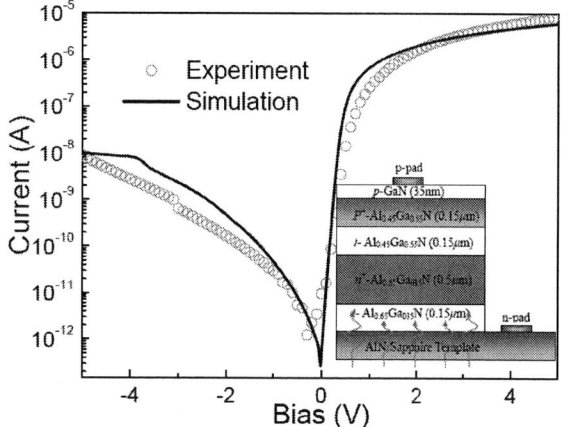

Figure 1. Dark current characteristics of AlGaN solar-blind *p-i-n* photodiode. The inset shows the device structure, the doping concentration of p^+, i, and n^+ layers are $2\times10^{18}cm^{-3}$, $3\times10^{16}cm^{-3}$, and $3\times10^{18}cm^{-3}$, respectively.

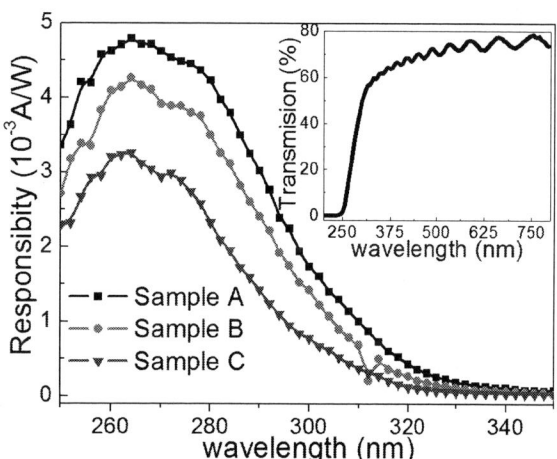

Figure 2. Spectral responsivity of three similar samples. The inset shows the spectral transmission of the wafer used in the fabrication of photodiode.

The spectral responsivity of the devices was measured using a light source consisting of a deuterium lamp and a xenon lamp, a monochromator, a chopper, and UV-grade focusing optics in a standard synchronous detection scheme. A calibrated, UV-enhanced Si detector was used to measure the illumination power density of the light source lamps over the measuring range of 200 to 500 nm. Since AlGaN solar-blind *p-i-n* photodiode often operates at small reverse bias, measuring voltage for spectral responsivity was set to -0.05V. Figure 2 presents the measured spectral responsivity of

three similar samples, the fabrication processing of three samples were completely same. Therefore, notable responsivity discrepancies are attributed to different epilayer qualities of three samples. In that sense, the sample A has the best crystalline quality, and sample C worst. However, all the spectral responsivity curves exhibit peak responsivities at 265nm corresponding to the cutoff wavelength of the $Al_{0.45}Ga_{0.55}N$ absorption layer. The inset of Fig. 2 is the transmission spectra of the wafer used in the fabrication. When the wavelength is shorter than 250nm, the transmission drops to nearly zero. This is because the *n*-type $Al_{0.65}Ga_{0.35}N$ layer contributes to the light absorption. Besides, clear interface interference can be observed as the wavelength is longer than 320nm.

IV. CONCLUSION

The dark current characteristics for $Al_{0.45}Ga_{0.55}N$ solar-blind *p-i-n* photodiodes have been reported. The simulated *I-V* curve is in good agreement with the experiment. It is found that the peak responsivity of 0.005A/W can be achieved at 265nm corresponding to the cutoff wavelength of the $Al_{0.45}Ga_{0.55}N$ absorption layer.

ACKNOWLEDGEMENTS

This work was supported by the National High Technology Research and Development Program of China (2011AA050508), Aviation Science Fund (20110190001), and Yao foundation.

REFERENCES

[1] M. Razeghi and A. Rogalski, "Semiconductor ultraviolet detectors," J. Appl. Phys., Vol. 79, pp. 7433, 1996.

[2] E. Ozbay, N. Biyikli, I. Kimukin, T. Tut, T. Kartaloglu, and O. Aytur, "High-performance solar-blind photodetectors based on $Al_xGa_{1-x}N$ heterostructures," IEEE J. Quantum Electron., Vol. 10, pp. 742, 2004.

[3] S. J. Pearton, J. C. Zolper, R. J. Shul, and F. Ren, "GaN: processing, defects, and devices," J. Appl. Phys., Vol. 86, pp. 1, 1999.

[4] J. C. Campbell, S. Demiguel, F. Ma, A. Beck, X. Guo, S. Wang, X. Zheng, X. Li, J. D. Beck, M. A. Kinch, A. Huntington, L. A. Coldren, J. Decobert, and N. Tscherptner, "Superconductivity and macroscopic quantum phenomena," IEEE J. Quantum Electron., Vol. 10, pp. 777, 2004.

[5] M. Akazawa and H. Hasegawa, "Sensing dynamics and mechanism of a Pd/AlGaN/GaN Schottky diode type hydrogen sensor," Phys. Status Solidi C, Vol. 4, pp. 2629, 2007.

[6] J. C. Carrano, T. Li, P. A. Grudowski, R. D. Dupuis, and J. C. Campbell, "Improved detection of the invisible," IEEE Circuits Devices Mag., Vol. 15, pp. 15, 1999.

[7] X. D. Wang, W. D. Hu, X. S. Chen, J. T. Xu, X. Y. Li, and W. Lu, "Photoresponse study of visible blind GaN/AlGaN *p-i-n* ultraviolet photodetector," Opt. Quant. Electron., Vol. 42, pp. 755, 2011.

[8] G. Parish, S. Keller, P. Kozodoy, J. P. Ibbetson, H. Marchand, P. T. Fini, S. B. Fleischer, S. P. DenBaars, and U. K. Mishra, "High-performance (Al, Ga)N-based solar-blind ultraviolet *p-i-n* detectors on laterally epitaxially overgrown GaN," Appl. Phys. Lett., Vol. 75, pp. 247, 1999.

[9] Synopsys: Synopsys Sentaurus Device user manual, USA 2008.

[10] X. D. Wang, W. D. Hu, X. S. Chen, J. T. Xu, L. Wang, X. Y. Li, and W. Lu, "Dependence of dark current and photoresponse characteristics on polarization charge density for GaN-based avalanche photodiodes," J. Phys. D: Appl. Phys., Vol. 44, pp. 405102, 2011.

Numerical simulation of high-efficiency InGaP/GaAs/InGaAs triple-junction solar cells grown on GaAs substrate

J. Liang[a], W. D. Hu[a,*], X. S. Chen[a,*], C. S. Xia[b], and L. W. Cheng[b]

[a]National Lab for Infrared Physics, Shanghai Institute of Technical Physics, Chinese Academy of Sciences, 500 Yu Tian Road, Shanghai, 200083, China
[b]Crosslight Software China, Suite 906, Building Jie Di, 2790 Zhongshan Bei Road, Shanghai 200063, China

Abstract

The structure parameters and illumination condition for InGaP/GaAs/InGaAs triple-junction solar cells grown on GaAs substrate have been numerically studied to search the optimal point of device performance. The dependences of conversion efficiency, I-V curves, and band diagram on the structure parameters and illumination condition have been investigated. Our work shows that the performances are largely dependent on the geometric design of device and the optimal parameters are extracted

I. INTRODUCTION

Multi-junction solar cell based on group III-V semiconductor material system use multiple subcell band gaps to divide the broad solar spectrum into smaller sections, each of which can be converted to electricity more efficiently. And it is a promising photovoltaic device to utilize solar energy not only because of its high conversion efficiency, but also it can be applied under cheaper concentrated sunlight optical system to achieve higher efficiency and reduce the cost per unit area at the same time[1].

In this paper, the InGaP/GaAs/InGaAs triple-junction solar cells on GaAs substrate are numerically investigated. The electrics characteristics such as I-V curves, efficiency and physical quantities like band diagram are analyzed under different illumination condition.

II. SIMULATION MODELS

The basic Poisson equation and drift-diffusion current continuity equations for electrons and holes are included in the simulator to calculate the electrical characteristics of the device. The ray-tracing model and optical generation-recombination term in the continuity equations are used for optical simulation[2]. Meanwhile, the flexible carrier mobility model and advanced Zener type tunneling model are introduced to simulate the behavior of the device for more accurate results compared to reality.

The structure discussed in this study is composed of three junctions, as shown in Figure 1. Each junction is consist of an n-type emitter layer and a p-type base (absorber) layer between the passivating window layer and back surface field layer (BSF, and the bottom junction has no BSF). Two tunnel junctions, which are all high doped and only 20 nm thick, are connecting

the junctions for better current transferred. The transparent InGaP grading layer is used for lattice match.

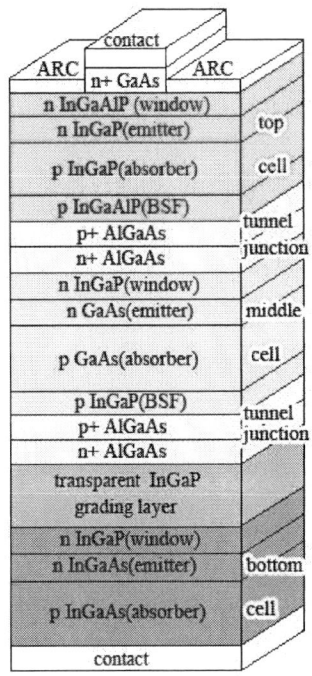

Figure 1. Schematic of InGaP/GaAs/InGaAs triple-junction solar cells

The thickness and doping of each layer is referred to Ref[3]. During the simulation, some crucial parameters, such as the thickness and doping of absorber layer are varied with illumination condition. Then the photoresponse curve and some other properties of the solar cells are calculated to study the performance of the device and find the optimal parameters for operating.

III. RESULT AND DISCUSSION

Figure 2 shows that the conversion efficiency changes with the thickness of p-type InGaAs(bottom cell) absorber and different illumination. The efficiency rapidly grows as the thickness of p-type InGaAs absorber increases at first. Then the efficiency stays at almost the same level regardless of the

---

*Corresponding author: wdhu@mail.sitp.ac.cn, xschen@mail.sitp.ac.cn,

change of thickness. The result of p-type GaAs absorber (not listed here) shows a similar trend. Meanwhile, before the concentration level reaches about 200 suns, more light power is provided to the device, more conversion efficiency is acquired. The reason of this phenomenon is that when the absorber layer is thick enough to absorb the incident light, the efficiency can reach its maximum values.

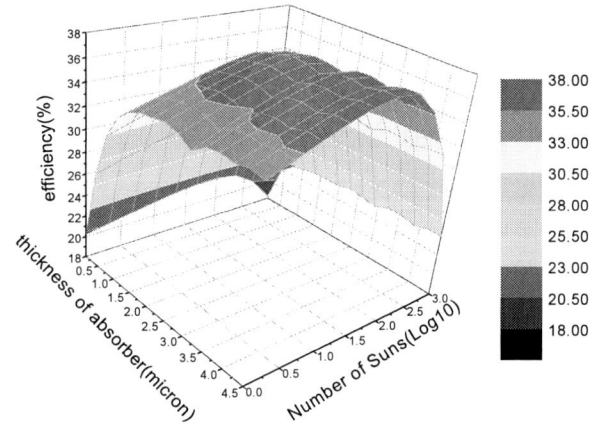

Figure 2 Conversion efficiency with the p-InGaAs absorber thickness changing from 0.1 to 4.5μm and concentrated condition from 1 sun to 1000suns (AM1.5G).

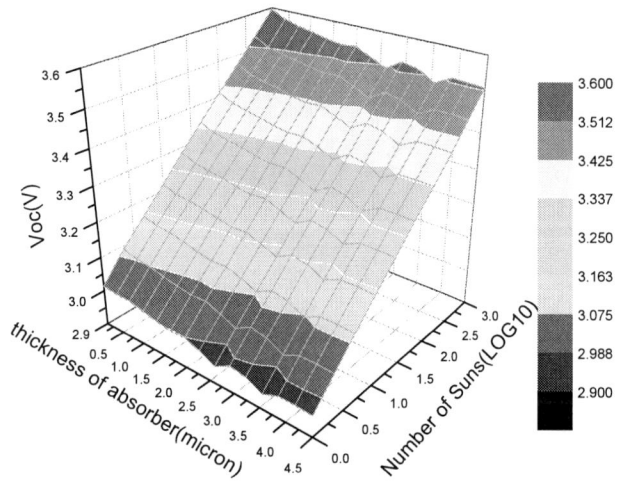

Figure 3 Open-circuit voltages with the p-GaAs absorber thickness changing from 0.1 to 4.5μm and concentrated condition from 1 sun to 1000 suns (AM1.5G).

Figure 3 shows that the open-circuit voltage goes down slightly with the augment of the thickness of middle cell p-type GaAs absorber and increases logarithmically with concentration. Similar results can be obtained with the different thicknesses of bottom cell p-type InGaAs or top cell p-type InGaP.

The whole open-circuit voltage is the sum of the individual values of each junction, which is the interval of quasi-Fermi levels for electron and hole respectively[4]. So the thickness of absorber layer has little effects on the electrical value reflects that the band diagram is not observably changed with the geometric structure. As the concentrated sunlight enhances, the generation rate in the junction increases and the interval of quasi-Fermi levels for electron and hole is widened. As a result, the open-circuit voltage grows logarithmically.

IV. CONCLUSION

The photoresponse and performance on the geometric design of device and illumination condition have been numerically simulated with the APSYS simulator from Crosslight Software Inc. The simulation results show that the thickness of absorber layer of each junction and concentrated sunlight have an important effect on both photoresponse and performance of the solar cells. In order to achieve the highest conversion efficiency, a illumination condition of 200~500 suns is optimal since higher concentration does not bring better performance. The p-type absorber layer is the most crucial part of each junction of the solar cell. Considering the process and cost of fabrications, the thicknesses of 2.7, 2.2, and 0.9μm for bottom, middle, and top junctions is thick enough to utilize the solar energy and no more materials are necessary.

ACKNOWLEDGEMENTS

This work was supported by the National High Technology Research and Development Program of China (2011AA050508), and Aviation Science Fund (Grant No. 20110190001).

REFERENCES

[1] R. King et al., "New horizons in III-V cell research", *21th European Photovoltaic Solar Energy Conference and Exhibition, Dresden, Germany, 2006.*

[2] Y. G. Xiao, Z. Q. Li, Z. M. Simon. Li "Modeling of GaInP/GaAs/Ge and the inverted-grown metamorphic GaInP/GaAs/GaInAs triple-junction solar cells" *High and Low Concentration for Solar Electric Applications III,* vol.7043, pp70430B, 2008

[3] J. F. Geisz, Sarah Kurtz, M. W. Wanlass, J. S. Ward, A. Duda, D. J. Friedman, J. M. Olson, W. E. McMahon, T. E. Moriarty, and J. T. Kiehl "High-efficiency GaInP/GaAs/InGaAs triple-junction solar cells grown inverted with a metamorphic bottom junction," *Appl. Phys. Lett* 91, 023502, 2007

[4] R. King et al, "Raising the efficiency ceiling with multi-junction III-V concentrator photovoltaics" *23rd European Photovoltaic Solar Energy Conference and Exhibition, Valencia, Spain, 2008,*

Structure Design of Refractive Index Sensor Based on LPFG with Double-layer Coatings

Zhengtian Gu, Tao Luo
Laboratory of Photo-electric Functional Films
College of Science, Uni. of Shanghai for Sci. and Tech.
Shanghai, China
E-mail: zhengtiangu@163.com

Kan Gao
Laboratory of Optical Fiber Sensors
No.23 Research Inst. of China Electronics Tech. Group,
Shanghai, China
E-mail: gaokan@siom.ac.cn

Abstract—**Based on the coupled-mode theory, the mode field distribution and mode transition of LP modes in an LPFG with double-layer coatings are studied theoretically. As a refractive index sensor, this LPFG structure is designed for high sensitivity by selecting a suitable thickness of the first coating. Date simulation indicates that the effective index of the LP cladding mode increases in steps with the first coating thickness, thus LP mode is guided to the overlay and the mode transition takes place, where the LP modes can be easily affected by the ambient. Thereby, the coated LPFG sensor should be designed to operate at the transition region. Further, for two types of LPFG with double-layer coatings, the transmittance spectrum, sensitivity and measurable dynamic range is analyzed. The sensitivity of the LPFG sensor is available to 10^3 with suitable overlay thickness.**

Keywords-Long-period fiber grating; LP mode; mode transiyion; effective index; sensitivity.

I. INTRODUCTION

The coupling of the long-period fiber gratings (LPFGs) happens between the forward-propagating core mode and the cladding modes. The effective index of the cladding modes of LPFGs is very sensitive to change of ambient refractive index (RI), so LPFGs have been wildly applied in fields of chemical and biological sensors. Ree et al. investigated firstly the behavior of an LPFG with Langmuir--Blodgett thin-film overlays[1]. It was found that the response of the LPFG with high refractive index overlays was sensitive to refractive indices higher than that of the cladding. Villar et al. studied the cladding mode transition and transmission spectrum of a coated LPFG based on the coupled-mode theory.[2] It is implied that a suitable overlay thickness can make some of the cladding modes transit into the overlay, and interact with the measurand. In view of the fact that the overlay index may be lower than that of the cladding, a novel LPFG with double-layer coatings is presented, which can be also acted as a refractive index sensor.

In this paper, a novel LPFG sensor is studied based on the coupled-mode theory described by Anemogiannis et al.[3] The transition of LP cladding modes is presented and the transmittance of this coated LPFG is discussed. Data simulation shows that the resonant wavelength shifts due to the minor variation of the overlay refractive index as the LPFG is located at the mode transition region. By selecting a suitable thickness of the first overlay, the high sensitivity of

this LPFG sensor can be obtained. Thereby, the sensing property and structure design for two types of LPFG with double-layer coatings are discussed, which will be guidelines for the practical design of LPFG sensors.

II. ANANLYSIS OF LP MODE FIELD IN COATED LPFG

A. Coupled-mode theory for LP modes

LP_{0j} modes are calculated based on the coupled-mode theory and transfer-matrix formulation. The transverse electric-field component propagating along the z-axis is given by

$$U_{0j,i}(r,\varphi,z) = \exp(-j\beta_{0j}z)A_{0j,i}J_0(r\gamma_{0j,i}) + B_{0j,i}Y_0(r\gamma_{0j,i}), \quad \beta_{0j}^2 > n_2^2 k_0^2,$$

$$U_{0j,i}(r,\varphi,z) = \exp(-j\beta_{0j}z)A_{0j,i}I_0(r\gamma_{0j,i}) + B_{0j,i}K_0(r\gamma_{0j,i}), \quad \beta_{0j}^2 < n_2^2 k_0^2,$$

$$(1)$$

where i=1, 2, 3 and 4 stands for the core, cladding and double-layer coatings, respectively; β_{0j} is the propagation constant of the LP_{0j} mode, $\gamma_{0j,i} = \sqrt{k_0^2 k_{n_i}^2 - \beta_{0j}^2}$ is the magnitude of the transverse wave number. We have the transfer matrix equation

$$m_{22}^{1,N+1}(\beta_{0j}) = 0. \quad (2)$$

After resolving Eq.(2), the effective indexes n_{eff} (or $\beta_{0j} = n_{eff}k_0$) of the LP_{0j} modes can be calculated.

Supposing that the input field amplitude $A_{01}(0)=1$, based on the coupled-mode theory, we can obtain the transmittance T_{LPFG} of the gratings with length L,

$$T_{LPFG}(L) = 10 \times \log_{10}[\text{Re}(A_{01}(L))^2 + \text{Im}(A_{01}(L))^2], \quad (3)$$

where $A_{01}(L)$ is the field amplitude at the end of the LPFG.

B. Mode transition regions of coated LPFG

Fig. 1 shows the effective refractive index of LP cladding modes versus the thickness of the first overlay thickness. As the thickness of the overlay increases to 0.5 μm, the LP_{02} mode is guided to the overlay. The effective refractive index is higher than 1.4447 as shown in Fig. 1. It means the cladding LP_{03} mode will become LP_{02}, and LP_{04} will become LP_{03}, and so forth. The phenomenon repeatedly emerges as the thickness increases to 2.0 μm, more modes are guided to the overlay and new re-organizations of the cladding modes takes place. The modes transit into the overlay, and they are more easily affected by the ambient. Thus a high sensitivity LPFG sensor can be designed by optimizing the overlay thickness.

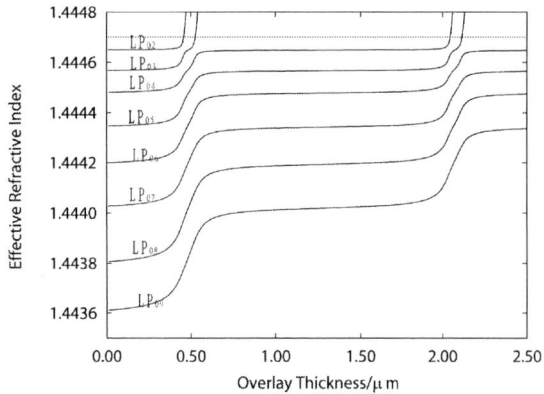

Fig.1 Relation of the effective refractive index with the overlay thickness

III. DESIGN STRUCTURE OF COATED LPFG SENSOR

A. First coating with higher refractive index

While the sensitive film index is lower than the cladding index, a coating with higher refractive index should be firstly coated on the cladding. For an LPFG sensor with double-layer coatings, the sensitive film index will be affected by the surrounding measurand. From Fig. 1, the thickness of the first coating must be located at the mode transition region. Fig. 2 shows the transmittance change of LPFG with the sensitive film index, in which the first coating thickness is selected to 0.5 μm according to Fig. 1. It is clear that the resonant wavelength shift obviously as a sensitive film index changes.

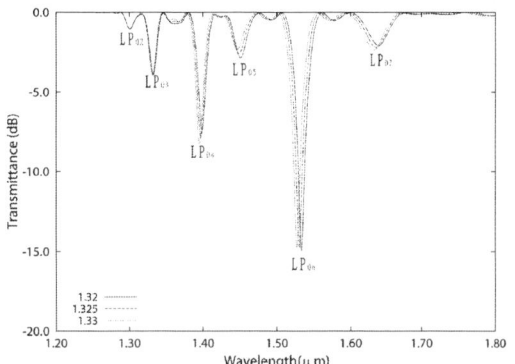

Fig.2 Transmittance of LPFG under different refractive index of the overlay

B. First coating with lower refractive index

For an LPFG coated with lower index overlay than the cladding, it can be also applied for a refractive index sensor by coating a sensitive film with higher index. Date simulation indicates the mode transits region shifts from lower index to higher index as the first coating thickness increases. Fig. 3 shows the dependence of transmittance on wavelength for

sensitive film refractive index. Fig. 4 shows the sensitivity and measurable dynamic range for sensitive film refractive index. For two observed wavelength 1610.0nm and 1633.5nm, the sensitivity are 4.6×10^3 and 1.5×10^3, while the measurable dynamic range of overlay index are 2.0×10^{-4} and 9.3×10^{-4}.

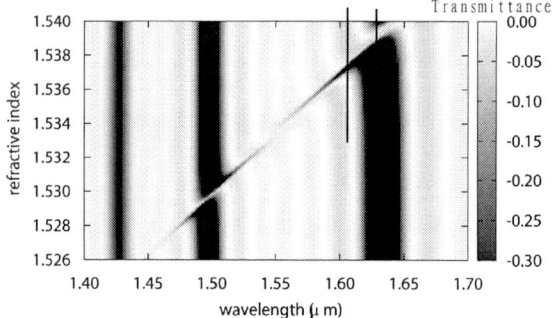

Fig.3 Dependence of *transmittance* on wavelength for overlay refractive index

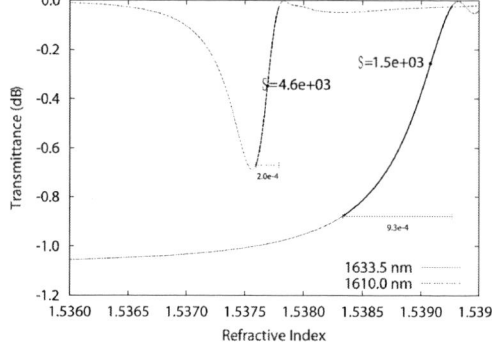

Fig.4 Sensitivity and measurable dynamic range for *overlay refractive index*

IV. CONCLUSION

An LPFG with double-layer coatings is presented and its sensing property and structure design is studied based on the coupled-mode theory. As the first coating thickness increases, the mode transition happens periodically. The transmittance of the LPFG located at the mode transition region shows that the resonant wavelength is very sensitive to the change of the refractive index of outside overlay. By selecting a suitable thickness of the first coating, the sensitivity of LPFG sensor is available to 10^3. This novel LPFG sensor with double-layer coatings overcomes the limitation of traditional coated LPFG in which the refractive index of sensitive coating must be higher than that of the cladding, and enlarges its scope of application, which allows it to be used far more widely.

REFERENCES

[1] Rees N D , James S W , Tatam R P , Ashwell G J, "Optical fiber long-period gratings with Langmuir–Blodgett thin-film overlays", Optics Letters, Vol. 27, pp. 686-688, 2002.

[2] Villar I Del Matias I R, Arregui F J and Achaerandio M, Nanodeposition of materials with complex refractive index in long-period fiber gratings", J. Lightwave Technology, Vol. 23, pp. 4192-4199, 2005.

[3] Anemogiannis E, Glytsis E N, Gaylord T K, "Transmission characteristics of long-period fiber gratings having arbitrary azimuthal/radial refractive index variations", J. Lightwave Technology , Vol. 21, pp. 218-227, 2003.

Experimental determination of minority carrier lifetime and recombination mechanisms in MCT photovoltaic detectors

Haoyang Cui, Naiyun Tang, Zhong Tang

School of computer and information engineering,
Shanghai University of Electric Power, 2103 Pingliang Road, Shanghai 200090, China, cuihy@shiep.edu.cn

Abstract—This paper presents an experimental study of minority carrier lifetime and recombination mechanisms in HgCdTe photodiode. The excitation light source is a wavelength-tunable pulsed infrared laser. A constant background illumination has been introduced to minimize the effect of the junction equivalent capacitor and the equivalent series resistance. The slow decay of the photo-generated voltage is recorded by a storage oscilloscope. By fitting the exponentially decay curve, the time constant has been obtained which is regarded as the photo-generated minority carrier lifetime of the HgCdTe photodiode. The experimental results show that the carrier lifetime is in the range of 18 ~ 407 ns at 77 K for the measured detectors of four compositions. It was found that the Auger recombination process is more effective for low Cd composition while the radiative recombination process became more important for high compensated materials. The Shockley-Read-Hall (SRH) recombination processes could not be ignored for all Cd composition.

Keywords- HgCdTe, minority carrier lifetime, open circuit photovoltage, Cd composition

I. INTRODUCTION

The minority carrier recombination lifetime has been known to be a based parameter of the semiconductor devices. Even though HgCdTe is used extensively for photovoltaic detectors, however, there is still a great deal of ambiguity issues, such as the minority carrier lifetime and the governing recombination mechanisms [1]. This is because of the instability of HgCdTe, which the material property may be changed during the process of the formation of *pn* junction. Therefore, the parameter of the raw material can not be applied to estimate the properties of *pn* junction devices. Moreover, there are great differences between the actual parameters and the design parameters such as trap concentration, carrier concentration, the junction depth, the junction width in conventional techniques. These factors have a lot of uncertainties effects on the *pn* junction. In order to determine the minority carrier lifetime, the measurements must be carried out on the actual devices then the extracted parameters are applied in devices design and simulate. Many measurements have been developed to determine the minority carrier lifetime such as: short-circuit current, open-circuit voltage decay (OCVD), pulse recovery technique etc [2,3]. However, the minority-carrier lifetimes of HgCdTe material are in nanosecond range, these methods not suit to measure such short lifetime.

The purpose of this paper is to measure the minority carrier lifetime using an improved photo-induced OCVD measurement technique which compensates the effects of the junction equivalent capacitor and the trap center on the measurements. The experiments results show that the carrier lifetime is in the range of 18 ~ 407 ns at liquid nitrogen temperature for the measured detectors of four compositions. From the experiment results, the governing recombination mechanisms are obtained by analyzing the free carrier recombination theory.

II. EXPERIMENTAL SETUP

The incident pulse laser having wavelength tuning range 2.3 ~ 10 μm was provided by an commercial optical parametric oscillator (OPG) and difference frequency generator (DFG) which were pumped by a picosecond Nd:YAG laser. The Laser delivered pulse of 30 ps in duration at a frequency 10 Hz. We could approximately these pulses as δ function illumination on the detector at $t = 0$, therefore, the influence of the falling time was avoided. In order to minimize the effect of the junction equivalent capacitor and the equivalent series resistance on the carrier lifetime measurements, we inducted an Oriel QTH lamp as the steady-state bias light source. A saturated steady-state of the *pn* junction output electric signal would be reached by turning the intensity of the bias light. Then the pulsed laser was illuminated on the sample and by recording the OCVD and fitting to the exponential decay curve, we could determine the minority carrier lifetime.

All HgCdTe samples were grown by MBE on GaAs substrates with CdTe buffer layers and an abrupt n^+-p structure were formed by the ion implantation of B^+ in p-type HgCdTe. As ZnS films were formed on the HgCdTe surface for passivation, the measured lifetime values were not influenced by the surface treatment. The structure of the HgCdTe photovoltaic detectors is shown in the inset of figure 1.The composition of $Hg_{1-x}Cd_xTe$ in our experiments are $x = 0.231$ ($\lambda_{Eg} \sim 8.6$ μm), $x = 0.305$ ($\lambda_{Eg} \sim 4.6$ μm), $x = 0.343$ ($\lambda_{Eg} \sim 3.7$ μm) and $x = 0.418$ ($\lambda_{Eg} \sim 2.9$ μm). The detectors were processed into 50×50 μm^2 area mesa structures. The photo-generated voltage on a *pn* junction has been recorded by a storage oscilloscope.

III. RESULTS AND DISCUSSION

A. minority carrier lifetime

Depending on the intensity of the excitation source, three different regions of the photovoltaic decay curve can be distinguished as high level injection, intermediate injection and low-injection [4]. The device actual is in low-injection condition. In this condition, where the excess minority carrier concentration is less than the equilibrium minority carrier concentration, the photovoltaic decay curve approaches exponential time dependence:

$$V_{\text{oc}} = \frac{kT}{q}\left[\exp\left(qV(0)/kT\right) - 1\right]\exp(-t/\tau) \tag{1}$$

which $V(0)$ is the open circuit voltage at the termination of excitation. Best fitting to the decay of the photovoltaic curve was realized with a second-order exponential decay function. The characteristic decay times were $\tau_1 \sim 2$ μs and $\tau_2 \sim 35$ μs. However, from the previous analysis, these long decay time constants were not the lifetimes of minority carriers but the presence of junction equivalent capacitor and trap energy level effects on excess carrier's relaxation.

The steady-state output photo-voltage of the *pn* junction would increase with the bias incident intensity increasing. The decay time constants of photovoltaic response induced by the pulsed laser were becoming shorter and the transient peak amplitude were decreasing with the incident intensity increasing. These phenomenons could be attributed to the junction equivalent capacitor and the trap center energy level effects on the decay curves became smaller when the bias light illuminated on the devices. The junction equivalent capacitor, equivalent series resistance and the trap center would be compensated or even cancelled under strong bias light condition. In this case, the photo-excited carriers annihilated by the recombination in base region and this photovoltaic decay time constant was related to the minority carrier lifetime. Since the values of the resistance and the carrier lifetime are much larger in the *p* region than in the n^+ region, and the photo-generated carriers in the emitter are about one percent of the carriers generated in the base, therefore, we can assume that the carriers stored in the base play a dominant role in the OCVD process [5].

The photo-excited OCVD decay curve was fitted with the expression in Eq (1), and the lifetime magnitude of the minority carrier in *p* region was determined to 190 ns. Using the method mentioned above, the minority carrier lifetime of the HgCdTe photodiode with different composition could be obtained. The results of the same composition came from different units of one array. There are distinctions between the different units of one array because the HgCdTe raw material is non-uniformity or the growing process can not be mastered. The excess carrier lifetimes extracted from our experiments are reasonable because of the lifetime magnitude consist with others results [2]. Generally speaking, there is a certain difficulty to precise measuring minority carrier lifetime. Even for the silicon material, the minority carrier's lifetime of accuracy scope is ±135% in different laboratory in American Society for Testing and Materials (ASTM) [6]. Therefore, it is acceptable that there is some divaricating in the lifetime experiment measuring of HgCdTe photodiode.

B. determination of recombination mechanism

The minority carrier lifetime τ obtained by the experiments are the present of radiative, Auger and SRH recombination mechanisms, therefore, the dominant recombination can be determined by analyzing the relationship of these mechanisms [7] which is given by:

$$\frac{1}{\tau} = \frac{1}{\tau_A} + \frac{1}{\tau_{Rad}} + \frac{1}{\tau_{SRH}} \qquad (2)$$

The radiative recombination lifetime τ_{Rad} and Auger recombination lifetime τ_A in HgCdTe material with different doping concentration can be obtained by theory calculation. There are four doping concentration HgCdTe material, only three of which are shown in Fig 1.

From Fig 1, one can descry that there are great divagates between the lifetimes obtained from the experiment and the theory calculation results which only considering the radiative and Auger recombination processes. This difference is the evidence that the SRH mechanism can not be neglected. For example, the radiative and Auger lifetime of the HgCdTe for $x = 0.231$ are 1200 and 170 ns (shown in Fig.1a) respectively while the measurement lifetime is about 70 ns. Using Eq (2),

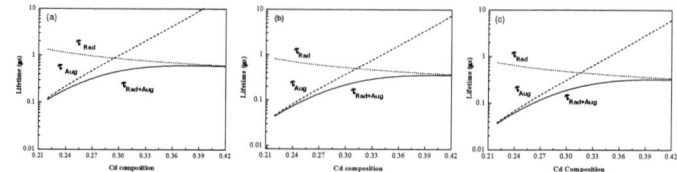

Fig 1 The calculated minority carrier lifetime as a function of composition for *p*-type $Hg_{1-x}Cd_xTe$ with the different doping concentration. (a) $4.96 \times 10^{15} cm^{-3}$ (b) $8.08 \times 10^{15} cm^{-3}$ (c) $8.87 \times 10^{15} cm^{-3}$. The lifetimes were assumed to be determined by Auger and radiative recombination mechanisms. Also shown are the calculated Auger and radiative lifetimes for these carrier concentrations.

the SRH lifetime is 130 ns. In this case, the effect of SRH and Auger process are the dominant mechanisms on carrier lifetime while the radiative process can be neglected. The lifetimes 510, 420 and 180 ns are radiative, Auger and SRH recombination respectively for $x = 0.305$; therefore, the only dominant recombination mechanism is SRH process. These results are consisted with the conclusion derived by Schacham [1] that the Auger process is more effective for low values while the SRH recombination are dominated for high compensated materials; The lifetimes 340, 6000 and 360 ns are radiative, Auger and SRH recombination respectively for $x = 0.418$, in this case, the radiative mechanism play more important role even can exceed the SRH process in the recombination process. For all samples, the SRH recombination has a full impact on minority carrier lifetime; therefore, it is an important recombination mechanism in minority carrier recombination process. In figure 4, our results also show that the carrier lifetime is in the range of 18 ~ 407 ns at liquid nitrogen temperature for the measured detectors of four compositions. With the composition increasing, the minority carrier lifetimes have an increasing tendency, and the lifetime of the short wavelength infrared detectors are the longest comparing to other detectors.

ACKNOWLEDGEMENT

Project supported by the National Natural Science Foundation of China (61107081), Innovation Program of Shanghai Municipal Education Commission of China (10YZ158), Shanghai Natural Science Foundation of China (10ZR1412300).

REFERENCES

[1] Schacham S E, Finkma E. Recombination mechanisms in *p*-type HgCdTe Freezeout and background flux effects. J Appl. Phys. 1985, 57(6):2001-9.

[2] Cui H Y, Li Z F, Quan Z J, Hu X N, Ye Z H, Lu W. Measurement of minority carrier lifetime in HgCdTe *p-n* junctions. Laser and Infrared. 2006, 36(11):1063-6,

[3] Khanna V K. Physical understanding and technological control of carrier lifetime in semiconductor materials and devices: A critique of conceptual development, state of the art and applications. Pro.gress in Quantum Electronics, 2005, 29, 59-163,

[4] Mahan J E, Ekstedt T W, Frank R I, Kaplow R. Measurement of minority carrier lifetime in solar cells from photo-induced open-circuit voltage decay. IEEE Transaction on Electron Devices, 1979, ED-26(5):733-9,

[5] Jain S C. Theory of photo induced open circuit voltage decay in a solar cell. Solid State Electronics 1981, 24(2):179-83,

[6] ASTM f28-91. Standard test method for minority-carrier lifetime in bulk germanium and silicon measurement of photoconductivity decay [S] 1997

[7] Lopes V C, Syllaios A J, Chen M C. Minority carrier lifetime in MCT Semicond Sci Technol 1993, 8:824-41

Polarization-Independent Self-Collimated Beam Splitting in Two-Dimensional Photonic Crystals

M. B. Yucel[1], O. A. Kaya[2], A. Cicek[3] and B. Ulug[1,*]

[1]Department of Physics, Faculty of Science, Akdeniz University, Campus 07058, Antalya/Turkey
[2]Department of Computer Education and Educational Technologies, Faculty of Education,
Inonu University 44280 Malatya/Turkey
[3]Department of Physics, Faculty of Arts and Sciences, Mehmet Akif Ersoy University, 15100 Burdur/Turkey

*bulug@akdeniz.edu.tr

Abstract- Polarization-independent splitting of self-collimated transverse-electric and transverse-magnetic polarized waves in a two-dimensional square photonic crystal are demonstrated. The beam splitting is facilitated by the existence of sharp edges in flat equifrequency contours.

I. INTRODUCTION

Photonic crystals (PC) are dielectric materials, whose refractive index is periodically modulated in space. They have been attracted a great deal of attention since they are able to manipulate light at the wavelength scale.

The beam splitter is one of the most crucial photonic components in which beam splitting in and outside the PC structure could be achieved by several methods, such as using line-defect PC waveguides [1], coupled-cavity PC waveguides [2] and directional coupling [3]. Beam splitters based on self-collimation effect, originating from complex dispersion of light in the PC have also been proposed [4]. The idea of using a directional band gap to split self collimated beams with large angular separation was suggested by Matthews et al [5] and experimentally demonstrated on the transverse-electric (TE) polarized beam [6].

In this work, polarization independent wide angular splitting of a self collimated beam inside a square PC is demonstrated and the influence of source width is discussed.

II. RESULTS and DISCUSSION

Two-dimensional (2D) PC is composed of alumina rods of radius r=1.55 mm and relative dielectric constant of 10.0 in a square lattice in air with a lattice constant of 3.5 mm. The Plane Wave Expansion (PWE) method, as implemented in the BandSOLVE software by RSoft Design Group is utilized to obtain the band structure (BS) and the equifrequency contours (EFC) for the infinite PC structure.

To show the self-collimation effect and the wave splitting, the EFCs of the transverse-magnetic (TM, electric field parallel to rods) and transverse-electric (magnetic field parallel to rods) modes are calculated for the 2nd and 3rd bands of the PC, as shown in Fig.1(a) and (b), respectively. The two bands overlap significantly for the TM case, whereas overlapping is negligible for the TE polarization. Besides, the slices (EFCs) at

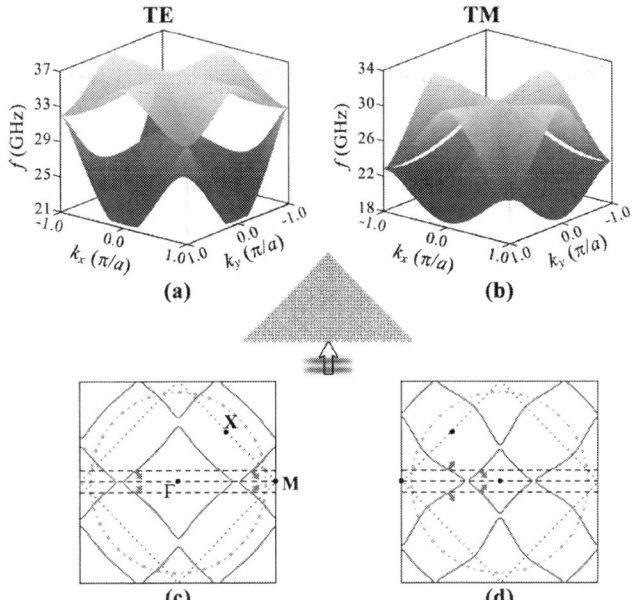

Fig.1-Band plots for the second and third TE **(a)** and TM **(b)** bands, accompanied by the corresponding EFCs at 27.2 GHz **(c)** and **(d)**, respectively. The inset depicts the beam splitter on which the waves are incident along the ΓM direction. The red-dotted squares and the green-dash-dotted circles in (c) and (d) represent the first BZ and the EFC at 27.2 GHz in air, respectively. The black-dashed horizontal lines and arrows in (c) and (d) represent the construction lines corresponding to source width in reciprocal space and the propagation directions of the beams splitting inside the PC, respectively.

27.2 GHz reveal that the contours are almost flat normal to the ΓX direction and own sharp corners along the ΓM direction. Thus, a wave of either the TE or the TM polarization incident at small angles along the ΓM direction is self-collimated within the PC with the refraction angle close to ±45°. However, a real source possesses finite transverse extent corresponding to a spread along the normal to the propagation direction in the reciprocal space, as suggested by the black dashed construction lines in Fig. 1(c) and (d). In this case, even if the wave is incident normally, a range of spatial modes are excited within the PC, most of which are self-collimated along the same direction, as demonstrated by the arrows in Fig. 1(c) and (d). Hence, a normally-incident beam of finite transverse extent is expected to split across the air-PC interface along the ΓM direction around 27.2 GHz, irrespective of its polarization, although the PC geometry is not disturbed by line defects, etc.

The EFCs for 27.2 GHz in Fig. 1(c) and (d) have two components centered on the Γ and M points. Thus, each construction line in Fig. 1(c) and (d) intersects the EFCs at two distinct points, resulting in further splitting of each beam to obtain more complicated devices, such as 1x4 splitters. However, the directions of velocity vectors in each splitting beam for both polarization in Fig. 1(c) and Fig. 1(d) are so close that differentiation of the further splitting of the beams becomes difficult, as in Fig. 2.

The propagation of the TE and TM-polarized waves in the 2D PC structure is simulated through the 2D Finite-Difference Time-Domain (FDTD) method, where the computational domain is surrounded by a perfectly matched layer absorbing boundary, as implemented in the FullWAVE software. The structure, on which the beams are incident along the ΓM direction, is visualized in the inset of Fig. 1. The output face of the PC is truncated normal to the ΓX direction so that a right isosceles triangular geometry is obtained and the waves inside the PC arrive at the PC-air interface normally, to minimize the reflection losses. The PC is excited by a plane wave at 27.2 GHz, whose transverse profile is Gaussian and the transverse width is varied between $2a$ and $10a$. The FDTD results are presented in Fig. 2.

As the source width increases, part of the EFCs remaining between the construction lines in Fig.1(c) and (d) becomes curved. Direction of the velocity vectors therefore deviates more and eventually, the beam width increases. Although this effect is evident in Fig. 2 for both polarizations it is much clear for the TE polarization in Fig. 2(a).

Polarization independent self collimated beam splitting could be achieved more efficiently in annular photonic crystals since they offer more flexibility to match the EFCs for the TE and TM polarizations in PC [7].

III. CONCLUSION

A 2D PC with high filling ratio of dielectric scatters in air is demonstrated to facilitate 50-50% splitting of incident beams with finite width at a specific frequency range upon refraction across the air-PC interface, irrespective of the source polarization.

TE and TM-polarized waves are split at close angles, while the transmitted beam width within the PC is higher in TE case. The splitting ability of TE waves is deteriorated as the source width is increased, whereas it is preserved for TM waves.

ACKNOWLEDGMENT

This study is supported by Akdeniz University Scientific Research Projects Unit.

REFERENCES

[1] P. Pottier, S. Mastroiacovo, and R. M. De La Rue, "Power and polarization beam-splitters, mirrors, and integrated interferometers based on air-hole photonic crystals and lateral large index-contrast waveguides", *Opt. Exp.*, Vol.14, pp.5617-5633, 2006.

[2] M. Bayindir, B. Temelkuran, and E. Ozbay, "Photonic-crystal-based beam splitters", *Appl. Phys. Lett.*, Vol.77, pp.3902-3904, 2000.

[3] I. Park, H.-S. Lee, H.-J. Kim, K.-M. Moon, S.-G. Lee, B.-H. O, S.-G. Park, and E.-H. Lee, "Photonic crystal power-splitter based on directional coupling", *Opt Exp.*, Vol. 12, pp.3599-3604, 2004.

[4] X. F. Yu and S. H. Fan, "Bends and splitters for self collimated beams in photonic crystals", *Appl. Phys. Lett.*, Vol. 83, pp. 3251-3253, 2003.

[5] A. F. Matthews, S. K. Morrison, Y. S. Kivshar, "Self-collimation and beam splitting in low-index photonic crystals", *Opt. Comm.*, Vol. 279, pp. 313-319, 2007.

[6] A. F. Matthews, "Experimental demonstration of self-collimation beaming and splitting in photonic crystals at microwave frequencies", *Opt. Comm.*, Vol. 282, pp. 1789-1792, 2009.

[7] A. Cicek and B. Ulug, "Polarization-independent waveguiding with annular photonic crystals", *Opt. Exp.*, Vol. 17, pp.18381-18386, 2009

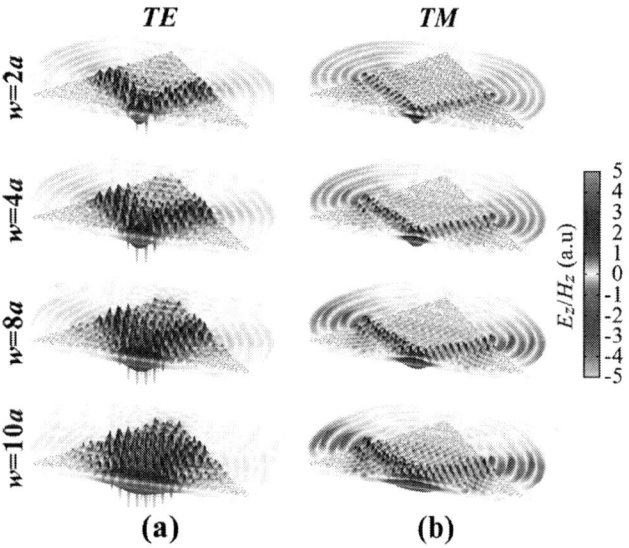

Fig. 2- FDTD simulations for the propagation of the beams at 27.2 GHz with TE **(a)** and TM **(b)** polarizations in 2D PC. The source width increases from $2a$ to $10a$ from top to down.

Portable 1.55μ m Terahertz Spectrometer and Imaging System

Sang-Pil Han[a], Namje Kim[a], Han-Cheol Ryu[a], Hyunsung Ko[a], Jeong-Woo Park[a], Min Yong Jeon[b], and Kyung Hyun Park[a]

[a] THz Photonics Creative Research Center, ETRI, Korea
[b] Department of Physics, Chungnam National University, Korea

Abstract—We demonstrate a portable terahertz (THz) spectrometer and imaging system. Absorption lines of water vapor in the free space are clearly observed by using the THz system. A THz imaging of a medical knife behind a poly-ethylene is finely measured by the same THz system as well.

I. INTRODUCTION

COMPACT, lightweight, and cost-effective terahertz (THz) spectroscopy and imaging systems have been gradually required for utilizing in the outdoors or moving situations such as the fields of security, non-invasive testing, food and agricultural goods quality control, and environment monitoring. Fiber-coupled THz systems can be one of solutions. They are lower cost, higher stability, portable THz systems as compared to free-space THz systems, since they have movable THz emitters and detectors [1]. Recently, compact THz emitting and detecting systems have been reported. Due to the utility of 1.55μm optical components, compact size, and cost-effectiveness, InGaAs-based fiber-coupled terahertz time-domain spectroscopy (THz-TDS) systems are considered promising [1]-[3]. In addition, compact and broadband continuous-wave (CW) THz optical beat sources, such as a monolithic dual-mode distributed feedback semiconductor laser and a 1.55 μm detuned dual-mode laser diode have been also developed to realize hand-held THz systems [4]-[7].

In this paper, we report a fiber-coupled InGaAs-based spectrometer and imaging system. Using the THz system, we present the experimental results of spectroscopy and imaging.

II. EXPERIMENTAL RESULTS

Our experimental setup of a fiber-coupled THz-TDS system is as in the following. It consists of a femtosecond laser with a pulse width of 70 fs, a computer-controlled delay line, an emitter module, a detector module, a 1×2 optical splitter, a dispersion-compensation fiber (DCF), a sine-wave function generator, and a lock-in amplifier. The DCF and SMF lengths in the system were tuned to compensate any pulse broadening [3].

The emitter (or detector) module depicted in Fig. 1 comprises a log-spiral antenna-integrated low-temperature grown (LTG) InGaAs photo-conductive antenna (PCA) chip on a printed circuit board (PCB), a hyper-hemi-spherical Si lens, and a fiber assembly. The fiber assembly is adjusted by a micromanipulator to couple the optical pulse signal to the active area of the PCA chip. In addition, a high-resistivity and collimating Si lens was used to reduce free-carrier absorption and to decrease the required number of THz components, such as parabolic mirrors and plastic lenses. In experiment, a

free-space distance between the emitter and the detector in the THz-TDS was 65mm. The THz output power of the emitter was about 130 nW when the bias voltage, the optical average pumping power, and the emitter photocurrent were 7.3 V, 9 dBm, and 0.8 mA, respectively. Lock-in integration time and measuring time on each delay step were 100 ms and 500 ms, respectively.

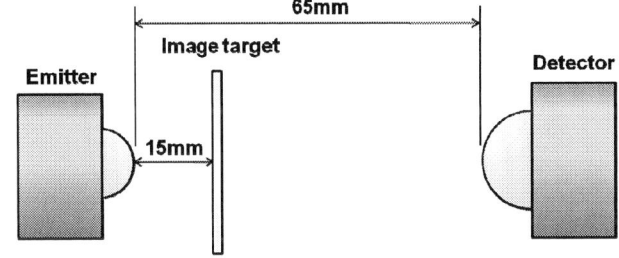

Fig. 1. Setup of THz-TDS module for measuring THz spectrum and imaging.

First of all, we measured water vapor in the free space for feasibility of the THz-TDS spectroscopy. Fig. 2 shows absorption spectrum of free space measured by using the THz-TDS system, where the time delay step and the frequency resolution are 0.1 ps and 2.44 GHz, respectively. As shown in Fig. 2, absorption lines of water vapor in the free space were clearly detected at 557, 752, 988, 1097, 1113, 1163, 1208, 1229, 1411, 1602, 1661, 1669, 1717, and 1762 GHz at the conditions of a relative humidity of about 10 % and room temperature. These results show that the THz radiation bandwidth of the THz-TDS system should be sufficiently higher than 2 THz.

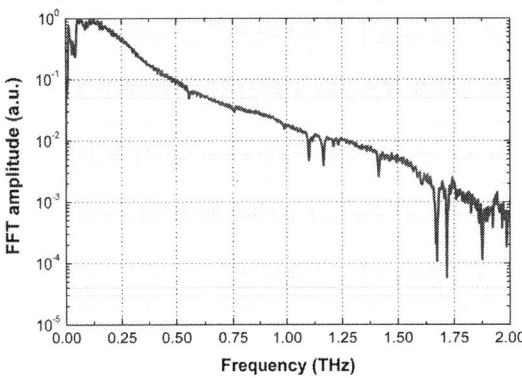

Fig. 2. Absorption spectrum of water vapor in the free space measured by using the THz-TDS system, where the time delay step and the frequency resolution are 0.1 ps and 2.44 GHz, respectively.

Then we measured the spot size of THz-wave beam along to THz radiation-direction by a knife-edge method to find an optimal position for THz imaging. With respect to the result as shown in Fig. 3, we chose an image target position of 15 mm far

978-1-4673-1602-6/12 $31.00 © 2012 IEEE

from the emitter.

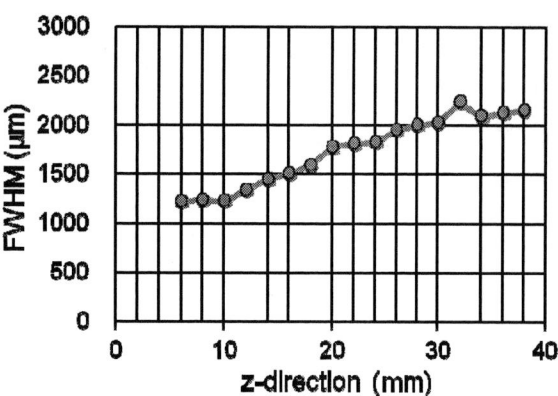

Fig. 3. Spot size of THz-wave beam along to THz radiation-direction.

Fig. 4 shows a THz imaging result of a medical knife behind a poly-ethylene with a thickness of 1 mm measured by using the THz-TDS system, where a signal-to-noise (SNR) of 300, a THz beam spot size of 1.5 mm, a cell size of 0.5×0.5 mm^2, a pixel resolution of 100×20 was set. The THz imaging of the medical knife was finely measured as in Fig. 4. We can see that the THz-TDS system should be enough an imaging system as well as a spectroscopy system.

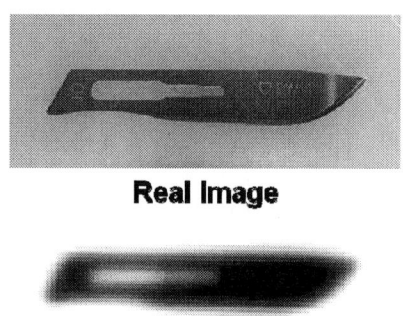

Fig. 4. Photograph of a medical knife, and its THz image measured by using the THz-TDS system.

III. CONCLUSION

We have successfully demonstrated a fiber-coupled THz-TDS system. The THz output power of the emitter was about 130 nW when the bias voltage, the optical average pumping power, and the emitter photocurrent were 7.3 V, 9 dBm, and 0.8 mA, respectively. Under the best alignment condition, absorption lines of water vapor in the free space were clearly measured by the THz-TDS system. Moreover, we have finely measured a THz imaging of a medical knife behind a poly-ethylene by using the THz-TDS system

ACKNOWLEDGEMENTS

This work was supported by the Public welfare & Safety research program through the National Research Foundation of Korea (NRF), by the Ministry of Education, Science and Technology-grant #2010-0020822.

REFERENCES

[1] B. Sartorius, H. Roehle, H. Künzel, J. Böttcher, M. Schlak, D. Stanze, H. Venghaus, and M. Schell, "All-fiber terahertz time-domain spectrometer operating at 1.5 μm telecom wavelengths," Optics Express 16, 9565-9570 (2008).

[2] H. Roehle, R. J. B. Dietz, H. J. Hensel, J. Böttcher, H. Künzel, D. Stanze, M. Schell, and B. Sartorius, "Next generation 1.5 μm terahertz antennas: mesa-structuring of InGaAs/InAlAs photoconductive layers," Optics Express 18, 2296-2301 (2010).

[3] S. -P. Han, H. Ko, N. Kim, H. -C. Ryu, C. W. Lee, Y. A. Leem, D. Lee, M. Y. Jeon, S. K. Noh, H. S. Chun, and K. H. Park, "Optical fiber-coupled InGaAs-based THz time-domain spectroscopy system," Optics Letters 36, 16, 3094-3096 (2011).

[4] N. Kim, J. Shin, E. Sim, C. W. Lee, D.-S. Yee, M. Y. Jeon, Y. Jang, and K. H. Park, "Monolithic dual-mode distributed feedback semiconductor laser for tunable continuous-wave terahertz generation", Opt. Express 17(16), 13851-13859 (2009)

[5] N. Kim, Y. A. Leem, M. Y. Jeon, C. W. Lee, S.-P. Han, D. Lee, and K. H. Park, "Widely Tunable 1.55 um Detuned Dual Mode Laser diode for Compact Continuous-Wave THz Emitter," ETRI Journal 33, 5, (2011).

[6] N. Kim, S.-P. Han, H. Ko, Y. A. Leem, H.-C. Ryu, C. W. Lee, D. Lee, M. Y. Jeon, S. K. Noh, and K. H. Park, "Tunable continuous-wave terahertz generation/detection with compact 1.55 μm detuned dual-mode laser diode and InGaAs based photomixer," Opt. Express 19, 15397 (2011).

[7] K. H. Park, N. Kim, H Ko, H. -C. Ryu, J. -W. Park, S. -P. Han, and M. Y. Jeon, "Portable terahertz spectrometer with InP related semiconductor photonic devices," Proc. SPIE Photonics West, Jan. (2012).

Optimization of InSb Infrared Focal Plane Arrays

Nan Guo[1], Wei-Da Hu[1,*], Xiao-Shuang Chen[1,*], Yan-Qiu Lv[2], Xiao-Lei Zhang[2], Jun-Jie Si[2] and Wei Lu[1]

[1]National Lab for Infrared Physics, Shanghai Institute of Technical Physics, Chinese Academy of Sciences, 500 Yu Tian Road, Shanghai, 200083, China

[2]Luoyang Optoelectronic Institute, Luoyang, Henan, 471009, China

Abstract

The quantum efficiency (QE) for mid-wavelength InSb infrared focal plane arrays has been numerically studied. Effects of the absorption length and thickness of p-region on device QE have been investigated. Our work shows that the optimum thickness of p-region is largely dependent on the absorption characteristics of the InSb.

I. INTRODUCTION

InSb which is a narrow-bandgap compound semiconductor has a response cutoff wavelength of 5.5μm at 77K. Due to its excellent absorption ability in the spectral range of 3μm-5μm and superior fundamental properties, InSb has been widely used in military and civil fields[1-8]. Therefore, it is particularly important to fully understand the photoresponse mechanisms of InSb to improve device performance[9]. In this paper, effects of the absorption length and thickness of p-region on the device QE are numerically investigated. The optimum thickness of p-region is largely dependent on the absorption characteristics of the InSb.

II. SIMULATION MODELS

For plain drift-diffusion simulation, the well known Poisson equation and continuity equations are self-coupled. The carrier generation-recombination process consists of SRH, Auger, and optical generation-recombination terms. Additionally, the tunneling effect is implemented in the continuity equations[10-13].

III. RESULT AND DISCUSSION

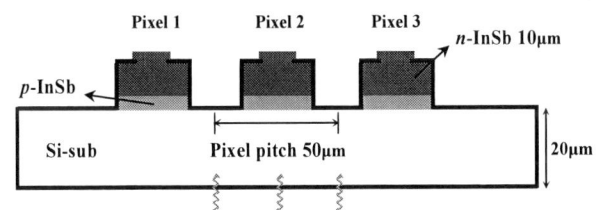

Figure 1. Schematic of linear InSb infrared focal plane arrays.

The InSb focal plane arrays discussed in this study are composed of three pixels, as shown in Figure 1. The n-region with the doping density of 10^{15}cm^{-3}, has a thickness of 10μm. The p-region is doped with 10^{17}cm^{-3} and its thickness is an adjustable parameter in the simulated process. The pixel pitch and filling factor are 50μm and 92% respectively. It should be noted that each element including an individual p-n junction forms an island on the 20-μm-thick Si substrate. During the simulation, only the center pixel is front-side illuminated using a 5μm incident light under 77K background, i.e., the optical energy is incident on the p-region. And the effect of antireflection coating is not taken into consideration. Finally the QE curve from pixel 2 is obtained.

For the conventional InSb detectors, the infrared radiation is incident on the n-type bulk InSb substrate. The photo-generated minority carriers will diffuse a long distance to p-n junction to be converted into electrical response. However, it is impossible to limit all inter-pixel migration of carriers. Some of them diffuse into neighboring junctions to form the crosstalk[14]. In our calculations, the light is directly incident on the thinner p-region instead of n-region. The carriers can diffuse to junction more easily with less recombination to lead a higher QE[15]. Moreover, all of the diodes are spaced from each other and the effect of crosstalk can be significantly reduced.

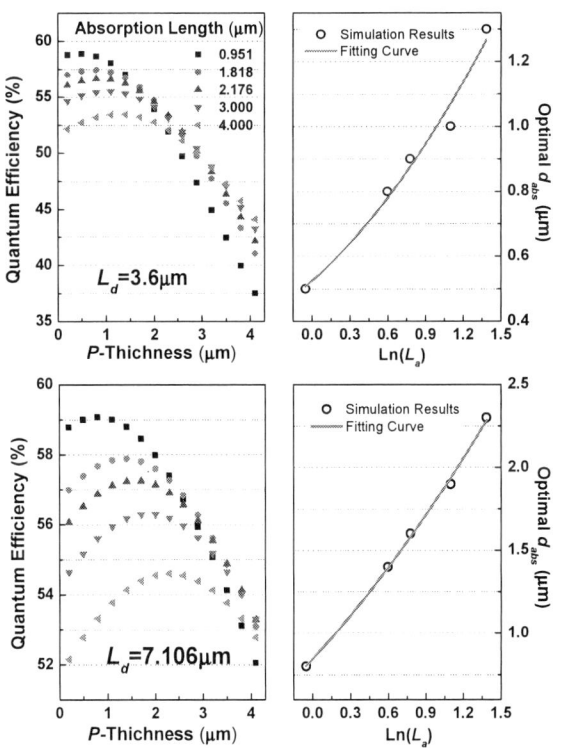

---

*Corresponding author: wdhu@mail.sitp.ac.cn, xschen@mail.sitp.ac.cn

Figure 2. QE vs. p-region thickness with L_a changing from 0.951 to 4.0μm for different L_d, i.e., 3.6, 7.106μm (left column). Fitting curve of the optimum thickness of p-region as a function of L_a (right column).

However the light can not be fully absorbed in the p-region due to its thinner thickness. Some of optical energy penetrates through the p-region into the n-region where still more minority carriers are generated[16]. Part of these additional carriers will diffuse back to the junction to contribute to the response. So the QE is dependent on not only carrier diffusion length L_d but also light absorption length L_a. In this paper, L_d refers to electrons diffusion length and that of holes is fixed at 81.2μm.

Figure 2 shows the simulated QE as a function of the p-region thickness with L_a changing from 0.951 to 4.0μm for different L_d, i.e., 3.6, 7.106μm (left column). By fitting the curve of the optimum p-region thickness d_{abs} as a function of L_a (right column), two empirical formulas which have the same polynomial format for different L_d are obtained:

$$d_{abs} = 0.52472 + 0.34239 \times \ln(L_a) + 0.13974 \times \ln^2(L_a) \quad \text{for } L_d\text{=3.6μm}$$

$$d_{abs} = 0.8449 + 0.83016 \times \ln(L_a) + 0.14969 \times \ln^2(L_a) \quad \text{for } L_d\text{=7.106μm}$$

IV. CONCLUSION

The quantum efficiency of mid-wavelength InSb infrared focal plane arrays has been numerically simulated with a two-dimensional simulator. Effects of the absorption length and thickness of p-region on device quantum efficiency have been investigated. The empirical formulas about the optimum thickness of p-region and the absorption length are obtained.

ACKNOWLEDGEMENTS

This work was supported by Aviation Science Fund (Grant No. 20110190001).

REFERENCES

[1] Y. T. Gau, L. K. Dai, S. P. Yang, P. K. Weng, K. S. Huang, Y. N. Liu, C. D. Chiang, F. W. Jih, Y. T. Cherng and H. Chang, "256 x 256 InSb focal plane arrays", *Proc. SPIE*, vol. 4078, pp.467, 2003.

[2] A. Rogalski, "Infrared detectors: an overview", *Infrared Physics & Technology*, 43, pp. 195-196. 2002.

[3] K. M. Chang, J. J. Luo, C. D. Chang and K. C. Liu, "Wet Etching Characterization of InSb for Thermal Imaging Applications", *Japanese Journal of Applied Physics*, 45, pp. 1477 2006.

[4] I. Kanno, S. Hishiki, O. Sugiura, R. Xiang, T. Nakamura, and M. Katagiri, "InSb cryogenic radiation detectors", *Nuclear Instruments and Methods in Physics Research A*, 568, pp. 416, 2006.

[5] I. Kimukin, N. Biyikli, and E. Ozbay, "InSb high-speed photodetectors grown on GaAs substrate", *Journal of Applied Physics*, 94, pp. 5414, 2003.

[6] I. Kimukin and N. Biyikli, "High-Speed InSb Photodetectors on GaAs for Mid-IR Applications", *IEEE JOURNAL OF SELECTED TOPICS IN QUANTUM ELECTRONICS*, 10, pp. 766, 2004.

[7] H. T. Pham, S. F. Yoon, D. Boning, and S. Wicaksono, "Molecular beam epitaxial growth of indium antimonide and its characterization", *J. Vac. Sci. Technol. B*, 25(1), pp. 11, 2007.

[8] W. J. Parrish, J. D. Blackwell, G. T. Kincaid and R. C. Paulson, "Low-cost high-performance InSb 256 x 256 infrared camera", *Proc. SPIE*, vol.

1540, pp. 274, 1991.

[9] N. Guo, W. D. Hu, X. S. Chen, C. Meng, Y. Q. Lv and W. Lu, "Optimization of Microlenses for InSb Infrared Focal-Plane Arrays", *Journal of Electronic Materials*, 40, pp. 1647-1650, 2011.

[10] W. D. Hu, X. S. Chen, Z. H. Ye, W. Lu, "A hybrid surface passivation on HgCdTe long wave infrared detector with in-situ CdTe deposition and high-density Hydrogen plasma modification", *Applied Physics Letters*, 99, pp. 091101, 2011.

[11] W. D. Hu, X. S. Chen, Z. H.Ye, C. Meng, Y. Q. Lv, W. Lu, "Effects of absorption layer characteristic on spectral photoresponse of mid-wavelength InSb photodiodes", *Optical and Quantum Electronics*, 42, pp. 801-808, 2011.

[12] W. D. Hu, X. S. Chen, F. Yin, Z. H. Ye, C. Lin, X. N. Hu, Z. J. Quan, Z. F. Li and W. Lu, "Simulation and design consideration of photoresponse for HgCdTe infrared photodiodes", *Optical and Quantum Electronics*, 40, pp. 1255, 2008.

[13] W. D. Hu, X. S. Chen, F. Yin, Z. H. Ye, C. Lin, X. N. Hu, Z. F. Li, W. Lu, "Numerical analysis of two-color HgCdTe infrared photovoltaic heterostructure detector", *Optical and Quantum Electronics*, Volume 41, pp. 699-704, 2009.

[14] M. Davis, M. Greiner, J. Sanders and J. Wimmers, "Resolution issues in InSb focal plane array system design", *Proc. SPIE*, vol. 3379, pp. 289, 1998.

[15] M. Davis and M. Greiner, "Indium antimonide large-format detector arrays", *Optical Engineering*, 50, pp. 061016, 2011.

[16] H. A. Timlin and C. J. Martin,"Electro-optical detector array", U.S.Patent No. 5,227,656, 1993.

A general transformation designing high gain lens antennas with homogeneous media

L. J. Huang[a], X. S. Chen[a, *], B. Ni[a], G. H. Li[a], Z. F. Li[a], and W. Lu[a, *]

[a] National Lab for Infrared Physics, Shanghai Institute of Technical Physics, Chinese Academy of Sciences,
500 Yu Tian Road, Shanghai, China 200083

Abstract

By employing finite embedded coordinate transformation method, we propose a general transformation to design a highly directive horn antenna with homogeneous and anisotropic media. Different from the layered lens antenna, our antenna consists of four triangle regions made of homogeneous and anisotropic media, which greatly reduces the difficulty of practical realization. Full wave simulation based on finite element method is performed to validate the performance of the antenna.

I. INTRODUCTION

In the past five years, transformation optics has triggered extensive interests all over the world since it provides unprecedented freedom and flexibility in manipulating the propagation of electromagnetic wave. Its main principle is based on the form invariance of Maxwell equations under coordinate transformation. The most fascinating application of transformation optics is the invisibility cloak [1-5], which is firstly introduced by Pendry and Leonhardt in 2006, respectively [1, 2]. An ideal invisibility cloak can guide electromagnetic wave smoothly flowing around the object inside it. The first free-space cloak with reduced material parameters was experimentally realized with split-ring resonators at microwave frequency by Schurig et al [3]. Besides the invisibility cloak, transformation optics is also used to design the other interesting devices, such as EM field concentrator [6], EM field rotator [7] and illusion optical device [8].

Recently, finite-embedded coordinate transformation method has been proposed to design beam shifter and splitter [9]. Later, it is extended to design the other interesting devices, such as beam compressor and expander [10], waveguide bends [10, 11], cylindrical-to-plane wave converter [12], and highly directive antenna [13]. However, most of devices reported are made of inhomogeneous and anisotropic materials, which greatly increase the difficulty of practical realization as well as limit the operation bandwidth. To remove the obstacles of material inhomogeneity and anisotropy, some novel transformations including quasi-conformal mapping are proposed. In this paper, we proposed a generalized transformation to design a highly directive antenna with homogeneous media. We just need to divide the antenna into four triangle blocks, and each block is made of homogeneous and anisotropic media. The simple material requirement makes the antennas relatively easy be

constructed by metamateirals. Full wave simulation based on finite element method is carried out to confirm the performance of the antenna under such a kind of transformation. The simulation results indicate that the radiation directivity of antenna is highly collimated and can be controlled arbitrarily. Thus, such a generalized transformation provides us more freedom and flexibility in designing the antenna with high direction.

II. TRANSFORMATION PRINCIPLE OF ANTENNAS

Fig. 1 shows the geometry structure of antenna. The rectangular area ABC_0D_0 in the virtual space is divided into four triangle areas AO_0B, BO_0C_0, $C_0O_0D_0$, and AO_0D_0, respectively. Simultaneously, the trapezoid area $ABCD$ in the physical space is also divided into four blocks AOB, BOC, COD, and AOD, respectively. Then, the four triangle regions AO_0B, BO_0C_0, $C_0O_0D_0$, and AO_0D_0 in the virtual space (x, y, z) are mapped into the four corresponding areas AOB, BOC, COD, and AOD in the physical space (x', y', z'). Here, four blocks AOB, BOC, COD and AOD are denoted by Region I, II, III and IV. The coordinate transformation function of above four mappings can be expressed as

$$
(a)\begin{cases} x' = a_{11}x + b_{11}y + c_{11} \\ y' = a_{12}x + b_{12}y + c_{12} \\ z' = z \end{cases} \quad (b)\begin{cases} x' = a_{21}x + b_{21}y + c_{21} \\ y' = a_{22}x + b_{22}y + c_{22} \\ z' = z \end{cases}
$$

$$
(c)\begin{cases} x' = a_{31}x + b_{31}y + c_{31} \\ y' = a_{32}x + b_{32}y + c_{32} \\ z' = z \end{cases} \quad (d)\begin{cases} x' = a_{41}x + b_{41}y + c_{41} \\ y' = a_{42}x + b_{42}y + c_{42} \\ z' = z \end{cases} \quad (1)
$$

$a_{i1}, b_{i1}, c_{i1}, a_{i2}, b_{i2}$, and c_{i2}, can be easily obtained and are found that they are only related to the coordinate value of A, B, C, D, O, C_0, D_0, and O_0, respectively. Here, i=1, 2, 3 and 4 represent Region I, II, III, and IV, respectively.

According to the principle of optical transformation, the material parameters in the four regions can be obtained by

$$
\overline{\overline{\varepsilon'}} = \overline{\overline{\mu'}} = \Lambda \bullet \Lambda^T / \det(\Lambda) \quad (2)
$$

in which $\Lambda_i = (a_{i1}, b_{i1}, 0; a_{i2}, b_{i2}, 0; 0, 0, 1)$, and $\det(\Lambda_i) = a_{i1}b_{i2} - a_{i2}b_{i1}$, i=1, 2, 3, 4, which represents Region I, II, III and IV.

III. SIMULATION RESULTS AND DISCUSSIONS

In order to verify the performance of the antennas, full wave simulation based on finite element method is performed. The boundary conditions surrounding computational region are set as perfect matched layer. The outer boundary of the antenna system except right boundary is defined as perfect electric conductor. A line current source with the magnitude of 1A is

---

* Corresponding author: xschen@mail.sitp.ac.cn, luwei@mail.sitp.ac.cn.

978-1-4673-1602-6/12 $31.00 © 2012 IEEE

located as the point (-0.1, 0.125). The working frequency of antenna is set as 5 GHz. For the sake of convenience and without loss of generality, we focus on the transverse magnetic (TM) incident wave with magnetic field polarizing along z axis. The coordinate values of A, B, C, D, C_0, and D_0, are (0, 0.05), (0, 0), (0.2, -0.125), (0.2, 0.175), (0.2, 0) and (0.2, 0.05).

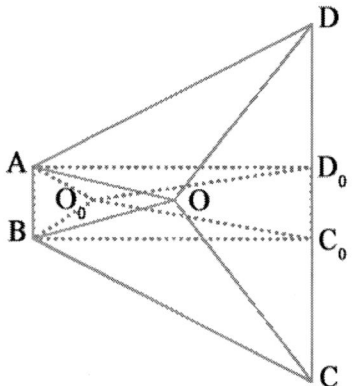

Figure 1(Color online) Schematic structure of high gain lens antenna.

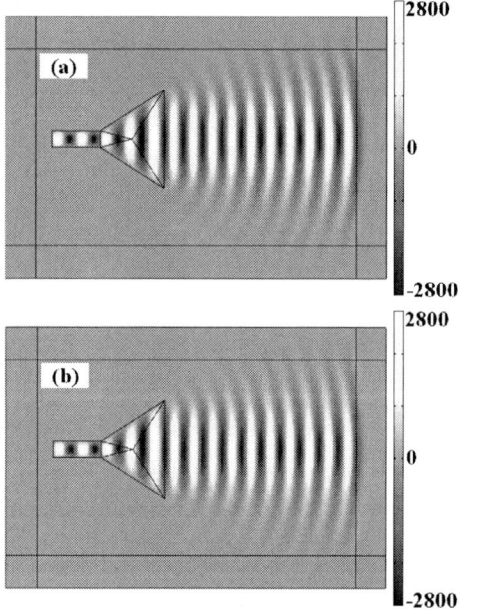

Figure 2 (Color online) Spatial distribution of magnetic field for antennas with different sets of material parameters.

Fig. 2 presents the magnetic field distributions of antennas. When the points O and O_0 are overlapped, and their coordinate values are set as (0.1, 0.025), as shown in Fig. 2(a), the magnetic field inside the antenna are successfully amplified to the whole trapezoid region and its distribution are symmetric. The output wave exiting antenna is highly collimated. Furthermore, since the triangle area ABO is mapped to itself, it is filled with air. Only other three blocks require homogeneous and anisotropic materials. If points O and O_0 do not locate at the same place and the coordinates of O and O_0 are (0.1, 0.025) and (0.1, 0.02), the material parameters of all four blocks are homogeneous and anisotropic. From Fig. 2(b), we find that the antenna stills performs well in radiating highly directive beam and the magnetic field outside the antenna is nearly the same as

that of Fig. 2(a) even though the field distributions inside the horn antenna are different. Thus, the general transformation provides us more freedom in designing the high gain lens antennas.

IV. CONCLUSION

In summary, we design a high gain lens antenna by using a general transformation. Such an antenna has one merit that it is made of homogeneous and anisotropic media, and thus is relatively easily constructed with optical metamaterials. Full wave simulation is carried out to confirm the performance of the antenna. According to the simulation results, it is found that the antenna perfectly fulfills the function of radiating EM beam with high directivity. Furthermore, the radiation direction of the antenna can be controlled artificially by varying the geometrical parameters.

ACKNOWLEDGEMENTS

This work was supported in part by the State Key Program for Basic Research of China grants (2011CB922004); the National Natural Science Foundation of China grants (61006090, 10725418, 10874196, 10990104, 11104299 and 60976092); the Fund of Shanghai Science and Technology Foundation grants (09DJ1400203, and 10JC1416100).

REFERENCES

[1] J. B. Pendry, D. Schurig, and D. R. Smith, "Controlling electromagnetic fields", Science **312**, 1780-1782 (2006)
[2] U. Leonhardt, "Optical conformal mapping", Science **312**, 1777-1780 (2006)
[3] D. Schurig, J. J. Mock, B. J. Justice, S. A. Cummer, J. B. Pendry, A. F. Starr, and D. R. Smith, "Metamaterial electromagnetic cloak at microwave frequencies", Science **314**, 977-980 (2006)
[4] L. J. Huang, D. M. Zhou, J. Wang, Z. F. Li, X. S. Chen, and W. Lu, "Generalized transformation for nonmagnetic invisibility cloak with minimized scattering", J. Opt. Soc. Ame, B **28**, 922-928 (2011)
[5] L. J. Huang, D. M. Zhou. J. Wang, Z. F. Li, X. S. Chen, and W. Lu, "Scattering characteristics of nonmagnetic invisibility cloak with minimized scattering", Opt. Communication **284**, 5523-5530 (2011)
[6] M. Rahm, D. Schurig, D. A. Roberts, S. A. Cummer, D. R. Smith, and J. B. Pendry, "Design of electromagnetic cloaks and concentrators using form-invariant coordinate transformations of Maxwell's equations", Photonics and Nanostructures: Fundamentals and Applications 6, 87-95 (2008)
[7] H. Y. Chen, and C. T. Chan, "Transformation media that rotate electromagnetic fields", Appl. Phys. Lett 90, 241105 (2007)
[8] L. J. Huang, D. M. Zhou, J. Wang, Z. F. Li, X. S. Chen, and W. Lu, "A generalized transformation to convert an arbitrary perfect electric conductor into another arbitrary dielectric object", Journal of Physics D: Applied Physics 44, 235102 (2011)
[9] M. Rahm, S. A. Cummer, D. Schurig, J. B. Pendry, and D. R. Smith, "Optical Design of reflectionless complex media by finite embedded coordinate transformations", Phys. Rev. Lett **100**, 063903 (2008)
[10] M. Rahm, D. A. Roberts, J. B. Pendry, and D. R. Smith, "Transformation-optical design of adaptive beam bends and beam expanders", Opt. Express **16**, 11555- 11567 (2008)
[11] W. X. Jiang, T. J. Cui, X. Y. Zhou, X. M. Yang, and Q. Cheng, "Arbitrary bending of electromagnetic waves using realizable inhomogeneous and anisotropic materials", Phys. Rev. E **78**, 066607 (2008)
[12] W. X. Jiang, T. J. Cui, H. F. Ma, X. M. Yang, and Q. Cheng, "Cylindrical-to-plane conversion via embedded optical transformation", Appl. Phys. Lett **92**, 261903 (2008)
[13] W. X. Jiang, T. J. Cui, H. F. Ma, X. M. Yang, and Q. Cheng, "Layered high-gain lens antenna via discrete optical transformation", Appl. Phys. Lett **93**, 221906 (2008)

The photocurrent of resonant tunneling diode controlled by the charging effects of quantum dots

Daming Zhou[a,b], Qianchun Weng[a], Wangping Wang[a], Ning Li[a], Bo Zhang[a], Xiaoshuang Chen[a,*], Wei Lu[a,*], Wenxin Wang[c], Hong Chen[c]

[a] National Lab for Infrared Physics, Shanghai Institute of Technical Physics, Shanghai, 200083, China
[b] Shanghai Advanced Research Institute, Chinese Academy of Sciences, Shanghai, 200083, China
[c] Institute of Physics, Chinese Academy of Sciences, Beijing 100080, China

Abstract

We experimentally studied the photocurrent of AlAs/GaAs/AlAs double barrier resonant tunneling diode (RTD), which is composed of an InAs layer of self-assembled quantum-dots (QDs) on top of the AlAs barrier layer. It is found that the charging InAs quantum dots can effectively modulate the carrier transport properties of the RTD. Moreover, we also found that the resonant tunneling current through a single energy level of an individual quantum dot is extremely sensitive to the photo-excited holes bound nearby the dot, and the presence of the holes lowers the electrostatic energy of the quantum dot state. In addition, it is also observed that the photocurrent can increase step by step with the individual photon pulse excitation when the illumination is low enough. The experiment results well demonstrated the quantum amplified characteristics of the device.

I. INTRODUCTION

Recently, QD based single photon detectors have attracted increasing attention mainly due to their single photon sensitivity and photon number resolving capability. It has been demonstrated that both field-effect transistor [1] and resonant tunneling diodes (RTD) [2, 3] modified which contains a layer of QD layer can detect single photons at temperature of 4.2K. The concept of RTD was first proposed by Tsu and Easki in 1973 [4]. In 2005, Blakesly first introduced the quantum dot concept into the conventional RTD. The experiment results showed that a single photon can be efficiently detected in the device. Physically speaking, when the RTD layers have the appropriate alignment voltage, a current can tunnel resonantly through the double barrier structure. It is indicated that the effective multiplication factor is in the order of 10^8. The very high multiplication factor is caused by the photo-excited holes being stored in the QD close to the RTD structure [5-8]. In this paper, we have investigated the photocurrent of AlAs/GaAs/AlAs double barrier resonant tunneling diode. It has been indicated how the photocurrent is influenced by the charging and discharging of the InAs dots. The Photo-excited holes reduce the negative charge in the dots, and thus lower the electrostatic energy of the tunneling channel of the active quantum dot. It is also found that the photocurrent increases in

step-like style with a laser pulse coming when the applied bias is fixed as constants.

II. DEVICE STRUCTURE

The structure was grown by molecular beam epitaxy on a (100) semi-insulating GaAs substrate. The detailed device structure is shown in Fig.1.

n+	GaAs	50 nm	top contact layer
i-	GaAs	150 nm	absorber layer
	GaAs	10 nm	cap layer layer
▲	InAs	QD ▲ ▲ ▲ ▲	
	GaAs	2 nm	spacer layer
	AlAs	3 nm	barrier layer
	GaAs	8nm	QW
	AlAs	3 nm	barrier layer
	GaAs	20 nm	spacer layer
n-	GaAs	430 nm	buttom contact layer layer
	AlAs	15 nm	etch stop layer
	GaAs	400 nm	buffer layer
GaAs (100)		substrate	

Fig. 1 Double barrier RTD structure with InAs dots being buried in the cathode side.

The fabricating progress contains three steps, that is the same as reference [7, 8]

III. RESULT AND DISCUSSION

Figure 2 shows the current-voltage (I-V) characteristics measured at 77K in the dark and under illumination by using laser light of wavelength $\lambda = 650nm$. There is a distinct difference in the resonant tunneling peaks at negative and positive collector-emitter voltages. It can be seen that the peak of the characteristic corresponds to the resonant tunneling through the double barrier structure for both forward and reverse biases. For reverse bias, we can find the explanation in reference [7]. Therefore, here we only discuss the case of forward bias. Notice that there is only light response at the voltage between 0.75V and 1.56V. When the voltage is greater than 1.5V, the device works in negative differential resistance region, and there is no light response.

---

* xschen@mail.sitp.ac.cn, luwei@mail.sitp.ac.cn

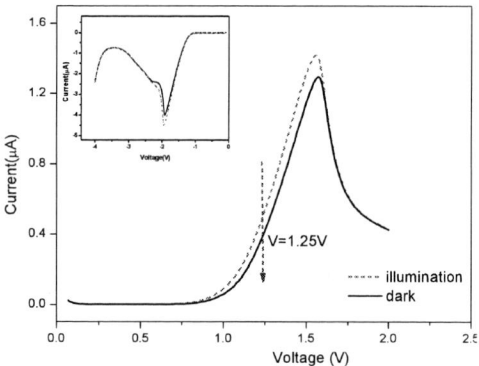

Fig.2 I-V curve at T=77K in the dark and under illumination with laser light ($\lambda = 650nm$). Under illumination the light response locates in the positive resistance region. Inset shows the light response with the reverse bias.

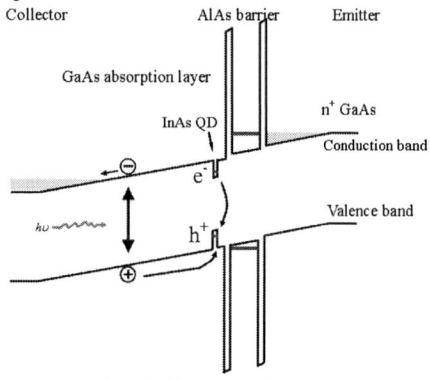

Fig.3 (a) Schematic band diagram of AlAs barrier QDRTD detector grown on a GaAs substrate shows the charge separation of photogenerated carriers in the absorption layer. The Photo-excited holes are attracted to the QDs and then are recombined with the electrons trapped inside.

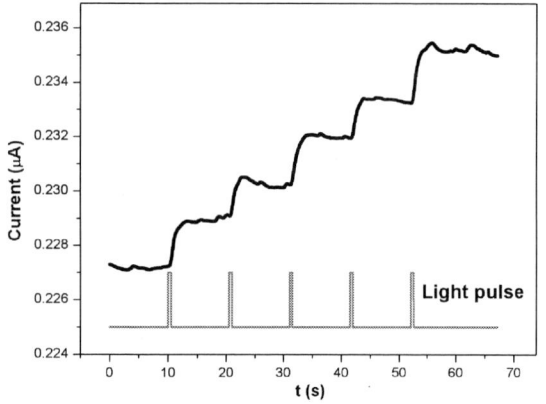

Fig. 3 (b) Time dependence of the tunnel current at V=1.25V with the excitation of laser pulse.

For conventional resonant tunneling diode, when the resonant energies between the double barriers approach the Fermi energy of the electrons at the emitter contact, the current maxima occurs. But here, the quantum dot layer in the structure has an important effect on the current-voltage characteristic of the QDRTD. The energy band diagram of the structure in forward bias is shown in Fig. 3(a). The response of the tunneling current to illumination is further illustrated in Fig.

3(b). Here we set the applied bias at V=1.25V, which is well below the resonant peak of I-V curve in the dark background. Since the conduction band level in the quantum dots has a lower energy than that in GaAs, therefore, each dot traps several excess electrons at the initial. For forward biased QDRTD with the light illumination, the photo-excited electrons are collected by the top contact, and the photo-excited holes are drifted to the InAs dot layer. Then the holes are captured by the QDs according to the external electrical field direction. When a quantum dot is depleted with a single photo-excited charge, the band bending can shift the energy levels further towards the resonance and cause the increase of the tunneling current. As shown in Fig.3(b), when a light pulse is incident, more holes can be captured by dots. As long as the life of holes is long enough, the current can keep the same value. When the next pulse comes, the current can increase step by step.

IV. CONCLUSION

In conclusion, we have indicated the photon detection based on a AlAs/GaAs/AlAs resonant tunneling diode containing a layer of InAs quantum dots. The Photo-excited holes reduce the negative charge in the dots, and thus lower the electrostatic energy of the tunneling channel of the active quantum dot. For a fixed applied bias, the photocurrent increases in step-like style with a laser pulse coming.

ACKNOWLEDGEMENTS

This work was supported in part by the State Key Program for Basic Research of China (2006CB921507), the National Natural Science Foundation of China (10725148, 10734090, 10990104, and 10874196), Key Fund of Shanghai Science and Technology Foundation (09dz2202200, 10JC1416100, and 10510704700).

REFERENCES

[1] Zhenghua An, T. Ueda, S. Komoyama and K. Hirakawa, Metastable excited states of a closed quantum dot with high sensitivity to infrared photons, Phys. Rev. B 75 (2007), 085417.
[2] J. C. Blakesley, P. See, A. J. Shields, B. E. Kardynal, P. Atkinson, I. Farrer, and D. A. Ritchie, Efficient single photon detection by quantum dot resonant tunneling diodes, Phys. Rev. Lett. 94 (2005), 067401.
[3] G. S. Buller and R. J. Collins, Single-photon generation and detection, Meas. Sci. Technol. 21 (2010), 012202.
[4] R. Tsu and L. Esaki, Tunneling in a finite superlattice, Appl. Phys. Lett. 22 (1973), 562.
[5] H. W. Li, B. E. Kardynal, P. See, A. J. Shields, P. Simmonds, H. E. Beere, and D. A. Ritchie, Quantum dot resonant tunneling diode for telecommunication wavelength single photon detection, Appl. Phys. Lett. 91 (2007), 073516.
[6] Y. Hou, W. P. Wang, N. Li, W. Lu and Y. Fu, Effects of series and parallel resistances on the current-voltage characteristics of small-area air-bridge resonant tunneling diode, J. Appl. Phys. 104 (2008), 074508.
[7] Wangping Wang, Ying Hou, Dayuan Xiong, Ning Li and Wei Lu, High photoexcited carrier multiplication by charged InAs dots in AlAs/GaAs/AlAs resonant tunneling diode, Appl. Phys. Lett. 92 (2008), 023508.
[8] W. P. Wang, Y. Hou, N. Li and W. Lu, W. X. Wang and H. Chen, Field effect enhanced quantum dot resonant tunneling diode for high dynamic range light detection, Appl. Phys. Lett. 94 (2009), 093511.

A bisection-function technique to characterize heat transport in high-power GaN-based light-emitting-diodes package

Liwen Cheng[1+], Yang Sheng[2], ChangshengXia[2],Weida Hu[1] and Wei Lu[1+]

[1]National laboratory for Infrared Physics, Shanghai Institute of Technical Physics, Chinese Academy of Science, 500 Yu Tian Road, Shanghai 200083,China
[2]Crosslight Software China, Suite 906, Building JieDi, 2790 Zhongshan Bei Road, Shanghai 200063, China

Abstract—The transient response of the junction temperature of packaged high-power GaN-based light-emitting diodes (LEDs) is numerically simulated. We found the heat transport in LEDs involves two evident processes and can be characterized by a bisection function. One process involves heat transfer from a LED chip to its slug submount, whereas the other involves the heat transfer from the slug submount to the ambient through the heat sink. The thermal time constant of the two processes are identifiable. The time constant of the first process is in millisecond order of magnitude, whereas that of the other process is in hundred-second order of magnitude. The thermal resistance in the two processes can be obtained by analyzing the transient response curve of the junction temperature.

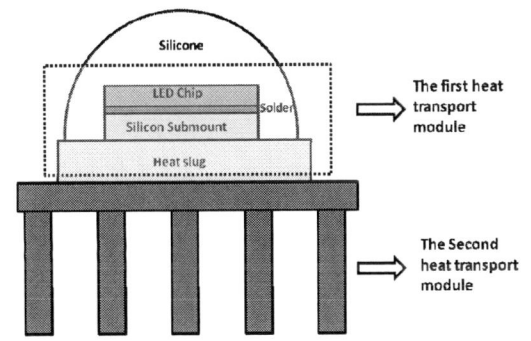

Fig. 1. Typical packaged structure of a high-power GaN LED.

I. INTRODUCTION

GaN-based light-emitting diodes (LEDs) have wide applications due to their many distinctive advantages, such as low-power consumption, long life span, and high brightness, among others[1]. High-power LEDs have the potential to replace traditional incandescent and fluorescent lamps. However, as the input power increases, additional heat is generated from the chip, resulting in a high junction temperature. As the junction temperature increases, the output power decreases and the wavelength shifts resulting in the degradation of efficiency and reliability. Therefore, an intensive understanding of the heat transport in high-power LED is becoming increasingly important. There were some reports about identity the thermal environment of a semiconductor chip by analyzing the transient response of the temperature. In the present paper, the heat transport of packaged LEDs is studied by simulation on the LED transient response of the junction temperature.

II. DEVICE STRUCTURE

The packaged structure of LEDs under study is shown in Fig. 1. A LED chip mainly consists of a 5 μm GaN active layer and a 75 μm sapphire substrate, and the chip size is 1 mm × 1 mm. The LED chip is connected to a 240 μm silicon submount by a 50 μm thick Pb/Sn solder. The silicon submount is connected to a 2.5 mm thick and 3 mm × 3 mm Cu heat slug. The slug is connected to an Al heat sink. The top of the LED chip is coated by silicone.

III. RESULTS AND DISCUSSIONS

In order to understand the thermal properties, this packaged structure is simulated to study the transient response of the junction temperature using FEM(Finite Element Method), which is a popular method be used to analyzing the thermal properties of LED chips and packages.

In the simulation, the electro-optical conversion efficiency is assumed to be 20%. The thermal boundary condition is natural convection, and the coefficient of convection is assumed to be 7.5 W/m °C. The ambient temperature is set to 25 °C. The simulation result is shown in Fig. 2.

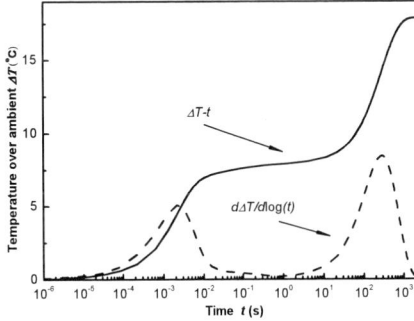

Fig. 2. Simulated curve of junction temperature and corresponding differential curve.

The transient response of the junction temperature involves two evident saturated processes, which can be obtained from the flat regions [$d\Delta T / d\log(t) = 0$] (Fig. 2). In addition, the first heat-saturating time is at millisecond level, whereas the second heat-saturating time is at hundred-second level. Considering the difference in the orders of magnitude for the two values, the slow heat-saturating process can be assumed as the stable heat environment compared with the fast heat-saturating process. Hence, the two heat transport processes can be approximated as two independent processes, and the dynamic equation of

+ email: lwcheng@crosslight.com.cn
+ email: luwei@mail.sitp.ac.cn

978-1-4673-1602-6/12 $31.00 © 2012 IEEE

thermal equilibrium can be described as

$$\frac{d\Delta T}{dt} = G - D\Delta T \qquad (1)$$

where ΔT is the junction temperature gradient above the ambient temperature, G is the ratio of heat generated, which includes heat generated inside and heat transferred from outside, and D is the ratio of heat transfer. Equation (1) of thermal equilibrium has a solution in exponential form

$$\Delta T = \Delta T_0[1 - \exp(-\frac{t}{\tau})] \qquad (2)$$

Hence, the transient response of the junction temperature can be described as a bisection-exponential function form:

$$\Delta T = \Delta T_1[1 - \exp(-\frac{t}{\tau_1})] + \Delta T_2[1 - \exp(-\frac{t}{\tau_2})] \qquad (3)$$

Then, the simulated transient response curve of the junction temperature is fitted using Equation (3) and the fitting result is shown in Fig. 3.The fitting equation is:

$$\Delta T = 4.47 \times [1 - \exp(-t/0.0018)] + 11.07 \times [1 - \exp(-t/307)] \quad (4)$$

According to the fitted curve (Fig. 3), the bisection-exponential function fits well the transient junction temperature response data.

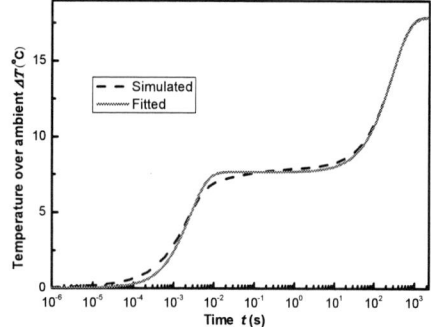

Fig. 3. Simulated data and fitting curve fitted by the bisection-exponential function.

As is known that thermal resistance is one of the important parameters used for evaluating the thermal performance of high-power LED packages. This parameter is defined as the ratio between the temperature difference of two points (or two areas) and the dissipated power. The symbol of thermal resistance is $R\Theta$ or R_{th}, and its unit is K/W or °C/W [2]. The relationship between the thermal resistance of the LED and the junction temperature can be expressed by the following equation:

$$R\theta_{\text{Junction-Ambient}} = \frac{T_{\text{Junction}} - T_{\text{Ambient}}}{P_d} = \frac{\Delta T_{\text{Junction-Ambient}}}{P_d} \qquad (1)$$

where T_{Junction} is the junction temperature of LED, and T_{Ambient} is the ambient temperature. P_d is the injected electrical power and given by $P_d = V_f * I_f$, where V_f is the operating voltage and I_f is the operating current.

For the packaged structure of LED (Fig. 1), the thermal transport path can be simplified into two modules of heat transfer: The first heat transport module is heat transferred from the chip to the heat slug, whereas the second heat transport module is heat conducted from the heat slug to the ambient by the packaged heat sink. This heat transport system can be represented by a series circuit model with two equivalent thermal resistances (Fig. 4).

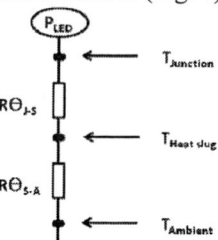

Fig.4. Schematic diagram of the packaged LED series circuit thermal model with two series resistances.

In Fig. 4, the overall thermal resistance of the packaged LED can be given by

$$R\theta_{\text{Junction-Ambient}} = \frac{T_{\text{Junction}} - T_{\text{Ambient}}}{P_d} = \frac{T_{\text{Junction}} - T_{\text{Slug}}}{P_d} + \frac{T_{\text{Slug}} - T_{\text{Ambient}}}{P_d}$$

And

$$R\theta_{\text{Junction-Ambient}} = R\theta_{\text{Junction-Slug}} + R\theta_{\text{Slug-Ambient}}$$

Considering that the fitted curve of the junction temperature has two terms of exponential functions [see Equation (3)], the first term of the exponential function can be attributed to the thermal resistance ($R\Theta_{J-S}$) contributed by the first heat transport module. The second term of the exponential function is attributed to the thermal resistance ($R\Theta_{S-A}$) contributed by the second heat transport module. Then, $R\Theta_{J-S}$ and $R\Theta_{S-A}$ can be obtained by $R\Theta_{J-S} = \Delta T_1/P_{LED}$=4.47 °C/W and $R\Theta_{S-A}= \Delta T_2/P_{LED}$=11.07 °C/W.

The overall thermal resistance of the packaged LED is given by $R\Theta_{J-A} = R\Theta_{J-S} + R\Theta_{S-A}$ =15.54 °C/W.

The thermal time constant for the two heat transport processes can be obtained from the fitted curve and the fitting parameters. The first thermal time constant for $R\Theta_{J-S}$ is 1.8 ms, which is determined by the first transport module. The second thermal time constant for $R\Theta_{S-A}$ is 307 s, which is determined by the second heat transport module.

IV. CONCLUSIONS

The transient response of the junction temperature of GaN LEDs with a typical packaged structure is examined. The results show that the heat transport in packaged LEDs is mainly influenced not only by heat transport from the chip to the heat slug, but also by heat transport from the heat slug to the ambient through the packaged heat sink. The thermal resistances and the corresponding thermal time constants can be obtained by measuring the transient junction temperature response curves of LEDs. Moreover, the heat transport in LEDs can be characterized by a bisection function. This technique can be useful in the analysis of thermal properties and in obtaining information on the thermal structure of packaged high-power GaN-based LEDs.

REFERENCES

[1]Tzer-En Nee, Jen-Cheng Wang, Hui-Tang Shen, et al. Joural of Applied physics, 102,033101 (2007)

[2]Lan Kim and Moo Whan Shin, IEEE Transactions on components and packaging technologies, vol.30, no. 4, pp. 632– 636, Dec. 2007

Running wave form Extended states in charactering the photocurrent of multiple quantum well and superlattice structured GaAs/AlGaAs solar cells

X. F. Yang[*], Y. S. Liu, and X. F. Jiang

Jiangsu Laboratory of Advanced Functional materials and College of Physics and Engineering, Changshu Institute of Technology, Changshu, China 215500

Abstract

The numerical simulations of the contribution of GaAs/AlGaAs multiple quantum well and superlattice to photocurrent in solar cells are presented. Effects of thickness of barrier and period on the extended and localized states, photocurrent have been investigated. Running wave method of calculation on the extended states of the two structures are adopt and compared.

I. INTRODUCTION

A combination of increased energy prices and fears over global warming are pushing up demand for photovoltaics (PVs), which drives many governments offering a lot support on the generation solar cells (SCs). Many theoretical and experimental works have been done to seek for a highly efficiency and lower price SCs [1-3], and with the improving manufacturing technology, the solar cells inserted with multiple quantum well (MQW) or superlattice (SL) have a optimistic market. By changing the width of barrier and period of quantum well, the SCs with low-dimensional structure present unique photocurrent characters. To improve the device performance, detail analysis of mechanisms of structure-related characteristics for SCs structured with MQW or SL is needed [5-12]. Different structural details such as layer thicknesses provide key insights into device simulation and design [4].

In this paper, the electrons and holes are assumed in running wave form to investigate the extend states and corresponding photocurrent in MQW or SL structured SCs. Effects of thicknesses of barrier and period on the performance of SCs are theoretically studied compared with single QW.

II. METHODS

The average optical transition rate in the multiple quantum well (QW) or superlattice (SL) from the initial state to the final state can be obtained by using the Fermi's golden rule [4]. The corresponding photocurrent (PC) spectrum from extended of valence band (VB) to extended states of conduction band (CB) is determined by the absorption coefficients and Fermi distribution of electrons and holes. Also is the PC spectrum

from the inverse transition. In this article, we suppose the carriers transporting in a running wave form as

$$\psi(z) = \begin{cases} e^{ikz} + re^{-ikz}, & z < -W/2 \\ Ae^{ikz} + Be^{-ikz}, & -W/2 \le z < W/2 \\ te^{ikz}, & z \ge W/2 \end{cases}$$

Here, we suppose the QW is along z direction with the width of QW being W. In the following calculation we suppose the electrons entering from one side while the holes from another side of the structure.

III. RESULT AND DISCUSSION

Figure 1 compares the extended states of electrons and holes at 2 and 10 periods when the well width being 5 nm and the barrier being 30 nm respectively.

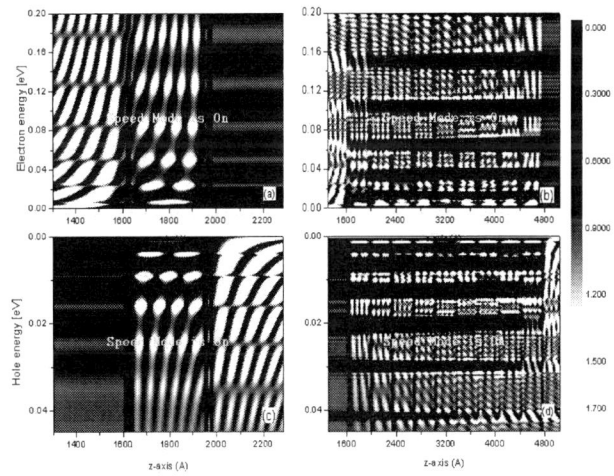

Fig. 1. The amplitude of extended state of electrons (upper) and holes (down) at 2 periods ((a) and (c)) and 10 periods ((b) and (d)) multiple QW.

Fig 1(a) and (c) show the amplitude of extended states of 2 periods QW after supposing the electrons entering from the left side and the holes entering from the right side. The extended states for 10 periods are presented in fig 1(b) and (d).

---

[*] Corresponding author: xflucky@mail.sitp.ac.cn

We can see that the holes are extensively reflected near the edge of QW. The sublevels of electrons and holes can be obviously observed at 10 periods QW. And the probability for electrons and holes to transit from the MQW is extensively reduced when the carriers' energy is very low, which will derive the PC in 10-periods MQW decreasing in turn.

Figure 2 shows the PC spectrum as a function of the thickness and periods of QW. In order to have a clearing view on the relationship of the barrier width and period of QW, the PC spectrum of single QW is inserted in every figure. Fig. 2(a), (b)and (c) present the PC spectrum at different period of 2, 3, 4, 5 and 10 with 2 nm barrier width, and we can see that it is a typical SL structure (the value of the barrier width is around that of the QW width). As shown in Fig. 2(a), (b) and (c), the PC spectrum is obviously improved comparing the single QW which agree well with the experimental results. However, for the 10 period SL, the PC is decreased because the electron is obviously reflected in the extended states.

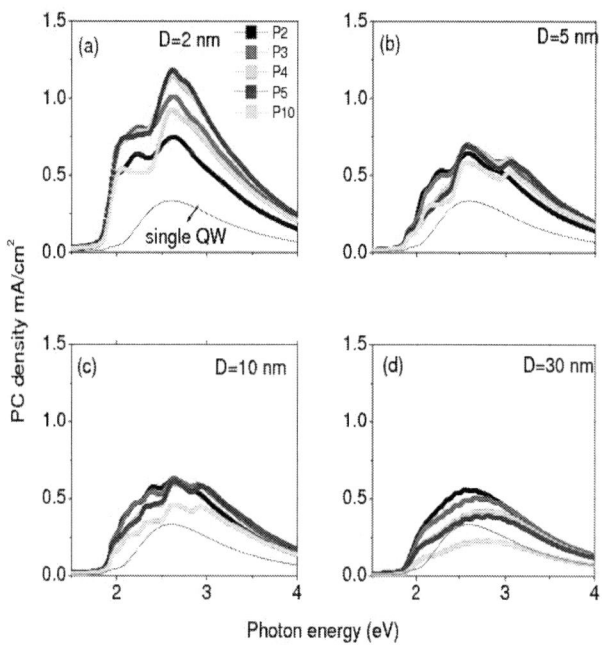

Fig. 2. PC spectrum as a function of barrier thickness (D) and periods (P). Here (a), (b), (c) and (d) correspond the PC at 2, 5, 10, 30 nm width barrier and P2 respects the period of QW being 2.

The PC spectrum as a function of period of MQW is shown in Fig. 2(d). Here the width of barrier is 30 nm which is 6 times of QW width, which means the structure is a typical MQW structure. The running wave assumption does not agree well with the experimental results. Usually, the carrier will relax in the barrier and reach a new balance, then transport to another QW. The lake of relax process in running wave assumption will introduce the decrease of PC due to extensively reflect of carriers as shown in Fig. 1(b) and (d). The Bloch wave assumption should be considered for the extended states distribution in MQW structure.

IV. CONCLUSION

The extended states of multiple quantum well (MQW) and superlattice (SL) structure in solar cells (SCs) are calculated in running wave assumption. The corresponding PC spectrum is simulated at different barrier and periods. The simulation results show that the carriers are extensively reflected at the bandage of MQW and SL structure, and the PC spectrum can be calculated for short period of SL when the carriers act in running wave form, while the running wave assumption can not work well in MQW and long period structure.

ACKNOWLEDGEMENTS

The authors thank the supports of the National Natural Science Foundation of China (NSFC) under Grants No. 11147162 and CSC-KTH program.

REFERENCES

[1] Lee, Y. J., Lee, M. H., Cheng, C. M., Yang, C. H., *Appl. Phys. Lett.*, vol. 98, pp. 263504, 2011.

[2] Liang, X. H., Zhang, Q. H., Lay, M. D., Stickney, J. L., *J. Am. Chem. Soc.*, vol. 133, pp. 8199, 2011.

[3] Yazawa, Y., Minemura, J., Tamura, K., Watahiki, S., Kitatani, T., Warabisako, T., *Solar Energy Materials and Solar Cells*, vol. 50, pp. 163, 1998.

[4] Y. F, *Superlattices and Microstructures*, vol. 30, pp. 69, 2001.

[5] A. Freundlich and A. Alemu, *Phys.stat. sol.*, vol. 2, pp. 2978, 2005.

[6] K. Tanaka, N. Kotera, and H. Nakamura, *J. Appl. Phys.*, vol. 85, pp. 4071, 1999.

[7] K. Tanaka and N. Kotera, Proc. 19th International Conference on Indium Phosphide and Related Materials, pp. 91–94, 2007.

[8] S. Kiravittaya, U. Manmontri, S. Sopitpan, S. Ratanathammaphan, C. Antarasen, M. Sawadsaringkarn, S. Panyakeow, *Solar Energy Materials and Solar Cells*, vol. 68, pp. 89, 2001.

[9] T. H. Loh, T. Miyamoto, T. Takada, F. Koyama, K. Iga, *J. Crystal Growth*, vol. 172, pp. 291, 1997.

[10] X. S. Jiang and P.K.L. Yu, *Appl. Phys. Lett.*, vol. 65, pp. 2536, 1994.

[11] M. C. Lynch, I. M. Ballard, D. B. Bushnell, J. P. Cconnolly, D. C. Johnson, T. N. D. Tibbits, K. W. J. Barnham, *J. Materials Sci.*, vol. **40**, pp. 1445, 2005.

[12] A. Alemu, J. A. Coaquira, and A. Freundlich, *J. Appl. Phys.*, vol. 99, pp. 084506, 2006.

A dual-band polarization insensitive metamaterial absorber with split ring resonator

B. Ni[a], X. S. Chen[a,*], L. J. Huang[a], J. Y. Ding[a], G. H. Li[a], and W. Lu[a,*]

[a] National Lab for Infrared Physics, Shanghai Institute of Technical Physics, Chinese Academy of Sciences
500 Yu Tian Road, Shanghai, China 200083

Abstract

A dual-band polarization insensitive absorber has been designed and studied in this work. Unlike previous dual band absorber composed of composite structures, only one square metal ring with a slit at the middle of each side has been designed to achieve dual-band absorption. The calculated results show two distinct absorption peaks of 0.96 at 10 GHz and 0.99 at 20 GHz. This dual-band absorber has many potential applications in scientific and technological areas because of its excellent absorption characteristics and concise structure.

I. INTRODUCTION

Metamaterials have attracted considerable interest due to their unique properties such as negative refraction, cloaking and superlensing. [1-3] Recently, metamaterials are used to design the perfect electromagnetic energy absorber. [4] After then, perfect absorber over wide range of frequencies, including microwaves, [5] THz, [6-7] IR,[8] and optical [9-10] have been investigated. Many characteristic absorber, for example, polarization insensitive absorption, [5] wide angle absorption [10] or dual-band absorption, [6-7] have achieved with various structures. It is worth paying more attention on dual-band absorber because of its potentially wide application areas such as transceiver system，spectroscopic imagers or detectors. Lately, polarization insensitive dual-band absorbers work at different frequency band have been reported. [11-12] These dual-band absorbers are designed by using composite square rings structure with different geometrical dimensions. A small ring is embedded in the bigger one. Each absorption peak corresponding to one of the composite rings. Moreover, due to the symmetric structure in the unit cell, the absorption peaks are insensitive to the polarization of the incident beam, which provides more efficient absorption for the nonpolarized incident beam. However, these structures are very complicated for both design and fabrication.

In this paper, we demonstrate a concise structured dual-band polarization insensitive absorber. Only one metal square ring simply with slits at the middle of each side are used to design dual band absorber. This split ring and metal plane separated by a dielectric spacer structure. Two distinct nearly 100% absorption peaks appear at 10 GHz and at 20GHz. The positions of the two peaks can be controlled by the width of the slits. In addition, both of the two peaks are also polarization insensitive.

II. Numerical model and simulations

Figure 1 shows the schematic structure of the dual-band absorber. The dual-band absorber consists of three layers. The top layer consists of an array of square copper ring with a slit at the middle of each side. The middle and bottom layer are FR4 and copper film, respectively. The parameters of the absorber, as shown in Figure 1, are a=10 mm, L=9.6 mm, w=0.5 mm, t_1=0.018 mm, t_2=0.78 mm and t_3=0.018 mm. g is the slit width and be set with different values in this letter. The electric conductivity of copper is σ=5.8×10^7 s/m and the electric permittivity of dielectric spacer is ε=$ε_1$+i$ε_2$=4+0.08i, which corresponds to a loss tangent tan(δ)=0.02. The incident wave propagates along z axis and the boundary conditions in the metal plane are set as periodic boundary conditions.

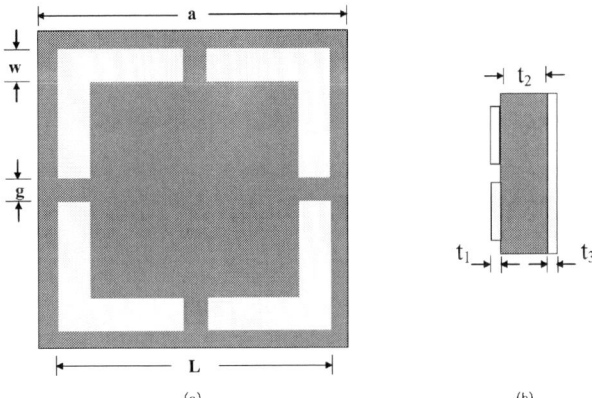

Figure 1. (Color online) Schematic structure of the dual-band absorber: (a) Top view and (b) side view of the absorber layers

To investigate the absorption spectra of the dual-band absorber, the finite-difference-time-domain (FDTD) method is employed. Assume that the absorption, reflectance and transmittance spectrums are A(ω), R(ω), and T(ω), respectively. The absorption A(ω) can be obtained by A(ω)=1-R(ω)-T(ω). Due to the metal ground plane, the transmission is zero, then the absorptivity A(ω)=1-R(ω).

Figure 2 presents the absorption spectra with the width of slits equals 1.1mm. The calculated results show two distinct absorption peaks of 0.96 at 10 GHz and 0.99 at 20 GHz. Compared to the absorber reported, our design are more simple

---

[*] Corresponding author: xschen@mail.sitp.ac.cn, luwei@mail.sitp.ac.cn.

978-1-4673-1602-6/12 $31.00 © 2012 IEEE

and easily fabricated. Besides, we also investigate the effect of the width (g) of slits on absorption spectra. It is found that the two absorption peaks are strongly influenced by the width of the slits, As shown in Figure 3, with g increasing from 0.5mm to 1.5mm, both two peaks are blue shift. The inner picture shows peak 2 around 20 GHz is more sensitive to the width changing compare to the other peak around 10 GHz. However, it should be emphasized that the absorption of the two peaks remain higher than 0.9 while g takes any value mentioned above.

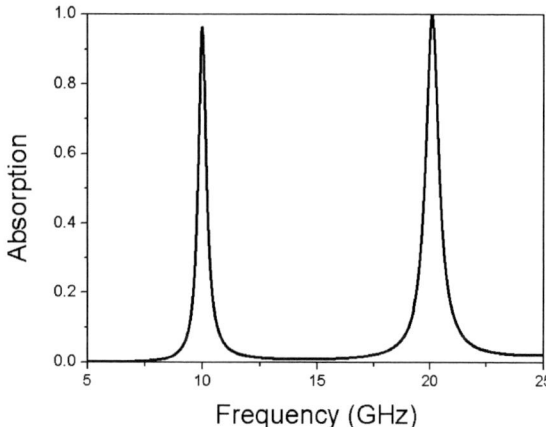

Figure 2 (Color online) Simulated absorption spectra of the dual-band absorber with a=10 mm, L=9.6 mm, w=0.5 mm, g=1.1 mm, t_1=0.018 mm, t_2=0.78 mm and t_3=0.018 mm.

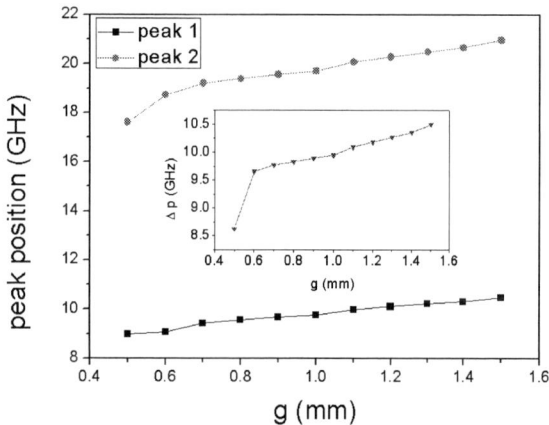

Figure 3 (Color online) The position of the absorption peaks as a function of the width of the slits. Inner: peak spacing of peak 2 and peak 1 with different slit width.

III. CONCLUSION

We demonstrate a dual-band polarization insensitive absorber in this letter. The calculated results show two distinct absorption peaks of 0.96 at 10 GHz and 0.99 at 20 GHz. In addition, the positions of the two peaks are found strongly influenced by the width of the slits. The peak 2 is more sensitive to the width changing compare to the peak 1. This dual-band absorber provided a promise application respect owning to its excellent characteristics and concise structure.

ACKNOWLEDGEMENTS

This work was supported in part by the State Key Program for Basic Research of China grants (2011CB922004); the National Natural Science Foundation of China grants (61006090, 10725418, 10874196, 10990104, 11104299 and 60976092); the Fund of Shanghai Science and Technology Foundation grants (09DJ1400203, and 10JC1416100).

REFERENCES

[1] D. R. Smith, Willie J. Padilla, D. C. Vier, S. C. Nemat-Nasser, and S. Schultz, "Composite Medium with Simultaneously Negative Permeability and Permittivity", Phys. Rev. Lett. 84, 4184–4187, 2000

[2] D. Schurig, J. J. Mock, B. J. Justice, S. A. Cummer, J. B. Pendry, A. F. Starr and D. R. Smith, "Metamaterial Electromagnetic Cloak at Microwave Frequencies", Science, Vol. 314, no. 5801 pp. 977-980, 2006

[3] J. B. Pendry, "Negative Refraction Makes a Perfect Lens", Phys. Rev. Lett. 85, 3966–3969, 2000

[4] N. I. Landy, S. Sajuyigbe, J. J. Mock, D. R. Smith, and W. J. Padilla, "Perfect Metamaterial Absorber", Phys. Rev. Lett, 100, 207402, 2008

[5] Na Liu, Martin Mesch, Thomas Weiss, Mario Hentschel, and Harald Giessen, "Infrared Perfect Absorber and Its Application As Plasmonic Sensor", Nano Lett, 10, 2342–2348, 2010

[6] Q.-Y. Wen, H.-W. Zhang, Y.-S. Xie, Q.-H. Yang, and Y.-L. Liu, "Dual band terahertz metamaterial absorber: Design, fabrication, and characterization", Appl. Phys. Lett. 95, 241111, 2009

[7] H. Tao, C. M. Bingham, D. Pilon, K. Fan, A. C. Strikwerda, D. Shrekenhamer, W. J. Padilla, X Zhang, and R. D. Averitt, " A dual band terahertz metamaterial absorber". J. Phys. D: Appl. Phys. 43, 225102, 2010.

[8] J. A. Mason, S. Smith, and D. Wasserman, "Strong absorption and selective thermal emission from a midinfrared metamaterial", Appl. Phys. Lett. 98, 241105, 2011

[9] J. Hao, J. Wang, X. Liu, W. J. Padilla, L. Zhou, and M. Qiu, "High performance optical absorber based on a plasmonic metamaterial ", Appl. Phys. Lett. 96, 251104, 2010.

[10] Chihhui Wu, Burton Neuner III, and Gennady Shvets , "Large-area wide-angle spectrally selective plasmonic absorber", Phys. Rev. B 84, 075102, 2011

[11] Yong Ma, Qin Chen, James Grant, Shimul C. Saha, A. Khalid, and David R. S. Cumming , "A terahertz polarization insensitive dual band metamaterial absorber", Optics Letters, Vol. 36, Issue 6, pp. 945-947, 2011

[12] Xiaopeng Shen, Tie Jun Cui, Junming Zhao, Hui Feng Ma, Wei Xiang Jiang, and Hui Li, "Polarization-independent wide-angle triple-band metamaterial absorber", Optics Express, Vol. 19, Issue 10, pp. 9401-9407, 2011

Physical model of an Optical Memory Cell with coupling quantum dots

L. Fan, F. M. Guo

Key Laboratory of Polar Materials and Devices, Ministry of Education,
East China Normal University, Shanghai, 200241, China
Email fmguo@ee.ecnu.edu.cn

Abstract: **The physical model was founded by Crosslight Apsys software for new type of photonic memory cell based on a quantum dot (QD)-quantum well (QW) hybrid structure. The physical mechanisms involved such as interband optical transition of quantum dots. The scan conditions and iterative algorithm was also set up to finish solving. Photon storage process has well proved based on I-V curve and transient time response obtained from the model. These are crucial in the signal readout-circuit design afterward.**

Key Words: **photon storage; quantum-dots; quantum-well; APSYS; physical model**

I. INTRODUCTION

There has been great interest in studying the storage and retrieval of the semiconductor optical memory cells [1-5]. In such devices, incoming optical signals are first stored as spatially separated electron-hole pairs and then retrieved by bringing stored electrons and holes to recombine at the same location radioactively. Generally speaking, a normal memorizer require over 10 million atoms for storage of a single bit of information while this kind of optical memory cell only needs a few thousand atoms involved to complete the same operation, which means smaller, cheaper and faster[4]. Crosslight Apsys package could be effectively applied to model performance of the optical memory cell and design.

This paper discusses the physical modeling of an Optical Storage Cell with coupling quantum dots in a well by using Crosslight Apsys software. First, an asymmetrical structure of quantum well-quantum dots was established by means of graphic user interface. Second, computer was programmed about the physical mechanisms involved such as interband optical Transition for quantum dots. The scan conditions and iterative algorithm was also set up to finish solving. Finally, some valuable diagrams are available including band, I-V and transient response, which give good agreements with measurements. Photon storage process is well proved based on I-V curve and transient time response obtained from the new model.

II. MODEL FO UNDED

Structure of the optical memory cell with quantum dots in well is shown in Fig.1 [5]. After a 1μm Si-doped (10^{18} cm^{-3}) GaAs buffer layer and an undoped 30 nm GaAs spacer, the undoped double barrier structure was deposited in the sequence of the first 25nm AlAs barrier, a 3nm GaAs interlayer, a 6 nm In0.15Ga0.85As QW, a 45 nm GaAs wide well, a 1.8 ML self-assembled InAs QD layer with a 5nm GaAs overlayer, and the second 25nm AlAs barrier. On the top, an undoped 30nm GaAs spacer and a Si-doped (10^{18}cm^{-3}) 30nm GaAs capping layer were overgrown. The ohmic contacts were separately made to the top and back contact layers. The top contacts was placed onto the cap layer with square aperture (50×50μm^2) left for the optical access.

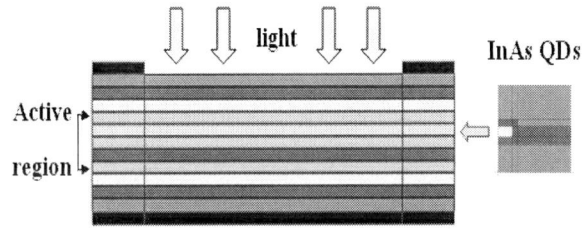

Fig.1 Structure of the memory optical cell

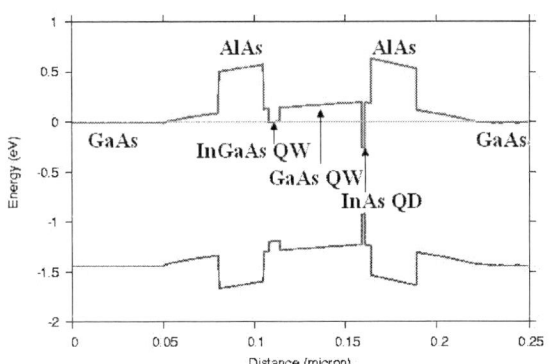

Fig.2 Band diagram profile along epitaxial direction

978-1-4673-1602-6/12 $31.00 © 2012 IEEE   57

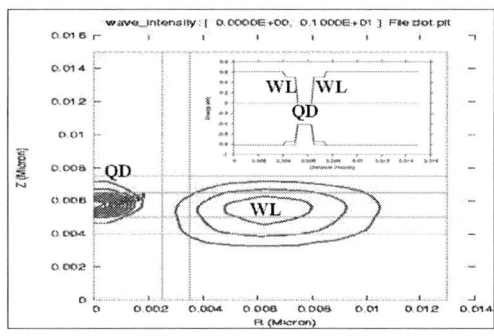

Fig.3 Band diagram profile of QD and its wave intensity

The band diagram of the device is depicted in Fig.2. Electron-hole pairs in the active region were first generated by laser pulse, and then dissociated by the reverse bias. Electrons generated in the wide well or in the QDs would drift or tunnel toward the lower potential energy and eventually would be captured by the thin InGaAs QW. The similar process would also occur to holes, except that they would drift in the opposite direction and would be trapped in the QDs. Due to the spatial separation of electrons and holes as well as the additional in-plane localization provided by the QDs, as the structure is reversely biased, excess electrons and holes can be stored at some time in the thin QW and QDs respectively. Fig.3 shows the band diagram profile of QD and its wave intensity.

III. RESULT AND DISCUSS

The photon storage characteristics of the device have been proved by its photoelectric response. Under the illumination of the wavelength 650 nm, the response current shows a step-like enhancement when reverse biased as shown in Fig.4. By contrast, whose photon energies were smaller than GaAs band gap, the step-like enhancement no longer appears. It is believed that the photovoltaic effect caused by the stored spatially separated electron-hole pairs pulls down the GaAs X minimum at the

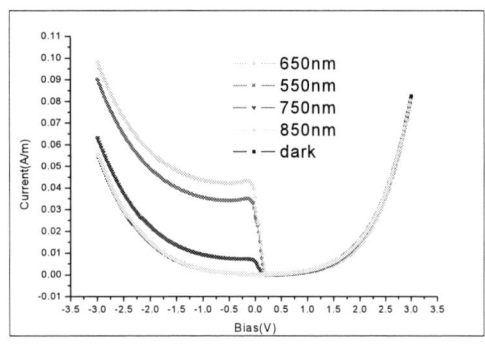

Fig. 4 Photocurrent & dark current versus bias voltage

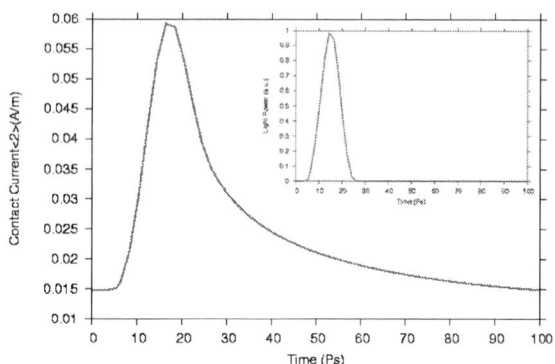

Fig.5 Transient response of the device at light pulse

outgoing interface of the AlAs emitter barrier, which eventually initiate the new Γ-X-X tunneling compared with that in the dark and result in the observed step-like photocurrent response.

As shown in Fig.5, the transient time response of the device demonstrates its photon storage characteristics. It still shows the stimulus of light pulse. The falling transient edge of the photocurrent as the photoexcitation turns off, mainly maps the decaying of electrons and holes, which were previously stored in the cell during the illumination. Its time constant is a measure of photonic memory time.

IV. CONCLUSION

An Optical Memory Cell with coupling quantum dots was successfully simulated. The photon storage process may well been proved by advanced numerical simulation.

ACKNOLEDGEMENTS

The work was supported by National Scientific Research Plan (2006CB932802, 2011CB932903) and State Scientific and Technological Commission of Shanghai (No. 078014194, 118014546).

REFERENCES

[1] Zimmermann S, Wixforth A, Kotthaus JP, et al. A Semiconductor-based photonic memory cell[J]. Science, 1999, 283 : 1292-1295.
[2] Tao S Q, Jiang Z Q, Yuan Q, Liu G Q and Xu M 2000 Chin. Phys. Lett. 17 675
[3] Li C D, Wang D L, Luo L, Yang H, Xia Z J and Gong Q H 2001 Chin. Phys. Lett. 18 541
[4] Lundstrom T, Schoenfeld W, Lee H, et al. Exciton storage in semiconductor self-assembled quantum dots[J]. Science, 1999.28 6: 2312-2314
[5] BIAN Song-Bao,TANG Yan, LI Gui-Rong, et al. Photon storage in optical memory cells based on a semiconductor quantum dots-quantum well hybrid structure [J], Chinese, Phys.Lett., 2003, 20: 1362-1365

General design of compact waveguide coupler with homogeneous media

L. J. Huang[a], X. S. Chen[a, *], B. Ni[a], G. H. Li[a], Z. F. Li[a], and W. Lu[a, *]

[a] National Lab for Infrared Physics, Shanghai Institute of Technical Physics, Chinese Academy of Sciences,
500 Yu Tian Road, Shanghai, China 200083

Abstract

Based on finite embedded coordinate transformation method, a general transformation is given to design a compact waveguide coupler. It removes the limitation of material inhomogeneity and makes the coupler be more easily realized. It also offers us much freedom and flexibility in choosing the appropriate material parameters to practically implement the device. Full wave simulation based on the finite element method is performed to confirm the performance of the waveguide coupler.

I.   INTRODUCTION

Recently, transformation optics has attracted tremendous attention in the world since it provides scientists and engineers with previously unavailable ability and flexibility to manipulate the propagating behavior of the electromagnetic (EM) wave and control the spatial distribution of EM fields. It fully takes the advantage of the form invariance of Maxwell equations under spatial coordinate transformations with the material parameters in the physical space being anisotropic and inhomogeneous. Transformation optics technique is firstly adopted to design an invisibility cloak [1-5], which can guide incident wave smoothly streaming around the object so that observer outside the cloak cannot detect it. The rapid development of metamaterial makes the practical realization of an invisibility cloak become possible. The first concept cloak with reduced material parameters was practically implemented with ten concentric split-ring resonators at microwave frequency [3]. Besides the invisibility cloak, it is also employed to design many novel devices, such as EM field concentrator [6], EM field rotator [7], and illusion optics device [8].

More recently, finite-embedded coordinate transformation has been introduced by Rahm and utilized in the design of beam shifter and splitter [9]. It also led to some other interesting device, such as waveguide expander and compressor [10], waveguide bender [10, 11], and waveguide coupler [12-14]. However, waveguide couplers shown in Refs [12-14] require anisotropic and inhomogeneous medium, which bring obstacles in practical realization of the device. To circumvent the limitation of the material inhomogeneity, Xu et al proposed a new transformation inspired by the design of the carpet cloak introduced by Chen et.al [15, 16]. In this work, a general transformation is introduced to achieve the compact waveguide coupler which is made of homogeneous medium

and the coordinate transformation employed by Xu is one special case of our transformation. The compact waveguide coupler can be realized with homogeneous uniaxial medium as long as the trapezoid region (connection part) is divided into several triangle blocks. Full wave simulation on the basis of finite element method is employed to verify the performance of the designed device. Simulation results indicate that the EM wave perfectly couples from one rectangular waveguide to another without causing any reflection. The coupler holds great promise in the application of integrated photonic circuit.

II.   TRANSFORMATION PRINCIPLE OF WAVEGUIDE COUPLER

The geometry structure of the waveguide coupler is presented in Fig. 1. Different from the coordinate transformation employed in Refs [12-14], we just need to map four triangular blocks AOB, BOM, MON, and AON in the virtual space (x, y, z) into four corresponding areas AEB, BEC, CED, and AED in the physical space (x′, y′, z′). As a result, only homogeneous and anisotropic medium are required to realize the coupler experimentally. If we define four blocks AEB, BEC, CED and AED as Region I, II, III and IV,   the corresponding transformation functions can be expressed as follows,

$$
\text{(I)}\begin{cases} x' = a_{11}x + b_{11}y + c_{11} \\ y' = a_{12}x + b_{12}y + c_{12} \\ z' = z \end{cases}
\quad
\text{(II)}\begin{cases} x' = a_{21}x + b_{21}y + c_{21} \\ y' = a_{22}x + b_{22}y + c_{22} \\ z' = z \end{cases}
$$

$$
\text{(III)}\begin{cases} x' = a_{31}x + b_{31}y + c_{31} \\ y' = a_{32}x + b_{32}y + c_{32} \\ z' = z \end{cases}
\quad
\text{(IV)}\begin{cases} x' = a_{41}x + b_{41}y + c_{41} \\ y' = a_{42}x + b_{42}y + c_{42} \\ z' = z \end{cases}
\tag{1}
$$

$a_{i1}, b_{i1}, c_{i1}, a_{i2}, b_{i2}$, and c_{i2}, can be easily derived. It can be found that they only depend on the coordinate value of A, B, C, D, E, M, N, and O, respectively. Here, i=1, 2, 3 and 4 represent Region I, II, III, and IV, respectively.

Then, based on the transformation principle, the material parameters in the four regions can be calculated

$$
\overline{\overline{\varepsilon'}} = \overline{\overline{\mu'}} = \Lambda \bullet \Lambda^T / \det(\Lambda)
\tag{2}
$$

in which $\Lambda_i = (a_{i1}, b_{i1}, 0; a_{i2}, b_{i2}, 0; 0, 0, 1)$, and $\det(\Lambda_i) = a_{i1}b_{i2} - a_{i2}b_{i1}$, i=1, 2, 3, 4, which represents Region I, II, III and IV. It is obvious that all constitute tensors are constants. Therefore, the material parameters are homogeneous and anisotropic.

III.   SIMULATION RESULTS AND DISCUSSIONS

In this part, full wave simulation is carried out to indicate the performance of the device. The boundary conditions on the left and right side of waveguide are set as perfect matched layer. The left and right boundary conditions are defined as scattering

---

* Corresponding author: xschen@mail.sitp.ac.cn, luwei@mail.sitp.ac.cn.

boundary condition. The other outer boundary conditions are set as perfect magnetic conductor. The incident wave propagates from left to right along x axis. The operating frequency of incident wave is 3 GHz. Without loss of generality, we only focus on the transverse electric (TE) incident wave with electric field polarized along z axis. The coordinate values of A, B, C, D, M, and N, are (0, 0.5), (0, -0.5), (0.5, -0.2), (0.5, 0.2), (0.5, -0.5), and (0.5, 0.5), respectively.

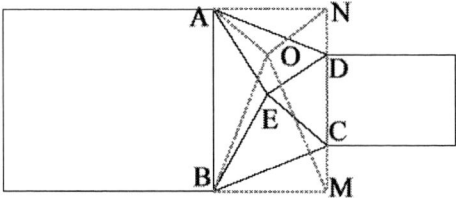

Figure 1 (Color online) Schematic structure of waveguide coupler.

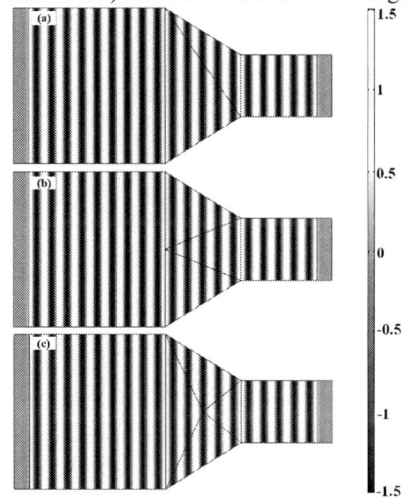

Figure 2 (Color online) Spatial distribution of electric field for different waveguide coupler when the trapezoid region is divided into two, three or four triangle blocks.

Electric field distribution of the device with different sets of material parameters are illustrated in Fig. 2. The trapezoid area consists of two, three, and four triangular regions, corresponding to the cases of Fig. 2(a) to (c). If the point E coincides with the point C, and simultaneously points O and M are overlapped, the coupler composes of two blocks. From Fig. 2(a), it can be seen that the EM wave perfectly couples from left waveguide to right waveguide without any reflection and energy loss. When the points O and E are overlapped and locate at the middle point of AB, as shown in Fig. 2(b), the device also performs well in transmitting EM field from left to right without causing any reflection. Here, it is worthy of pointing out that the device is degenerated into the case presented in Ref [15] when O (or E) locates at the side CD. Fig. 2(c) presents the electric field distribution of coupler when the points O and E are still overlapped but located in region ABCD. The device also has superior performance. Besides, the EM field inside the trapezoid area can be compressed and stretched without changing the functionality of the device as long as the coordinate values of M and N are varied.

IV. CONCLUSION

In summary, we introduce a general transformation to design the compact waveguide coupler. With the transformation, the coupler can be realized with homogeneous and anisotropic medium, which greatly reduces the difficulty of practical implementation of the device. Full wave simulation base on finite element method is performed to consolidate the good performance of the coupler.

ACKNOWLEDGEMENTS

This work was supported in part by the State Key Program for Basic Research of China grants (2011CB922004); the National Natural Science Foundation of China grants (61006090, 10725418, 10874196, 10990104, 11104299 and 60976092); the Fund of Shanghai Science and Technology Foundation grants (09DJ1400203, and 10JC1416100).

REFERENCES

[1] J. B. Pendry, D. Schurig, and D. R. Smith, "Controlling electromagnetic fields", Science **312**, 1780-1782 (2006)

[2] U. Leonhardt, "Optical conformal mapping", Science **312**, 1777-1780 (2006)

[3] D. Schurig, J. J. Mock, B. J. Justice, S. A. Cummer, J. B. Pendry, A. F. Starr, and D. R. Smith, "Metamaterial electromagnetic cloak at microwave frequencies", Science **314**, 977-980 (2006)

[4] L. J. Huang, D. M. Zhou, J. Wang, Z. F. Li, X. S. Chen, and W. Lu, "Generalized transformation for nonmagnetic invisibility cloak with minimized scattering", J. Opt. Soc. Ame, B **28**, 922-928 (2011)

[5] L. J. Huang, D. M. Zhou, J. Wang, Z. F. Li, X. S. Chen, and W. Lu, "Scattering characteristics of nonmagnetic invisibility cloak with minimized scattering", Opt. Communication **284**, 5523-5530 (2011)

[6] M. Rahm, D. Schurig, D. A. Roberts, S. A. Cummer, D. R. Smith, and J. B. Pendry, "Design of electromagnetic cloaks and concentrators using form-invariant coordinate transformations of Maxwell's equations", Photonics and Nanostructures: Fundamentals and Applications 6, 87-95 (2008)

[7] H. Y. Chen, and C. T. Chan, "Transformation media that rotate electromagnetic fields", Appl. Phys. Lett 90, 241105 (2007)

[8] L. J. Huang, D. M. Zhou, J. Wang, Z. F. Li, X. S. Chen, and W. Lu, "A generalized transformation to convert an arbitrary perfect electric conductor into another arbitrary dielectric object", Journal of Physics D: Applied Physics 44, 235102 (2011)

[9] M. Rahm, S. A. Cummer, D. Schurig, J. B. Pendry, and D. R. Smith, "Optical Design of reflectionless complex media by finite embedded coordinate transformations", Phys. Rev. Lett **100**, 063903 (2008)

[10] M. Rahm, D. A. Roberts, J. B. Pendry, and D. R. Smith, "Transformation-optical design of adaptive beam bends and beam expanders", Opt. Express **16**, 11555-11567 (2008)

[11] W. X. Jiang, T. J. Cui, X. Y. Zhou, X. M. Yang, and Q. Cheng, "Arbitrary bending of electromagnetic waves using realizable inhomogeneous and anisotropic materials", Phys. Rev. E **78**, 066607 (2008)

[12] P. H. Tichit, S. N. Burokur, and A. Lustrac, "Waveguide taper engineering using coordinate transformation technology", Opt. Express **18**, 767-772 (2010)

[13] K. Zhang, Q. Wu, F. Y. Meng and L. W. Li, "Arbitrary waveguide connector based on embedded optical transformation", Opt. Express 18 17273-17279 (2010)

[14] X. F. Zang, and C. Jiang, "Manipulating the field distribution via optical transformation", Opt. Express **18**, 10168-10176 (2010)

[15] H. Y. Xu, B. Zhang, G. Barbastathis, and H. D. Sun, "Compact optical waveguide coupler using homogeneous uniaxial medium", J. Opt. Soc. Ame, B **28**, 2633-2636 (2011)

[16] X. Z. Chen, Y. Luo, J. Zhang, K. Jiang, J. B. Pendry, and S. Zhang, "Macroscopic invisibility cloaking of visible light", Nat. Commun **2**, 176 (2011)

Simulation of Resonant Tunneling Structures: Origin of the *I-V* multi-peak and Plateau-like Behaviour

J. Wen, Q.C. Weng, L. Li, D.Y. Xiong[*]

Key Laboratory of Polar Materials and Devices, Ministry of Education, East China Normal University,
Shanghai 200241, People's Republic of China

Abstract-**Plateau-like behavior and multi-peak of the *I-V* curves of an AlAs/GaAs/AlAs double-barrier resonant tunneling diode combined with a layer of InAs QDs (QD-RTD) are simulated. Our simulation results show that the coupling between the energy level in the emitter QW (QD) and that in the central quantum well is the key point in understanding the origin of the *I-V* multi-peak and plateau-like structure. The embedded designed QD layer at the emitter spacer can enhance this effect. The effects of device temperature on the *I-V* characteristics are obtained. Our results provide the physical basis for understanding and utilizing the plateau-like behavior of *I-V* curves in designing resonant tunneling devices.**

I. INTRODUCTION

With the interest in electronic phenomena in quantum tunneling structure growing over the past two decades, the double barrier resonant tunneling structure (RTS) has been extensively studied due to its novel physical mechanism and device applications[1-6]. In this paper, plateau-like and multi-peak behaviours of the *I-V* curves of an AlAs/GaAs/AlAs double-barrier resonant tunneling diode (RTD) and that combined with a layer of InAs QDs (QD-RTD) have been simulated and analyzed in detail.

II. SIMULATION RESULTS

The samples in our experiment and simulation were grown by molecular beam epitaxy on semi-insulated GaAs (100) substrate. The material layers, from top to bottom, were piled as follows: a 50 nm highly n-doped GaAs (2×10^{18}cm^{-3}) as an emitter layer, a 150 nm *i*-GaAs spacer layer, a layer of self-assembled InAs QDs capped by 10 nm GaAs (for sample *B*), a 3 nm AlAs barrier, an 8 nm GaAs quantum well (QW), a 3 nm AlAs barrier, a 20 nm *i*-GaAs spacer layer and a 430 nm graded n-doped GaAs (from 1×10^{16}cm^{-3} to 1×10^{18}cm^{-3}) as a collector layer (denoted as sample A and B and showed in Figure1 and 2, respectively).

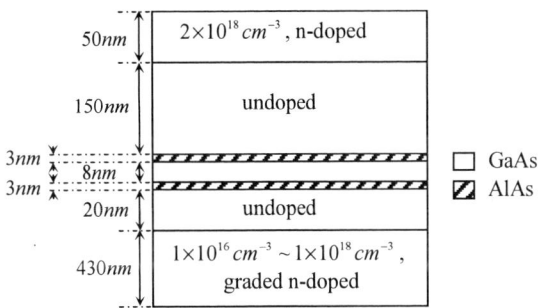

FIG.1. Structure of sample *A*

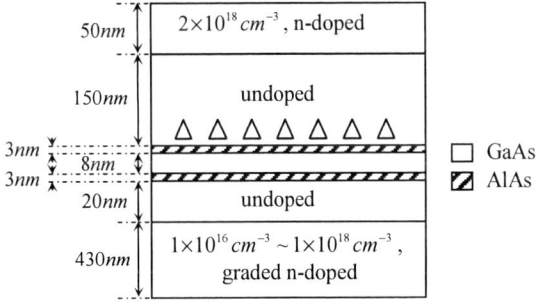

FIG.2. Structure of sample *B*

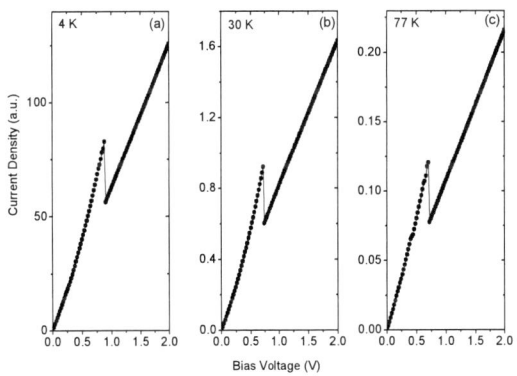

FIG. 3. Comparison of the current densities for sample *A*

---

*Correspondence Email: dyxiong@ee.ecnu.edu.cn

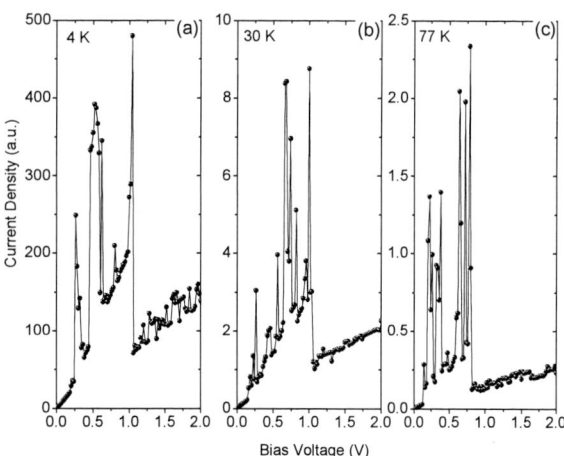

FIG. 4. Comparison of the current densities for sample *B*

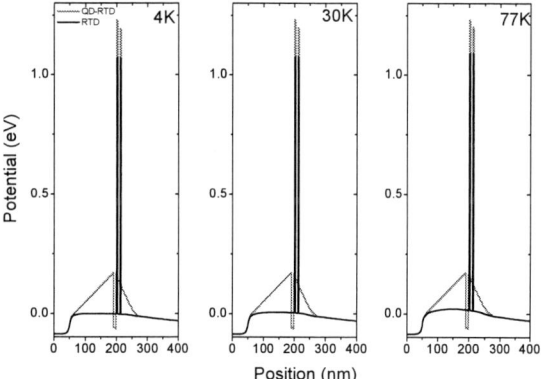

FIG. 5. Comparison of the potential between sample *A* and *B*

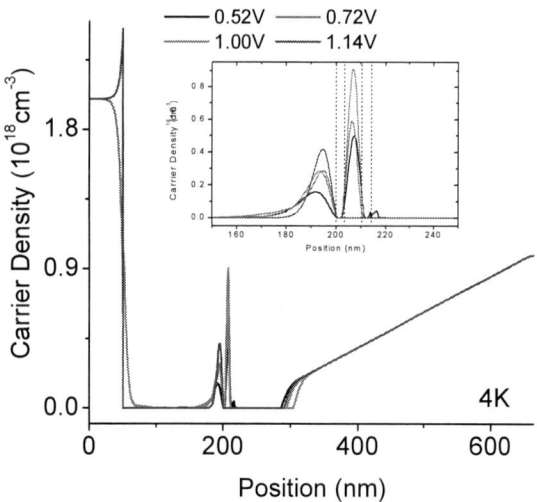

FIG. 6. Comparison of carrier density diagram of sample A
at bias of 0.52V, 0.72V, 1.00V and 1.14V at 4K

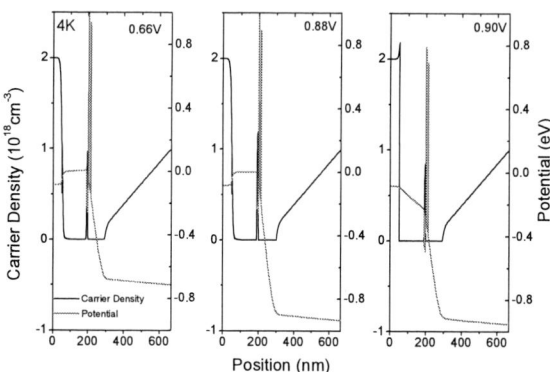

FIG. 7. Comparison of carrier density diagram of sample B with their potential diagram

The InAs QD layer has been treated as the InAs QW layer. Figure 5 indicates that the potential near the charged dots is clearly pushed up, which result in the multi-peak in *I-V* curves. The results show that the coupling between the energy level in the emitter QW (QD) and that in the central quantum well is the key point in understanding the origin of the *I-V* plateau-like and multi-peak behaviour and the embedded designed QD layer at the emitter spacer would enhance this effect. The effects of device temperature on the *I-V* characteristics are presented. The simulated results provide the physical basis for utilizing the plateau-like behavior of *I-V* curves in designing quantum resonant tunneling devices.

ACKNOWLEDGEMENTS:

This work was partially supported by the National Nature Science Foundation of China (Grant No. 61106092)

REFERENCE

[1] L. L. Chang, L. Esaki, and R. Tsu, Appl. Phys. Lett. 24, 593 (1974).

[2] J. O. Sofo and C. A. Balseiro, Phys. Rev. B. 42, 7292 (1990).

[3] L. Worschech, S. Reitzenstein, A. Forchel, Appl. Phys. Lett. 77, 3662 (2000).

[4] Kasturi Mukherjee, N. R. Das, J. Appl. Phys. 109, 053708 (2011).

[5] Peiji Zhao, H. L. Cui, D. Woolard, K. L. Jensen, F. A. Buot, J. Appl. Phys. 87, 1337 (2000)

[6] Wangping Wang, Ying Hou, Dayuan Xiong, Ning Li, Wei Lu, Wenxing Wang, et al. Appl. Phys. Lett. 92, 023508 (2008)

An Equivalent Circuit Model for the Long-Wavelength Quantum Well Infrared Detectors

L. Li,   Q.C. Weng,   J. Wen,   D. Y. Xiong[*]

Key Laboratory of Polar Materials and Devices, Ministry of Education, East China Normal University,
Shanghai 200241, China

Abstract-**We present an equivalent circuit model for AlGaAs/GaAs long wavelength quantum well infrared photodetectors (LW-QWIPs). Bias dependence of the dark current and photocurrent is described with the aid of analogue circuit modelling technique in TINA software. This model can be integrated with the readout circuit for the whole device circuit optimization further. The temperature dependence of dark current has also been incorporated into this circuit model. The designed parameters of the LW-QWIPs can be fed into this model as user-defined inputs to simulate the detector performance. The obtained results agree well with the experimental measurements.**

I. INTRODUCTION

The development of quantum well infrared photodetectors (QWIPs) has prompted its application in large area focal plane arrays[1-5]. Such applications require the integration of readout circuit with the photodetectors. For the optimization of the readout circuit, it is necessary to simulate the circuit along with the QWIPs. The operation of the QWIPs is usually represented by a set of equations based on its semiconductor device physics. However, these physical equations cannot be used directly when the photodetectors are connected with there readout circuit for the whole circuit simulation. To overcome this difficulty and make it applicable for a wide range of device operating conditions, this paper is trying to employ a few current sources in parallel with the device resistance and provide the bias and temperature dependence of the dark current.

II. Circuit design

The operation of AlGaAs/GaAs long wavelength QWIPs (LW-QWIPs) can be represented by two equations that take into account the dark current and photocurrent as a function of bias across the device and working temperature. These equations have been incorporated for circuit simulation using the analogue circuit modelling technology in TINA software. This model can be convenient integrated with the readout

circuit for the whole circuit simulations further. Circuit model has been simulated in TINA software. Transient and DC current-voltage(I-V) analysis has been performed. Simulated results from this model are in good agreement with the experimental data in a wide range of operating conditions verifying the validity of the model.

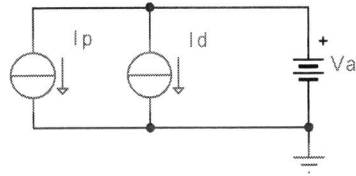

FIG. 1. Current composition of a QWIP

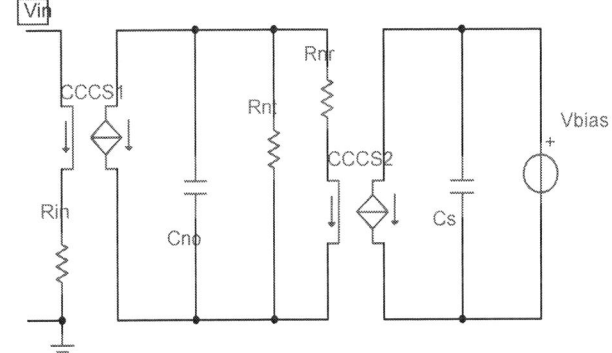

FIG. 2.   Photocurrent circuit model of QWIP

Figure 1 shows the current composition of the LW-QWIPs where I_p is photocurrent and I_d is dark current. The transient equivalent circuit has been shown in Figure 2. Figure 3 presents the photocurrent circuit model with input signal. Figure 4 shows the total current response of the pulse signal input where VM2 is the equivalent input voltage and AM1 is the response photocurrent in figure.3. Figure 4 shows the pulse signal input and current response. Figure 5 and 6 present the complete equivalent circuit. And figure 7 give a comparison of the simulated dark current and the experiment

*Correspondence Email: dyxiong@ee.ecnu.edu.cn

data at different working temperatures.

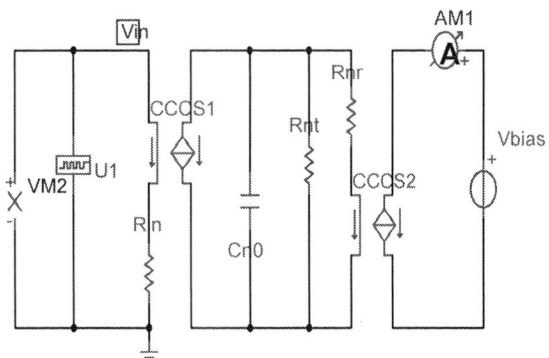

FIG. 3. Photocurrent circuit model with input signal

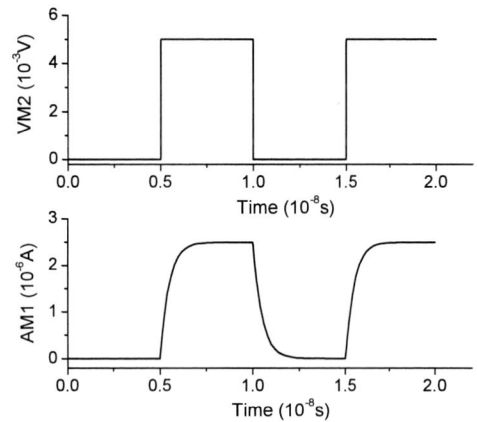

FIG. 4. Pulse signal input and current response

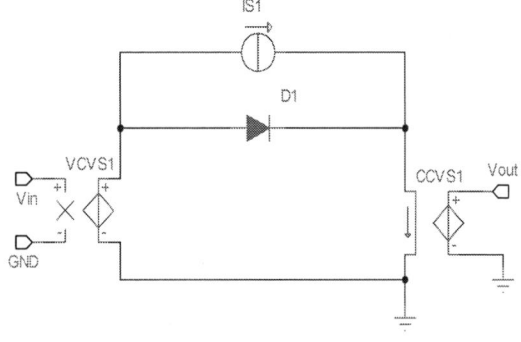

FIG. 5.   Equivalent circuit of $n(V)$

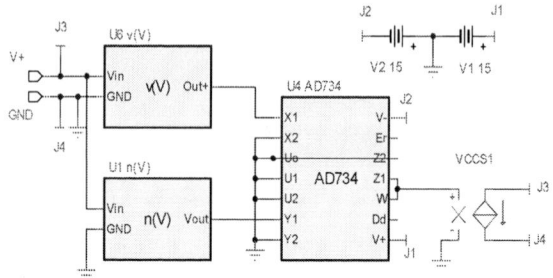

FIG. 6.   Dark current equivalent circuit

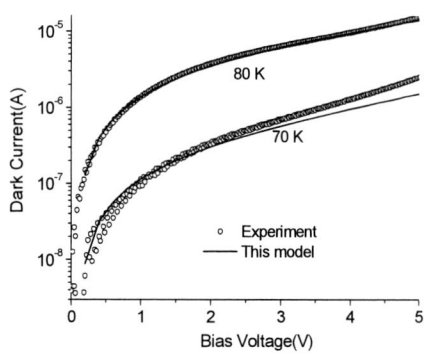

FIG. 7.   Comparison of the simulated dark current and the experiment data

Acknowledgements:

This work was partially supported by the National Nature Science Foundation of China (Grant No. 61106092)

REFERENCE

[1] B. F. Levine. Quantum well infrared photodetectors. J. Appl.Phys.,1993, 74: 1-81.

[2] H. C. Liu, B. F. Levine, J. F. Andersson. Quantum Well Intersubband Transition Physics and Devices. Dordrecht[Netherlands]:Kluwer, 1994.

[3] S. D. Gunapala, H. C. Liu, H. Schneider. Proceedings of Workshop on Quantum Well Infrared Photodetectors QWIP 2000. Holland: Elsevier, 2000.

[4] SHEN S C. Comparison and competetion between MCT and QW structure material for use in IR detectors.Microelectronics Journal.1994,25(8):713-739

[5] Y. Fu, W. Lu, Physics of Semiconductor Quantum Devices,BeiJing, Sciencep, 2005

Analytic solution of the nonlinear equation

Hyuk-jae Lee, Seok Lee, Deok Ha Woo, Taik Jin Lee, and Jae Hun Kim*

Korea Institute of Science and Technology
Seoul, 130-650, Korea, jaekim@kist.re.kr

Abstract- **We analytically solve the nonlinear wave equation of the beam, which travels through the nonlinear Kerr medium. The tanh function method, a powerful method solving the traveling wave equation, is applied to the self-guiding light.**

I. INTRODUCTION

The electromagnetic fields are satisfied with the Maxwell's and wave equations. In the case that light travels through a nonlinear medium, we should consider nonlinear effect from the nonlinear medium. However, it is very hard to analytically solve the nonlinear equation. Numerical approach has been developed to solve the nonlinear traveling electromagnetic field [1]. The split-step Fourier method is well known as the approach understanding the nonlinear effects in the optical fibers [2]. However, this method takes a long time for obtaining the results. Thus, several different finite-difference schemes were suggested as other methods to analyze nonlinear effects [3]. The inverse scattering method was applied to nonlinear partial differential equation for analytic solution [4]. However, the applications of this method are limited to some specific cases.

Recently, the tanh function method, applicable to various cases, is introduced as a new and most effective method for finding exact solutions of a nonlinear wave equation satisfied with a traveling wave through the nonlinear medium [5]. Furthermore, some studies showed that the tanh function method could be extended to the spatial types of nonlinear equations [6]. The reason that these methods were possible is due to powerful software that can deal with the tedious algebraic computation.

In this paper, we applied the generalized tanh function method to find the analytic solution of the nonlinear equation describing the traveling wave through the nonlinear Kerr medium.

II. SOLVING THE NONLINEAR PARTIAL DEFFERENTIAL EQUATION

If we consider the response of the optical beam traveling through the substantial thickness of nonlinear homogeneous medium to a harmonic electric field, the self-guiding of light is described mathematically by Helmholtz equation. For simplification [7], we assume that the electric field varies slowly in the z-direction and does not vary in the y-direction. Using the approximation and the inverse Fourier transformation in time, the Helmholtz equation becomes

$$\frac{\partial^2 E}{\partial x^2} - i\alpha \frac{\partial E}{\partial z} + \beta \frac{\partial^2 E}{\partial \partial z} + \chi(\mid E \mid^2 E) = 0. \quad (1)$$

For the simplification, we used the constants α, β and χ for complex constants.

For applying to the tanh function method, the first step is to introduce the wave transformation $E(x,z,t) = U(\xi)$, $\xi = x + \eta z + \lambda t$, and changes (1) to an ordinary differential equation. The next step is to introduce a new variable $T = T(\xi)$, which is the solution of the Riccati equation, $T' = k + T^2$. Hence all derivatives of T with respective to ξ are polynomial in T as

978-1-4673-1602-6/12 $31.00 © 2012 IEEE

$T'' = -2T(k + T^2)$.

To find the solution, we propose the polynomial solution $U(T) = \sum_{v=0}^{M} a_v T^v$, where the positive integer M can be determined by balancing the highest derivative term with nonlinear terms. Here we can obtain $M = 1$ after simple calculation. We try the polynomial solution as

$$U(T) = a_0 + a_1 T . \qquad (2)$$

After we substitute (2) into the ordinary differential equation of (1), the constants $k, \eta, \lambda, a_0, a_1$ are obtained as setting all coefficients of T^v to zero. Then we get the algebraic equations of the constants. a_0, a_1 and k are given by solving the algebraic equation,

$$a_0 = 0, \quad a_1 = -\frac{\sqrt{2(1+\eta\lambda\beta)}}{\sqrt{\chi}}, \quad k = \frac{(a\eta - b\eta\beta)^2}{8(1+\eta\lambda\beta)} \quad (3)$$

and the others are arbitrary.

The final solution of (1) is given by

$$E(x,z,t) = \frac{\sqrt{2}\sqrt{k(1+\eta\lambda\beta)}}{\sqrt{\chi}} e^{i(ax + k_0 z + bt)}$$
$$\times \tanh(\sqrt{k}(x + \eta z + \lambda t), \quad (4)$$

where $a = \frac{1}{2}(a\eta - b\eta\beta)$ and b, c, η, λ are arbitrary constants. Figure 1 shows the wave propagation based on (4).

III. CONCLUSION

We showed that the generalized tanh function method can be applied to the nonlinear partial differential equation describing the traveling light through nonlinear homogeneous medium. This shows that the generalized tanh function method is more applicable than the original tanh function method. Equation (2) describes the traveling wave that is similar to soliton solution that travels steadily without a variation.

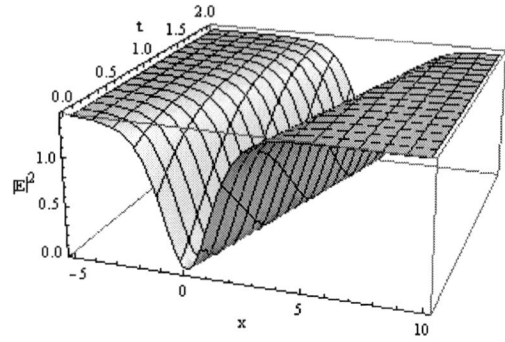

Fig. 1. Propagation of the square of (4) after fixing the z coordinate and the constants

ACKNOWLEDGMENT

This work was partially supported by the public welfare & safety research program (2010-0020796) and the Pioneer Research Center Program (2010-0019313) through the National Research Foundation of Korea (NRF) funded by the Ministry of Education, Science and Technology (MEST).

REFERENCES

[1] Govind and P. Agrawal, *Nonlinear Fiber Optics*, 3rd ed., Academic Press, San Diego, CA, 2001.

[2] R. A. Fisher and W. K. Bischel, "The role of linear dispersion in plane-wave self-phase modulation," App. Phys, Lett. Vol. 23, pp661-663, December 1973.

[3] T. R Taha and M. I. Ablowitz, "Analytical and numerical aspects of certain nonlinear evolution equations. II. Numerical, nonlinear Schrödinger equation," J. Comput. Phys. vol. 55, pp203-230, Agust 1984.

[4] W. Malfliet, "Solitary wave solutions of nonlinear wave equations," *Amer. J. Phys.* vol. 60, pp. 650-654, July 1992.

[5] E. Parkers, "Eaxt solutions to the two-dimensional Korteweg-de Vries-Burgers equation," *J. Phys.* vol. A 27, pp. L497-L501, July 1994.

[6] S. A. El-Wakil and M. A. Abdou, "Modified extended tanh function method for solving nonlinear partial differential equation," Chaos, Solitons & Fractals, vol. 31, pp321-330, 2007.

[7] B. E. A Saleh and M. C. Teich, *Fundamentals of Photonics*. A Wiley-interscience publication, New York, pp754, 1991

An Evaluation of Photoresist Thickness for Semiellipsoid Microlens Fabrication before Thermal Reflow Using the Prolate Spheroid Approximation

Shih-Yu Hung
Department of Automation Engineering
Nan Kai University of Technology
Nantou, Taiwan, syhung@nkut.edu.tw

Chien-Hsin Hung
Graduate Institute of Precision Engineering
National Chung Hsing University
Taichung, Taiwan

Abstract—**We present a new semiellipsoid microlens fabrication method using the lift-off and alignment exposure processes. The lift-off method is used to create an elliptical copper base before the thermal reflow process. During the photoresist thermal reflow process, the elliptical base can precisely define the bottom shape of the liquid photoresist. The prolate spheroid approximation method is developed to estimate the thickness of elliptic photoresist column required by the semiellipsoid microlens of a certain height, with the error being controlled within ±3%.**

I. INTRODUCTION

Due to the surface tension, only the microlens in spherical cap shape can be produced by the traditional thermal reflow method [1, 2]. In this work, an elliptical copper base is formed on a wafer by using the lift-off process. Based on geometric analysis, we aim to provide a theoretical approach for designing a semiellipsoid microlens shape. Prolate spheroid approximation method will be developed to estimate the thickness of the elliptical photoresist column for the semiellipsoid microlens of a certain height. After the photoresist coating process, exposure should be conducted through mask alignment. And after development, the elliptical photoresist column on the copper base can be obtained. Due to their high surface tension, liquid photoresist tend to minimize their surface trying to achieve a semiellipsoid shape during the photoresist thermal reflow process. The elliptical base can change the contact area and contact mode between the liquid photoresist and copper base, and thus change the surface tension of the liquid photoresist accordingly. During the photoresist thermal reflow process, the elliptical copper base can precisely define the bottom shape of the liquid photoresist.

II. THE PRINCIPLE OF DESIGN-PROLATE SPHEROID APPROXIMATION

To obtain the desired ellipsoid microlens geometric shape, the volume of elliptical photoresist column must be precisely estimated before beginning the thermal reflow process. Fig. 1. illustrates a ellipsoid microlens model and its coordinate system. The general equation for the ellipsoid is:

$$\frac{x^2}{A^2} + \frac{y^2}{B^2} + \frac{z^2}{C^2} = 1 \qquad (1)$$

where A is the length of semi-axis in the x-direction, B is the length of semi-axis in the y-direction and C is the length of semi-axis in the z-direction. Consider $K(0, 0, k)$ a point on the height of the ellipsoid such that $0 \leq k \leq C$. The plane parallel to x-y plane going through the point K will intersect the ellipsoid and the filled cap is the volume we want to calculate. The volume of the resulting ellipsoidal cap is given by

$$V_{\text{ellipsoidal cap}} = \pi AB \left(\frac{2C}{3} - k + \frac{k^3}{3C^2} \right) \qquad (2)$$

A prolate spheroid can be formed by rotating an ellipse about its major axis; in other words, an ellipsoid with two equal lengths of semi-axes in x and z directions. A prolate spheroid is a spheroid in which the polar axis (y-direction) is greater than the other axes, like a rugby ball. With $k = C - h$ and $C = A = R$, the volume of the prolate spheroid cap can be derived as below

$$V_{\text{prolate spheroidal cap}} = \pi \frac{B}{R} \left(Rh^2 - \frac{h^3}{3} \right) \qquad (3)$$

where the equatorial radius R of a prolate spheroid is

$$R = \frac{h^2 + a^2}{2h} \qquad (4)$$

where h is related to the height of microlens and a is the length of minor semi-axis for the elliptical base as indicated in Fig. 2. The equation of an ellipse for y-z plane is

$$\frac{y^2}{B^2} + \frac{z^2}{C^2} = 1 \qquad (5)$$

b is the length of major semi-axis for the elliptical base as indicated in Fig. 2. Since the ellipse goes through the point $(0, b, k)$ and $C = R$, so

$$\frac{b^2}{B^2} + \frac{k^2}{R^2} = 1 \qquad (6)$$

From (6) with $k = R - h$, it yields

$$B = \frac{b}{\sqrt{2 \left(\frac{h}{R} \right) - \left(\frac{h}{R} \right)^2}} \qquad (7)$$

Before melting, the volume of elliptical photoresist column is

$$V_{\text{elliptical column}} = \pi abH \qquad (8)$$

where H is the thickness of elliptical photoresist column. We assume that the photoresist volume does not change during the thermal reflow process based on mass conservation, that is

$$\pi abH = \pi \frac{B}{R}\left(Rh^2 - \frac{h^3}{3}\right) \qquad (9)$$

The thickness of elliptical photoresist column can be evaluated using the prolate spheroid approximation as

$$H = \frac{B}{R}\frac{h\left(3a^2 + h^2\right)}{6ab} \qquad (10)$$

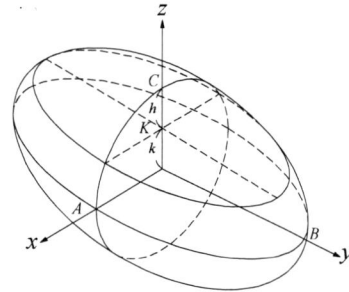

Fig. 1. The coordinate system defined for a semiellipsoid microlens model.

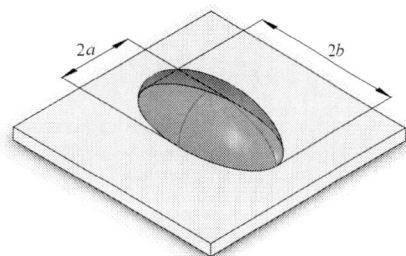

Fig. 2. a and b are the lengths of minor and major semi-axes for a semiellipsoid microlens, respectively.

III. RESULTS AND DISCUSSION

The relationship between theoretical height (h) of the semiellipsoid microlens and experimental results with respect to various the thickness (H) of elliptical photoresist column is depicted in Fig. 3. For example, the lengths of minor and major semi-axes for an elliptical base are 45μm and 90μm, respectively. The semiellipsoid microlens with a height of 20μm can be obtained using the elliptical photoresist column with H=11μm as shown in Fig. 3. According to the experimental result using the elliptical photoresist column with a thickness of 11μm, the height of the semiellipsoid microlens was 20.5μm after thermal reflow process. There were the errors varied within ±3% between the experimental results and the calculated values using (10). This shows that the approximate method is valid. The prolate spheroid approximate solution

using (10) is proven feasible and can be a basis for designing ellipsoid microlens in various sizes.

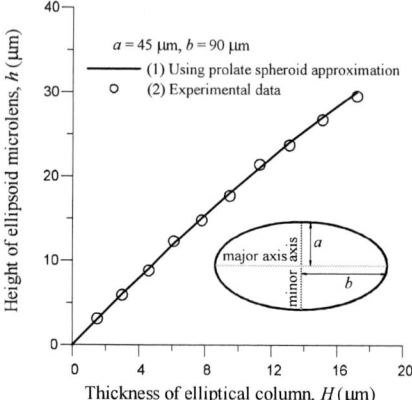

Fig. 3. The height of ellipsoid microlens computed using prolate spheroid approximate method compared with the experimental results.

IV. CONCLUSION

A semiellipsoid microlens can be placed onto the tip of a single-mode fiber end to improve the power coupling efficiency from a laser diode. The semiellipsoid microlens allows increasing the fiber spot size and numerical aperture. We presented a robust and reliable fabrication method for a semiellipsoid microlens array [3, 4, 5].

Before thermal reflow process, prolate spheroid approximation method was developed to estimate the required thickness of the elliptical photoresist column for the semiellipsoid microlens of a certain height. The variations between the actual photoresist thickness and theoretical calculation for making the semiellipsoid microlens were within ±3%. This proved that the theoretical model is feasible.

ACKNOWLEDGMENT

This work is supported by the National Science Council of Taiwan, through Grant No. NSC-100-2221-E-252-009- MY3.

REFERENCES

[1] T. H. Lin, H. Yang and C. K. Chao, "Concave microlens array mold fabrication in photoresist using UV proximity printing," *Microsyst. Technol.*, vol. 13, pp. 1537-1543, 2007.

[2] T. H. Lin, H. Yang, R. F. Shyu and C. K. Chao, "New horizontal frustum optical waveguide fabrication using UV proximity printing," *Microsyst. Technol.*, vol. 14, pp. 1035-1040, 2008.

[3] J. Y. Hu, C. P. Lin, S. Y. Hung, H. Yang and C. K. Chao, "Semi-ellipsoid microlens simulation and fabrication for enhancing optical fiber coupling efficiency," *Sens. Actuators. A: Phys.*, vol. 147, pp. 93–98, 2008.

[4] C. K. Chao, J. Y. Hu, S. Y. Hung and H. Yang, "Theoretical Prediction of Fiber Coupling for Ellipsoidal Microlens," *J. Mech.*, vol. 26, pp. 29-36, 2010.

[5] C. H. Hung, S. Y. Hung, M. H. Shen and H. Yang, "Semiellipsoid microlens fabrication method using UV proximity printing," *Appl. Opt.*, vol. 51, pp. 1122-1130, 2012.

Self-Switching Using SOA-Assisted Sagnac Interferometer

Vahid Ahmadi[a], Morteza Jamali[a] and Mohammad Razaghi[b]

[a] Dept. of Electrical and Computer Eng., Tarbiat Modares University, Tehran, Iran. Email: v_ahmadi@modares.ac.ir
[b] Dept. of Electrical and Computer Eng., University of Kurdistan, Sanandaj, Iran. Email: m.razaghi@uok.ac.ir

Abstract—**We propose and investigate self-switching mechanism by Sagnac interferometer based semiconductor optical amplifier (SOA) for subpicosecond pulses. The various switch characteristics such as phase differentiation between propagated pulses, SOA gain and switch extinction ratio all in time domain are shown.**

I. INTRODUCTION

The switching characteristics of SOA as a nonlinear component in all-optical networks have been of large attention in recent researches. Sagnac switch based on SOA is one of favorite structures to achieve this goal. By using modified nonlinear Schrödinger equation (MNSLE) considering main which includes nonlinear effects, we study gain and phase dynamics of SOA. The analysis is based on improved finite difference beam propagation method (IFD-BPM) [1]. In Sagnac loop, cross phase modulation (XPM) technique is used by employing SOA switching nonlinearities [2,3]. The input light pulse enters the first loop through input port of input coupler and splits unequally into two counter-propagating pulses, (u and v), with $\pi/2$ phase difference. In our proposed scheme, SOA is offset from the center of the loop using optical delay line (ODL). This will cause SOA to operate in different regimes for two input pulses. This phenomena, if other Sagnac switch parameters are tuned suitably can lead to sufficient phase difference between SOA's output counter-propagating pulses required for proper switching mechanism. Nonlinear effects of SOA become more important in subpicosecond regime rather than in the picoseconds regime. The main nonlinear effects are: self phase modulation (SPM), two photon absorption (TPA), carrier heating (CH), spectral hole burning (SHB), Kerr effects and gain dispersion. In subpicosecond regime, besides SPM effect which is the dominant phenomenon for picosecond pulses, the effects of SHB and CH phenomena on pulse shape and spectrum are more noticeable. Due to these effects the switch output pulse is broadened. In comparison to previous works [2,3] in our proposed scheme we don't need additional pump pulse for switching purpose besides using two independent input and output couplers, which leads to more tunability. Furthermore the double Sagnac structure with symmetric output coupler (coupling ratio 0.5) can be used as a pattern effect compensator [4].

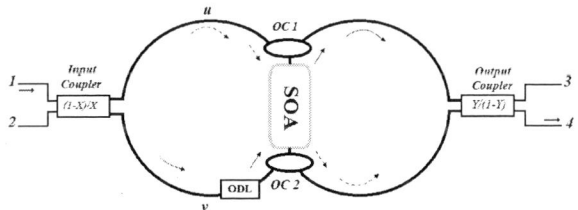

Fig.1. Schematic of Sagnac-based switch with unequal power distribution due to 2×2 couplers. OC: optical circulator, ODL: optical delay line.

II. THEORY

A. Self-switching Operation Perinciple

The operation of the self- switching can be described with the help of the schematic diagram shown in Fig. 1. The switch consists of two optical loops formed by the joint input and output ports of two independent 2×2 couplers and a SOA that can be offset from the midpoint of the loops. When an input pulse enters the loop through one of the input ports of input coupler, it splits asymmetrically to two counter-propagating pulses, (u and v). Each coupler induces a $\pi/2$ phase difference between its outputs. The low power optical pulse is injected several picoseconds before the high power optical pulse. These delays cause changes in both gain and refractive index of SOA for two counter-propagating pulses and therefore, propagating pulses experiences different dynamic states. As a result, due to SOA nonlinearities the phase difference occurs between these two pulses.

B. Theory of proposed self-switched scheme

The optical input pulse injected in one of the input ports [e.g., Port 1, see Fig. 1], is distributed unequally to u and v pulses. The power splitting ratios in input and output coupler are X and $1-Y$, respectively. The u and v pulse powers after passing through input coupler are $p_u = XP_{in}$ and $p_v = (1-X)P_{in}$ respectively, where P_{in} is input power. The high power optical pulse arrives 7 ps (the time needed for a pulse propagated through the SOA cavity) after the low power optical pulse. As mentioned before, SOA induces a nonlinear phase shift ($\Delta\varphi_{NL}$) to optical pulses. When $\Delta\varphi_{NL} = \pi$, maximum extinction ratio between switch outputs (Port 3 and Port 4) can be reached. The basic interferometric equations that describe the output pulses at the output ports (P3 and P4 respectively), can be written as

$$P3 = YP_u(t) + (1-Y)P_v(t) + 2 \times (\sqrt{Y(1-Y)P_u(t)P_v(t)} \times cos[\varphi_u(t) - \varphi_v(t)]) \quad (1)$$

and

$$P4 = (1 - Y)P_u(t) + YP_v(t) - 2 \times (\sqrt{Y(1 - Y)P_u(t)P_v(t)} \\ \times \cos[\varphi_u(t) - \varphi_v(t)]) \qquad (2)$$

where $P_u(t)$ and $P_v(t)$ are powers of the u and v pulses after passing through the SOA and $\varphi_u(t)$ and $\varphi_v(t)$ are the corresponding phase shifts. These two parameters are related to $\Delta\varphi_{NL}$ by

$$\Delta\varphi_{NL} = \varphi_u(t) - \varphi_v(t) = -\frac{\alpha_N}{2} ln\left(\frac{G_u(t)}{G_v(t)}\right) \qquad (3)$$

where $G_u(t)$ and $G_v(t)$ are the SOA gain sensed by the u and v pulses and α_N is the linewidth enhancement factor associated with the gain changes due to carrier depletion.

C. SOA Model

To take into account all nonlinear effects in the SOA, a set of MNLSEs are solved numerically. The analysis is based on central difference approximation in time domain and trapezoidal integration technique for spatial steps [1].

III. RESULTS

In our simulation, the SOA has a length of 500 μm with operating wavelength at 1550 nm. The asymmetric couplers are obtained by using an unbalanced multi-mode interference (MMI) devices with a coupling ratio $X = Y = 0.3$. The input pulse shape is *sech²* and Fourier transform limited with 200 fs full width at half maximum (FWHM) and input energy is equal to 1 pj. The variation of phase difference and its cosine with time at $g_0 = 78.5$ cm⁻¹ (which is related to amplifier bias current [1]) are shown in Fig. 2. It is shown that for specific time spans, $\Delta\varphi_{NL}$ between two pulses reaches π. Fig. 3 shows the time evaluation of SOA's gain sensed by each pulse, for the same g_0. The solid line is SOA gain variation at u pulse input facet and dashed line is SOA gain variation at v pulse input facet. The two dip structures shown in this figure are due to time delay between u and v pulses. The shallow dip in solid line curve is related to input u pulse and the deep dips corresponds to output v pulse. In dashed line curve the sallow hole is due to output u pulse and the deep dip corresponds to input v pulse. The switching characteristic of our structure is plotted in Fig. 4. The effect of g_0 on extinction ratio (P4/P3 by use of (1) and (2)) is illustrated in Fig. 4. For g_0 less than 40

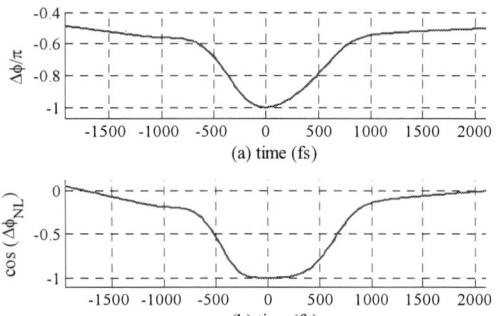

Fig.2 (a) Variation of phase difference between two pulses, (b) Time variation of cos(Δφ_NL).

is transmitted to port 4 (switching port) and the switching function will be achieved.

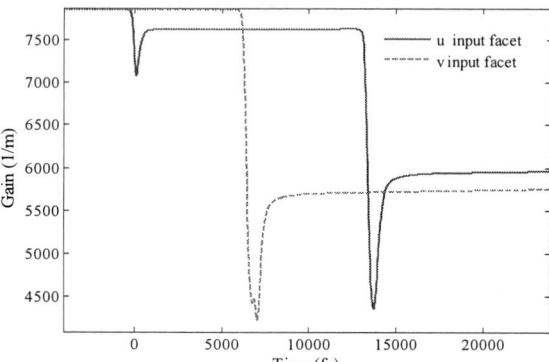

Fig. 3 Gain variation versus time for each facet of SOA.

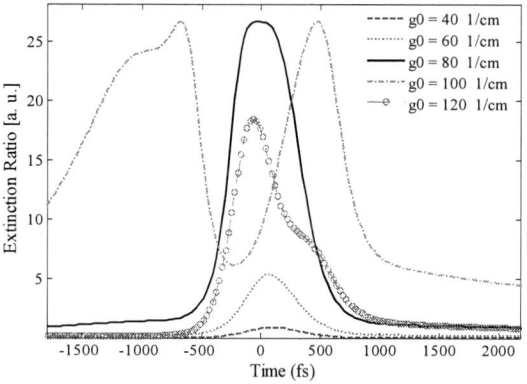

Fig.4 Extinction ratio for various values of small signal gain (g₀) versus time.

IV. CONCLUSION

We numerically analyzed self switching mechanism in SOA-assisted Sagnac interferometer by unequal distribution of input optical pulse. We showed that the switching function in our proposed scheme can be accomplished by varying the SOA's bias current.

ACKNOWLEDGMENT

This work was supported by Iran Telecommunication Research Center (ITRC).

REFERENCES

[1] M. Razaghi, V.Ahmadi and M. J. Connelly, "Comperehensive Finite-Difference Time Dependent Beam propagation Model of Counter propagation Picosecond Pulses in a Semiconductor Optical Amplifier," *IEEE J. Lightwave Technol.*, vol. 27, pp. 3162–3174, 2009.

[2] K.E. Zoiros, J. Vardakas, T. Houbavlis and M. Moyssidis, "Investigation of SOA-assisted Sagnac recirculating shift register switching characteristics," *Optik*, vol. 116, pp. 527–541, 2005.

[3] G. Papadopoulos and K.E. Zoiros, "On the design of semiconductor optical amplifier-assisted Sagnac interferometer with full data dual output switching capability" *Optics & Laser technol.*, vol. 43, pp.697–710, 2011.

[4] E. A. Patent, J. J. G. M. van der Tol, N. Calabretta, and Y. Liu, "A pattern effect compensator," in *Proc. IEEE/LEOS Symp. (Benelux Chapter)*, Brussels, Belgium, pp. 233–236, 2001.

Study on the structure characteristics of HgCdTe infrared detector using laser beam-induced current

X. K. Hong*, H. Lu, D. B. Zhang

College of Physics and Electronic Engineering, Changshu Institute of Technology, Changshu, Jiangsu, China 215500

Abstract

The structure characteristics of typical *n+-on-p* HgCdTe infrared detector have been studied by laser beam-induced current (LBIC). The dependence of LBIC on laser wavelength, junction depth and localized leakage has been presented. The spreading length of minority carrier of p-type region (*Lsp*) is extracted by the exponential decay fitting of the curve of LBIC. It is found that the peak magnitude of LBIC and junction depth approximates to a linear relationship for practical values of device fabrication. The *Lsp* monotonously increases with junction depth. A notable shift of LBIC profile is observed when localized leakage exists. This provides a powerful explain for LBIC applying to characterize the structure and process uniformity of HgCdTe infrared detector.

I. INTRODUCTION

HgCgTe infrared detectors are becoming one of the most important dectectors due to its superior performance [1-3]. However, seeking a high-yield and low cost technology is still challenging for infrared dectectors' continuous development. At present, amounts of works have been done on the growth of high quality materials and fabrication techniques of devices. Characterization of individual devices is also an important procedure for practical device applications. Traditionally, individual devices within these large 2-D arrays have been characterized after flip-chip hybrid bonding to the silicon readout circuitry. However, in comparatively low-yield technologies such as large infrared focal plane arrays, significant cost savings would be expected to result if nondestructive yield and uniformity characterization were to be undertaken at an earlier stage in the fabrication process. As an attractive candidate, laser beam induced current (LBIC) is a qualitative manner, which has been used for a number of years [4-8]. In this technique a focused laser beam is scanned across the surface of the semiconductor, and the current flowing in the external circuit between two nominally shorted ohmic contacts on either side of the scan area is measured, as shown in Fig. 1. the remote ohmic contacts can be removed when finish the measurement. Hence, nondestructive characterization is performed because no electrical contacts to individual devices are needed.

Since the induced current is highly dependent on large number parameters, analysis of LBIC measurements is still a challenging task. Correlations of LBIC signal on various device parameters are needed for the technique to gain wider practical applications.

This paper presents a system study of the dependence of LBIC signals of typical *n+-on-p* HgCdTe infrared detector on laser wavelength, junction depth and localized leakage.

II. DEVICE STRUCTURE AND SIMULATION MODEL

The core structure of HgCdTe infrared detector is *p-n* junction. This paper focuses on the special case of a single, isolated, ideal *p-n* junction diode structure, where the length of *p*-type region is $200\mu m$, the thickness is $8\mu m$. The initial junction depth is $1\mu m$, the n-type region is $20\mu m$ in length. The typical *n+-on-p* device structure is then formed with doping density of $N_a=1\times10^{16}cm^{-3}$ and $N_d=1\times10^{17}cm^{-3}$. In addition, only front-side illumination is considered. Simulations of LBIC on $Hg_{0.78}Cd_{0.22}Te$ long-wavelength infrared photodiodes at 120K is performed using a commercial semiconductor modeling package.

For plain drift-diffusion simulation the well known Poisson equation and continuity equations are used. The optical generation rate can be expressed by:

$$G^{opt}(z) = J(x,y,z_0) \cdot \alpha(\lambda,z) \cdot \exp\left[-\left|\int_{z_0}^{z}\alpha(\lambda,z)dz\right|\right] \quad (1)$$

where λ is the wavelength, $J(x,y,z_0)$ is the optical beam spatial variation of intensity over a window where rays enter the device, z_0 is the position along the ray where absorption begins, and $\alpha(\lambda,z)$ is the absorption coefficient along the line[7, 9].

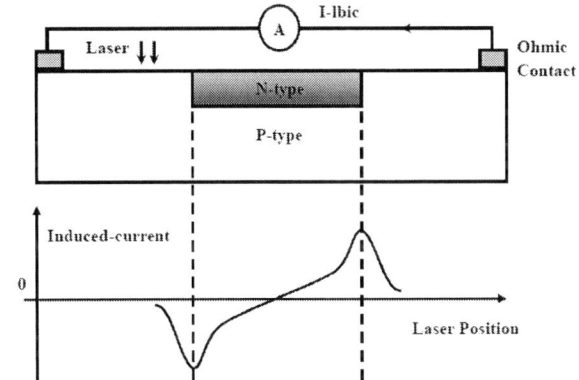

Fig. 1 LBIC measurement configuration and the typical bipolar LBIC profile

III. RESULTS AND DISCUSSION

The dependence of LBIC profile on laser wavelength is shown in Fig. 2. Almost no change is found for wavelength ranging from $0.40 \mu m$ to $1.3 \mu m$, which indicates the depth of laser penetration is shallow compared with the depth of junction. That's to say, the light absorption only exists in the surface of our device.

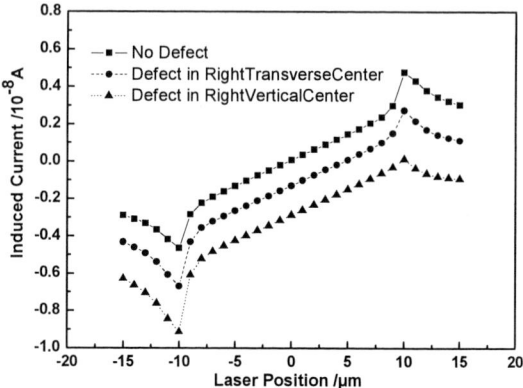

Fig. 4 Simulated LBIC profiles of devices with and without localized leakage caused by metal contaminations

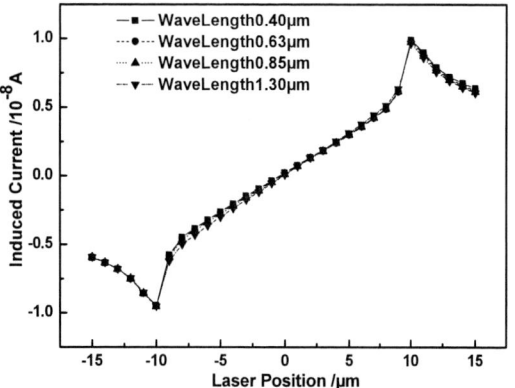

Fig. 2 Dependence of LBIC profile on laser wavelength

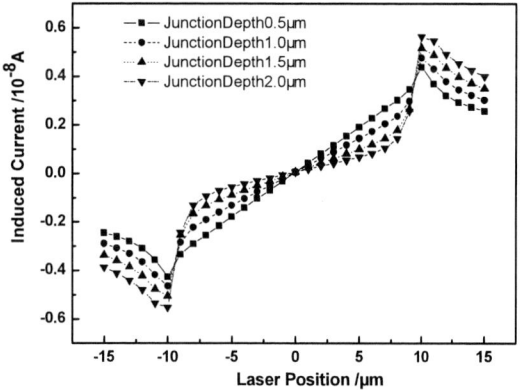

Fig. 3 LBIC profile with different p-n junction depth

Fig. 3 shows the LBIC profile with different *p-n* junction depth. It is found that the peak magnitude of LBIC and junction depth approximates to a linear relationship for practical values of device fabrication. An exponential decay model is used to extract the theoretical spreading length of minority carrier by fitting the LBIC curve in *p*-type region. The *Lsp* monotonously increases with junction depth. Then the value of *Ls* can be used to provide a direct qualitative indication of process uniformity.

Fig. 4 gives simulated LBIC profiles of devices with and without localized leakage caused by metal contaminations. The leakage was modeled by including a small piece of metal that was ohmic to both sides of the junction. The side length of square metal is about $0.2 \mu m$. Asymmetry is apparent with localized leakage. The shift in zero point of LBIC current is clearly linear with the leakage position.

IV. CONCLUSION

The dependence of LBIC on laser wavelength, junction depth and localized leakage has been presented to depict the structure characteristics of typical n^+-*on-p* HgCdTe infrared detector. The spreading length of minority carrier of *p*-type region is extracted by the exponential decay fitting of the curve of LBIC. It is found that the peak magnitude of LBIC and junction depth approximates to a linear relationship for practical values of device fabrication. The *Lsp* monotonously increases with junction depth. A notable shift of LBIC profile is observed when localized leakage exists. This provides a powerful explain for LBIC applying to characterize the structure and process uniformity of HgCdTe infrared detector.

REFERENCES

[1] A. Rogalski, [Infrared Detectors], Gordon and Breach Publishers, Amsterdam, 391(2000).
[2] S. C. Shen, "Comparison and competition between MCT and QW structure material for use in IR detectors," Microelectronics J. 25, 713-739(1994).
[3] Jun Wang, Xiao-Shuang Chen, Wei-Da Hu*, Lin Wang , Wei Lu , Faqiang Xu , Jun Zhao , Yanli Shi , Rongbin Ji, Amorphous HgCdTe infrared photoconductive detector with high detectivity above 200 K, Applied Physics Letters, 99, 113508-1-113508-3 (2011).
[4] D. A. Redfern, E. P. G. Smith, C. A. Musca, J. M. Dell, and L. Faraone, "Interpretation of current flow in photodiode structures using laser beam-induced current for characterization and diagnostics," IEEE Transactions on Electron Devices, VOL. 53, No. 1, Jan. 2006.
[5] D. A. Redfern, W. Fang, K. Ito, C. A. Musca, J. M. Dell, and L. Faraone, "Low temperature saturation of p-n junction laser beam induced current signals," Solid State Electron., vol 48, pp. 409-414, 2004.
[6] F. Yin, W. D. Hu, B. Zhang, Z. F. Li, X. N. Hu, X. S. Chen and W. Lu, Simulation of laser beam induced current for HgCdTe photodiodes with leakage current, Optical and Quantum Electronics, 41, 805-810(2009).
[7] Fei Yin, Weida Hu, Zhijue Quan, Bo Zhang, Xiaoning Hu, Zhifeng Li, Xiaoshuang Chen, Wei Lu "Extraction of electron diffusion length in HgCdTe photodiodes using laser beam induced current" ACTA PHYSICA SINICA, 58, 7884-7890 (2009)
[8] J. Bajaj and W. E. Tennant, "Remote contact LBIC imaging of defects in semiconductors," J. Cryst. Growth 103, 170-178(1990).
[9] W. D. Hu, X. S. Chen, F.Yin, Z. J. Quan, Z. H. Ye, X. N. Hu, Z. F. Li and W. Lu, "Analysis of temperature dependence of dark current mechanisms for long-wavelength HgCdTe photovoltaic infrared detectors," J. Appl. Phys. 105, 104502-1-104502-8 (2009).

Gap in pagination due to withheld paper.

Pages 73-74

Optimization of Detector Arrays and Circuits Targeted for Precision Calculation in Infrared Laser Interferometer

Xiaojie Sun

Shanghai Institute of Technical Physics of the Chinese Academy of Sciences
500# YuTian Road
Shanghai, 200083 China, sxiaojie@mail.sitp.ac.cn

Abstract—A low noise transimpedance amplifier (TIA) is used in a wide band PIN (Positive Intrinsic Negative) laser detector arrays to transform the photo current produced by an infrared interfering laser power to an output voltage with a specified amplitude and frequency response. In this paper we consider the specifications of a PIN detector array coupled with a TIA circuit. Then the following issues that influence high precision calculation results will be investigated: low noise performance of the detector array and TIA pre-amplifier; fluctuation effects of amplitude caused by frequency modulation interfering signal; photosensitive area (PA) of the detector and physical distance between detectors.

We find that noise performance related with signal to noise ratio (SNR) defines the minimum calculation error. And PA related with sensitivity has influence on junction capacitance and amplitude fluctuation. Meanwhile bias circuit mode related with dark current also influence amplitude response. Based on these issues, a PIN detector array pair is constructed of five sensors arranged in cross. The center response wavelength of the detector array is around 850 nm according to the requirement of the referential laser interferometer.

I. INTRODUCTION

Fourier Transform Infrared Spectrometer (FTIRS) is a main load in the next generation geostationary meteorological satellite. In a FTIRS with traditional structure, laser beams reached orthogonal placed detector arrays in Fig 1 [1]. Photocurrent responses from detector arrays are then amplified and transformed into voltage signal through the TIAs. Tilts of moving mirror appeared in scan motion will exist a dynamic phase difference between the voltage signals on sensors, e.g. HL and HR, or HL' and HR', see EQ (1) to EQ (4).

$$T_{HL} = I_{fm} + I_{sm} + 2*M(\alpha)*\sqrt{I_{fm} + I_{sm}}* \\ \cos(2*\pi*v_{OPD}*t/\lambda + \theta_{\delta HL} + \Delta\theta(\alpha)) \quad (1)$$

$$T_{HR} = I_{fm} + I_{sm} + 2*M(\alpha)*\sqrt{I_{fm} + I_{sm}}* \\ \cos(2*\pi*v_{OPD}*t/\lambda + \theta_{\delta HR} - \Delta\theta(\alpha)) \quad (2)$$

$$R_{HL} = I_{fm'} + I_{sm'} + 2*M(\alpha)*\sqrt{I_{fm'} * I_{sm'}}* \\ \sin(2*\pi*v_{OPD}*t/\lambda + \theta'_{\delta HL} + \Delta\theta(\alpha)) \quad (3)$$

$$R_{HR} = I_{fm'} + I_{sm'} + 2*M(\alpha)*\sqrt{I_{fm'} + I_{sm'}}* \\ \sin(2*\pi*v_{OPD}*t/\lambda + \theta'_{\delta HR} - \Delta\theta(\alpha)) \quad (4)$$

Where: I_{fm} represents the laser intensity (or laser power) from the reflection of fixed mirror; I_{sm} represents the laser intensity from the reflection of scanning mirror; $M(\alpha)$ represents the interference efficiency of the referential laser signal [2][1][3], λ represents the wavelength of the laser; $\theta_{\delta HL}$ represents the static phase difference from standard cosine laser signal; v_{OPD} represents the velocity of OPD (Optical Path Difference); $\Delta\theta(\alpha) = 4*\pi*d*\alpha/\lambda$ represents the dynamic phase error induced by the tilt angle(α) [4][5]; d represents distance between detectors in horizontal or vertical side.

With normalization and perpendicular compensation [5], $\Delta\theta(\alpha)$ is calculated through inverse trigonometric functions. By comparing the phase differences between $\Delta\theta(\alpha)$ s, the tilt α between fixed mirror and scanning mirror can be calculated, see EQ (7).

$$\alpha = 4*d*\Delta\theta/\lambda \quad (7)$$

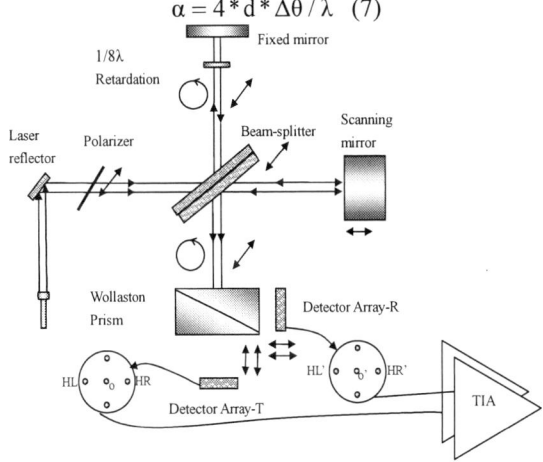

Fig. 1. Detectors and optic path of laser interferometer

II. METHODOLOGIES

A. The requirement of SNR

To simplify the circuit structure and cut down the difficulty for space based application, a common JFET amplifier (LF356) is used as the main TIA. Fig 2 is the TIA circuit coupled with the photodiode. Meanwhile the maximum measurable physical tilt angle α is limited by SNR of combination of detector and TIA, see EQ (6) [5]. When λ and d is fixed for system configuration, SNR defines the measurement dynamic range. Noise modeling of combination of photodiode and TIA is shown in Fig 3 [6]. Measurement of output noise is listed in table I.

$$\alpha_{max} = \lambda / (2 * d * SNR) \quad (6)$$

B. The requirement of amplitude fluctuation

Normalization process of a trigonometric function is sensitive to signal amplitude variation. Yet it's a frequency modulation signal at detector side caused by interferometer specification. So fluctuation of signal is seen as noise when calculating tilt angle. Fig 4 is the fluctuation noise equivalent tilt angle. Hence the bandwidth of the detector defines the sensitivity of tilt measurement. Fig5 (a) and (b) is the comparison of voltage fluctuation between previous version and after optimization of detector and TIA.

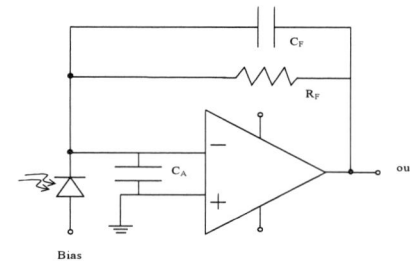

Fig. 2. Scheme of detector and TIA

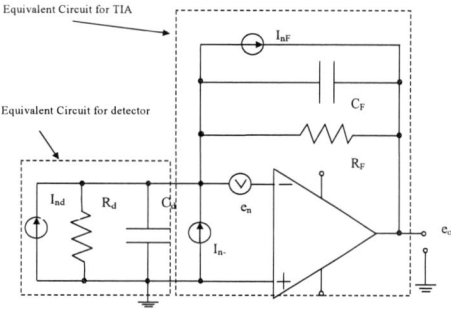

Fig. 3. Noise model of detector and TIA

TABLE I

MEASUREMENT OF OUTPUT NOISE

RMS noise (μV) / Detector No.	Detector & TIA	Sampling System	TIA	Detector	3dB BW (kHz)
T_{HL}	751.6	145	56	735.4	745
T_{HR}	731.4	145	56	714.7	682
R_{HL}	661.8	145	56	643.3	856
R_{HR}	753.6	145	56	737.4	623

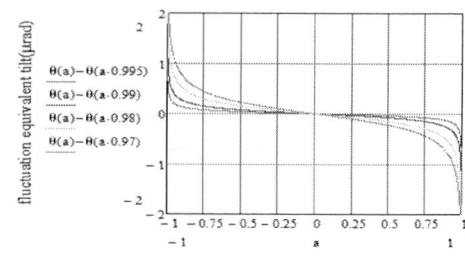

Fig. 4. Voltage equivalent tilt angle

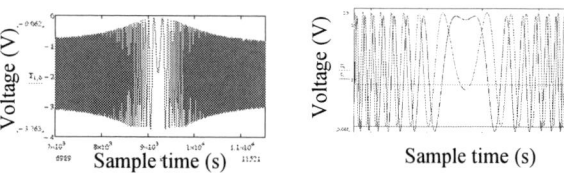

Fig. 5. Voltage fluctuation VS time variable modulation signal

III. CONCLUSIONS

This paper proposes a five element PIN detector array with cross ranged placement. The detector array is fabricated with common anode process and biased properly. Through analysis of direct noise caused by detectors and circuits, calculation noise caused by amplitude fluctuation, an optimized combination of parameter for photosensitive area, distance between detectors, transimpedance resistor and capacitor is derived. With a familiar JFET worked as TIA to suppress the dark current noise, the tilt measurement accuracy reached infra one micro-radian.

ACKNOWLEDGEMENT

This work is supported by the Knowledge Innovation Program of Shanghai Institute of Technical Physics (SITP), Chinese Academy of sciences.

REFERENCES

[1] Huandong Wei, Jianwen Hua and Zuoxiao Dai et al., "Optimization of the referential laser signal to Fourier Transform Spectrometer," *SPIE, Proc.* Vol. 7382, 73820W, pp. 10.1117-12.835069, 2009.

[2] Louis W. Kunz and David Goorvitch, "Combined Effects of a Converging Beam of Light and Mirror Misalignment in Michelson Interferometry," *Applied Optics, Proc.* Vol. 13, n° 5, pp. 1077-1079, May 1974.

[3] Charles S. Williams, "Mirror Misalignment in Fourier Spectroscopy Using a Michelson Interferometer with Circular Aperture," *Applied Optics, Porc.* Vol. 5, n° 6, pp. 1084-1085. June 1966.

[4] MACOY N. H and BROBERG H, "Dynamic alignment design and assessment for scanning interferometers," *SPIE, Proc.* Vol. 2832, pp.126-154. 1996,

[5] Douglas L. Cohen, "Performance degradation of a Michelson interferometer when its misalignment angle is a rapidly varying, random time series," *Applied Optics, Proc.* Vol. 36, No.18, pp. 4034-4042, 1997.

[6] C. D. Motchenbacher and J. A. Connelly, *Low-noise Electronic System Design.* John Viley & Sons, INC. NewYork, 1993.

[7] Bahram Zand, Khoman Phang, and David A. Johns, "Transimpedance Amplifier with Differential Photodiode Current Sensing," *IEEE Transl.* Vol. 2, pp. 624-627. 1999.

The investigation of the transient photovoltage in HgCdTe infrated photovoltaic detectors

Haoyang Cui, Naiyun Tang, Zhong Tang

School of computer and information engineering,

Shanghai University of Electric Power, 2103 Pingliang Road, Shanghai 200090, China, cuihy@shiep.edu.cn

Abstract—The changed polarity of transient photovoltage (TPA) from negative to positive induced by ultra fast lasers illumination is studied in the HgCdTe p-n junction photovoltage detector. The negative photovoltaic-response decrease obviously and even disappear by blocking the laser beam with an aperture to limit the illumination area of the linear array detectors. A combined theoretical model of p-n junction and Schottky contact can explain this new phenomenon well. Using the TPA technique and the combined model, the characters of p-n junction and Schottky contact will be distinguished. Therefore, it could be used in characterizing the Ohmic contact of the detectors electrodes, and its sensitivity is expected to be much higher than the steady states methods.

Keywords- HgCdTe, transient photovoltage, polarity change, Schottky contact

I. INTRODUCTION

Since the first synthesis in 1958, HgCdTe infrared detectors have been intensively developed over the past fifty years [1, 2]. In a pixel of the linear array of HgCdTe photovoltaic detectors, the photo-generated carriers separated by the p-n junction will be injected in the readout circuit from HgCdTe-metal interface. Therefore, for the n⁺-on-p backside illuminated hybrid linear array detectors, the electrodes preparation is one of the most important issues influencing the detectors performance. Generally, a coupled of methods have been devoted to characterizing the Ohmic contact of the detectors electrodes [3, 4]. Most of the methods are using steady-state electric testing techniques, such as *I-V*, *C-V* test. However, the signals of HgCdTe-metal interface and p-n junction are in superposition with each other in the electric steady-state testing. As a powerful tool to investigate the electric property of the detectors, the *I-V* testing hardly distinguishes the photoelectric signals of the HgCdTe-metal interface from p-n junction, and the information from the interface will be captured difficulty. Moreover, the electrical *I-V* testing can only indirectly reflect the influence of HgCdTe-metal interface on the devices photoelectric properties. It is an indirect measurement method.

In this paper, an advantageous technique, transient photovoltage (TPV) [5-7], could largely complement this subject. For example, when irradiated with a picosecond pulsed laser, the transient photo response show the changed polarity of TPV, that is, there is an apparent negative valley first then it evolves a positive peak. By considering the electrode and the pixels configurations in the linear array of the detectors, the negative and the positive photovoltages can be attributed to the HgCdTe-metal interface barriers behavior and p-n junction respectively. Thus, the photovoltaic responses could be distinguished from the time domain and polarity.

Such a changed polarity of TPV phenomenon gives us an insight into observing the photo-generated carriers immigrate and could be used in charactering the Onmic contact of the electrode.

II. EXPERIMENTAL RESULTS

A. Experimental setup

The HgCdTe n⁺-on-p photodetector was grown by MBE on (100) GaAs substrate with a buffer layer of CdTe. The p-n junction was formed by B⁺ implantation into the p-type HgCdTe layer, resulting in an abrupt n⁺-on-p structure. The acceptor and donor concentration are 8×10^{15} and 1×10^{17} cm⁻³, respectively. The sample was mounted in a liquid nitrogen-cooled Dewar for measurement.

The incident laser pulse was provided by a commercial optical parametric oscillator and difference frequency generator pumped with a picosecong Nd:YAG laser (EXSPLA PG401/DFG). The pulse duration was 30 ps and the repetition rate was 10 Hz. Comparing to the shortest rising time of tens of nonoseconds in the pulsed response profile of the HgCeTe photodiode, the laser pulse can be approximated as a δ function in our experiment. A small portion of the laser beam was reflected by a beam splitter and measured using an energy detector in order to monitor the exciting energy. The pulsed photo-response of the HgCdTe detector was measured from the voltage drop across a 50 Ω load-resistor. Both signals from the energy detector and the HgCdTe detector were fed into an oscilloscope to monitor and record the puls profiles. An average of 500 pulsed profiles was recorded to eliminate the pulse-to pulse fluctuation and improve the signal-to noise ratio. An aperture was used to limit the illumination area of the linear array detectors by blocking the laser beam.

B. Experimental results

In the ideal case, the photo-response of a photodiode irradiated by laser should show a simple rapid increase and slow decay process due to the relative large value of capacitance and resistance of p-n junction. However, the photo-response profiles of the HgCdTe photodiode excited by pulsed laser are very different. When irradiate with the laser, the time profile of the responding voltage shows an apparent negative valley during the first 15 ns, and then it evolves a positive peak. By changing the excitation laser intensity the transient photo-response of the detector show the similar time evolution profiles, no matter for the case of single photon absorption transition that the photon energy is larger than the bandgap or for the case of two-photon absorption transition that the photon energy is smaller than the bandgap. The contradictions of the ideal situation and the experimental results means that a photovoltaic mechanism different with p-n junction may exist.

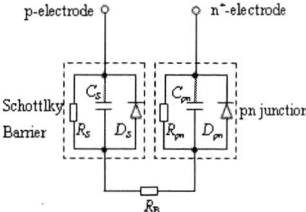

Fig 1 The equivalent circuit of the HgCdTe photodiode

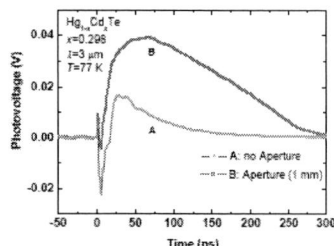

Fig 2 Transient photovoltage of the HgCdTe
photodiode with no aperture and 1 mm aperture.

To explain the polarity change in TPV, a combined theoretical model including p-n junction and Schottky barrier are proposed, the equivalent circuit of the HgCdTe photodiode is shown in Fig1. For n⁺-on-p type HgCdTe optoelectronic device, it is generally believed that n⁺ electron interface is a good Ohmic contact. While there is a big work-function between the p-HgCdTe and metal, therefore, p electrod interface is not easy to form Ohmic contacts, and it is even possible to form a Schottky contact. The opposite built-in electric filed of the Schottky barrier contact comparing to p-n junction provides the possibility for the generation of negative valley and positive peak in the TPV, while the high-frequency character of the Schottky barrier provide the possibility for the generation of negative photovoltage prior to positive photovoltage.

Two aspects of evidence are discussed in order to verify the correctness of the above theoretical model. (a) Considering the common p-electrode configurations surround around all pixels in the linear array of the detectors and the laser spot size is much larger than the pixel size, the common p-electrode coverage area will be illuminated inevitable, even constitute the main part of the photoexcited area. Thus, if the p-electrode interface formed a Schottky barrier, the electrode interface will generate an apparent negative photovoltage. The experimental results (Fig 2) show that the negative photovoltaic-response decrease obviously and even disappear by blocking the laser beam with an aperture to limit the illumination area of the linear array detectors. (b) In the following, we estimated the concentration of the photo-generated carrier, and the results show that the concentration is high enough to form an obvious photovoltage in the p-electrode interface no matter for the case of single-photon and two-photon absorption transition. The photoexcited carriers density within the absorption region induced by one-photon absorption processed

are expressed as: $\Delta n = \dfrac{W \cdot \lambda \cdot \alpha}{1.24 \cdot S} \cong 10^{14}\text{-}10^{16} \text{ cm}^{-3}$, the concentration decline by

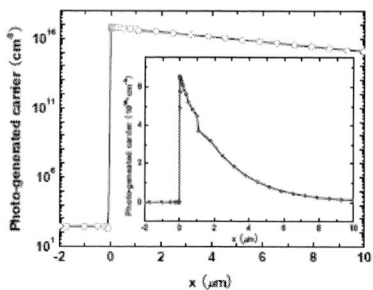

Fig 3 The photo-excited carriers density distribution after the laser illumination. The inset is the semi logarithmic coordinates.

about one order of magnitude near the electrode away from the incidence plane (Fig 3). For the case of two-photon absorption transition, the photo-excited carriers density in the irradiated region of the device are uniformly distributed, which is

expressed as: $\Delta n = \dfrac{\beta I^2}{2\hbar\omega}\Delta t \cong (0.5\text{-}30)\times10^{12} \text{ cm}^{-3}$. The above calculations show

that there are relatively high concentrations of photo-excited carriers at the electrode interface which is far from the plane of incidence. Therefore, if the p-electrode interface formed a Schottky barrier instead of Ohmic contact, the photo-excited carriers at the electrode interface will be separated and formed a negative photovoltage.

ACKNOWLEDGEMENT

Project supported by the National Natural Science Foundation of China (61107081), Innovation Program of Shanghai Municipal Education Commission of China (10YZ158), Shanghai Natural Science Foundation of China (10ZR1412300) .

REFERENCES

[1] W. Lu, L. He, X.S. Chen, L. Ding, S.L. Sun, J. Tang, J.H. Su, The development of HgCdTe infrared detector technology in China,Proc. of SPIE Vol. 7298,2009, 72982Z

[2] J. Shao, X. Lü, S.L. Gou, W.Lu, Impurity levels and bandedge electronic structure in as-grown arsenic-doped HgCdTe by infrared photoreflectance spectroscopy, Physical Review B, 2009, 80(15):155125-1-155125-11

[3] J. V. Mikhelashvili, G. Eisenstein, V. Garber, S. Fainleib, G. Bahir, D. Ritter, M. Orenstein, ON theextraction of linear and nonlinear physical parameters in nonideal diodes. J. Appl. Phys. 1999, 85(9):6873-83

[4] M. Y. Lee, Y. S. Lee, H. C. Lee, Behavior of elemental tellurium as surface generation recombination centers in CdTe/HgCdTe interface, Appl. Phys. Lett., 2006,88(20):204101-3

[5] Y. Yao, S. X. Yu, B. F. Ding, D. L. Li, X. Y. Hou, C. Q. Wu, A combined theoretical and experimental investigation on the transient photovoltage in organic photovoltaic cells, Appl. Phys. Lett., 2010,96(20):203306-8

[6] H. Y. Cui, Z. F. Li, Z. L. Liu, C. Wang, X. S. Chen, X. N. Hu, Z. H. Ye, W. Lu, Modulation of the two-photon absorption by electric fields in HgCdTe photodiode, Applied Physics Letters,2008,92(2):021128-30

[7] H.Y. Cui, Z.F. Li, F.J.Ma, X.N.Hu, Z.H.Ye, W.Lu, Negative photovoltaic-responses in HgCdTe infrared photovoltaic detectors irradiated with picosecond pulsed laser, J. Infrared Millim. Waves, 2009, 28(3):161-164

Simulation of carrier dynamics in graphene on a substrate at terahertz and mid-infrared frequencies

N. Sule*, K. J. Willis[‡], S. C. Hagness, I. Knezevic

Department of Electrical and Computer Engineering,
University of Wisconsin-Madison, Madison WI 53706, USA
*Email: sule@wisc.edu

Abstract—**We calculate the complex conductivity of graphene in the terahertz (THz) to mid-infrared (mid-IR) frequency range using a numerical simulation that couples the two-dimensional (2D) ensemble Monte Carlo technique (EMC) for carrier transport, the three-dimensional (3D) finite-difference time-domain (FDTD) technique for electrodynamics, and molecular dynamics (MD) for short range Coulomb interactions. We demonstrate the effect of the typically used silicon-dioxide substrate on the high-frequency carrier dynamics in graphene and show good agreement between recent experimental results and our numerical simulations.**

I. INTRODUCTION

Graphene is a promising material for novel electronic and optoelectronic device applications [1]–[3]. Although considerable work has been done on the fundamental electronic properties [4] and low-field *dc* transport [5] in graphene, high-frequency carrier dynamics in this material has not been extensively studied [6]. From a technological perspective, the application of graphene in optoelectronic devices, such as transparent conductors, photodetectors, and metamaterials, requires a deeper understanding of the interaction of high-frequency electromagnetic waves with the carriers in this two-dimensional system. Moreover, the effect of a substrate and the impurities therein cannot be ignored for device applications.

II. NUMERICAL MODEL

Here, we calculate the frequency-dependent conductivity of graphene and demonstrate the effect of a typical SiO_2 substrate on the carrier dynamics. Our numerical simulation combines the ensemble Monte Carlo (EMC) technique for solving the Boltzmann transport equation for carrier transport in the two-dimensional (2D) graphene layer with the three-dimensional (3D) finite-difference time-domain (FDTD) technique [7] for solving Maxwell's curl equations throughout the domain. The short-range Coulomb forces exerted on the carriers by impurity ions are calculated using molecular dynamics (MD). The EMC, FDTD, and MD solvers are coupled in each time step, such that the combined FDTD-MD fields accelerate the carriers and the EMC carrier motion produces the sourcing current density. This coupled EMC-FDTD-MD technique has previously been used to accurately calculate the *ac* conductivity of doped silicon [8], [9].

[‡] Current address: AWR Corporation, 11520 North Port Washington Road, Mequon, WI 53092, USA

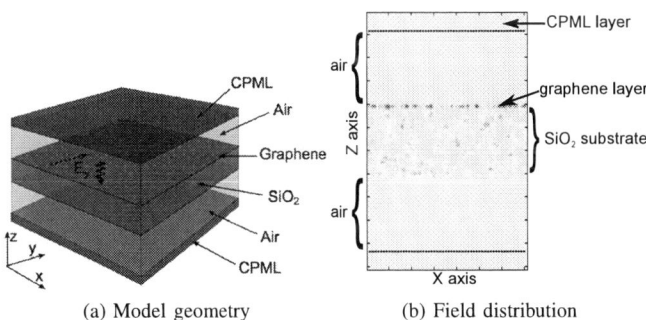

(a) Model geometry (b) Field distribution

Fig. 1. (a) 3D EMC-FDTD model geometry. The wavy arrow denotes the direction of propagation of the wave, while the direction of the electric field is shown by the dotted arrow. The domain boundaries perpendicular to x and y are terminated by periodic boundary conditions. (b) Visualization of the spatial distribution of the electric field (E_y) stemming from the influence of the carriers in the graphene sheet and the charged impurities in the SiO_2 substrate at the start of the simulation.

The simulation domain, as shown in Fig. 1a, consists of a monolayer graphene sheet with air on top and a SiO_2 substrate at the bottom, with dielectric constants of $\epsilon_g = 2.5$, $\epsilon_a = 1.0$, and $\epsilon_s = 3.9$, respectively. We use periodic boundary conditions for the bounding planes perpendicular to the plane of carrier transport (the xz and yz planes in Fig. 1a), in order to simulate a very large sheet of graphene. The simulation domain is terminated with convolutional perfectly matched layer (CPML) boundary conditions [7] on the top and bottom boundaries. The initial electric field distribution due to the carriers and charged impurities is shown in Fig. 1b.

The 2D carrier transport is limited by the Coulomb fields of the impurities in the substrate, in addition to scattering with longitudinal acoustic (LA) and optical (LO) phonons, as well as polar surface optical (SO) phonons due to the substrate. The LA and LO scattering rates are calculated using the tight-binding Bloch (TBB) wave functions and are fitted to *ab-initio* electron-phonon scattering rates to deduce the deformation potentials [10]. The SO scattering rates are calculated based on the interaction Hamiltonian from Ref. [11], and the TBB wave functions.

III. RESULTS

Ensemble averages of the drift velocity, carrier density, and energy are calculated at each time step. Fig. 2 shows the time evolution of the carrier velocity for a 13 THz plane wave

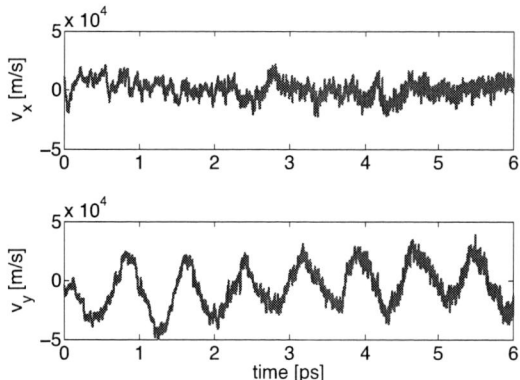

Fig. 2. Time evolution of the average carrier velocity for a 13 THz plane wave excitation with electric field along the y-direction (as shown in Fig. 1). Carrier density is $6.3 \times 10^{12}\,\mathrm{cm}^{-2}$ and the ensemble contains around 10,000 carriers. Transport in this case is limited by LO, LA, and SO scattering in addition to the Coulomb impurity interaction.

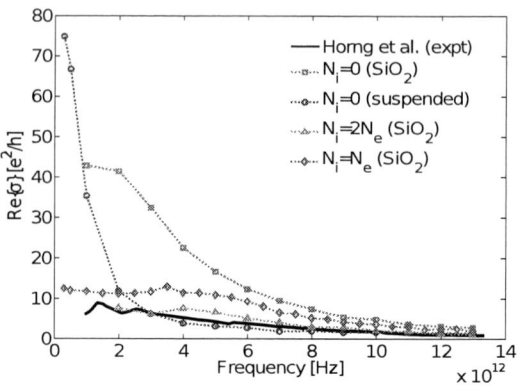

Fig. 3. The real part of the complex conductivity of graphene as a function of frequency at the carrier density of $N_e = 6.3 \times 10^{12}\,\mathrm{cm}^{-2}$, which is the same density for a gate voltage of 50 V in Ref. [6]. Blue circles denote suspended graphene without impurities, red squares denote graphene on SiO_2 substrate without impurities, green diamonds and pink triangles denote different impurity levels in supported graphene, and the black solid curve corresponds to the experimental data from Horng et al. (Ref. [6]). Good agreement with experimental data is found for impurity density exceeding the electron density twofold ($N_i = 2N_e$.)

excitation, with electric field along the y-direction. Current density and electric field phasors are calculated using fast Fourier transform after a steady state has been reached, and are used to calculate complex conductivity based on the formula $\hat{\sigma}(\omega) = \frac{E^*(\omega)J(\omega)}{|E(\omega)|^2}$. In Fig. 3, we show the real part of the complex conductivity as a function of frequency for supported and unsupported (suspended) graphene with a carrier density of $6.3 \times 10^{12}\,\mathrm{cm}^{-2}$. We contrast transport in a suspended graphene sheet, where the intrinsic LA and LO phonon-electron scattering limits carrier dynamics, with transport in supported graphene, where the inclusion of SO phonon-electron and impurity scattering is necessary. A significant effect on the conductivity of graphene due to the SO phonons (red squares) as well as the impurities (green diamonds and pink triangles) is clearly seen in Fig. 3. For example, the conductivities of suspended graphene (blue squares) at $4\,\mathrm{THz}\,(\mathrm{Re}\{\sigma\} = 3.8\,\mathrm{e}^2/\mathrm{h})$ and $10\,\mathrm{THz}\,(\mathrm{Re}\{\sigma\} = 1.6\,\mathrm{e}^2/\mathrm{h})$ are lower than those of supported graphene without impurities (red squares) by factors of 6 and 3, respectively. In contrast, at frequencies below 1 THz, the conductivity of suspended graphene exceeds that of supported graphene. We find good agreement with the experimental data [6] (black curve) corresponding to the same carrier density tuned via a gate voltage of 50 V, for a simulation with an impurity density that is twice the carrier density.

IV. CONCLUSION

In conclusion, we have calculated the complex conductivity of graphene in the THz to mid-IR frequency range by simulating the high frequency carrier dynamics using a coupled EMC-FDTD-MD numerical solver. We demonstrate that the substrate has a significant influence on carrier transport at low frequencies, while its influence becomes less pronounced at frequencies greater than 10 THz. Our results show good agreement with recent experimental data [6] for practically relevant carrier and substrate impurity densities. Moreover, we can independently tune the impurity density and also predict the frequency dependent conductivity of graphene supported

on an ideally clean SiO_2 substrate. Such a characterization of the high-frequency conductivity of supported and gated graphene might provide useful for photonic and optoelectronic device applications.

ACKNOWLEDGMENT

This work has been supported by the NSF through the University of Wisconsin MRSEC (award DMR-0520527), and by the AFOSR (awards FA9550-09-1-0230 and FA9550-11-1-0299).

REFERENCES

[1] Ph. Avouris, Z. Chen and V. Perebeinos. "Carbon-based electronics," *Nature Nanotech.* 2, pp. 605-615 (2007).

[2] A. K. Geim, and K. S. Novoselov. "The rise of graphene," *Nature Mater.* 6, pp. 183-191 (2007).

[3] F. Bonaccorso, Z. Sun, T. Hasan, and A. C. Ferrari. "Graphene photonics and optoelectronics," *Nature Photon.* 4, pp. 611-622 (2010).

[4] A. H. Castro Neto, F. Guinea, N. M. R. Peres, K. S. Novoselov, and A. K. Geim. "The electronic properties of graphene," *Rev. Mod. Phys.* 81, pp. 109-162 (2009).

[5] S. Das Sarma, Shaffique Adam, and E. H. Hwang. "Electronic transport in two-dimensional graphene," *Rev. Mod. Phys.* 83, pp. 407-470 (2011).

[6] J. Horng et al. "Drude conductivity of Dirac fermions in graphene," *Phys. Rev. B* 83, 165113 (2011).

[7] A. Taflove and S. C. Hagness. *Computational Electrodynamics: The Finite-Difference Time-Domain Method*, 3rd ed., Artech House, 2005.

[8] K. J. Willis, S. C. Hagness and I. Knezevic. "Terahertz conductivity of doped silicon calculated using ensemble Monte Carlo/finite-difference time-domain (EMC/FDTD) simulation technique," *Appl. Phys. Lett.* 96, 062106 (2010).

[9] K. J. Willis, S. C. Hagness and I. Knezevic. "Multiphysics simulation of high-frequency carrier dynamics in conductive materials," *J. Appl. Phys.* 110, 063714 (2011).

[10] N. Sule and I. Knezevic. "Phonon-limited electron mobility in graphene calculated using tight-binding Bloch waves," submitted to *Phys. Rev. B* (2012).

[11] A. Konar, T. Fang and D. Jena. "Effect of high-κ dielectrics on charge transport in graphene-based field effect transistors," *Phys. Rev. B* 82, 115452 (2010).

Hydrogenated Graphene: Structures and Surface Work Function

N. Jiao, Chaoyu He, C. X. Zhang and L. Z. Sun*

Laboratory for Quantum Engineering and Micro-Nano Energy Technology, Xiangtan University, Xiangtan 411105, China

lzsun@xtu.edu.cn

Abstract: The structures and surface work functions of graphanes with five fundamental configurations are systematically studied with the density functional theory. We find that, from the point of view of energy, hydrogenated graphene prefer forming the chair graphane than the other ones. The work function and layer thickness of the five structures vary with the hydrogenation, providing important theoretical data for experimental identifying the configurations of graphanes by STM and AFM.

Kewwords: Graphene, hydrogenation, graphane, Surface work function.

Introduction

Since the first discovery of grapheme in 2004 [1], graphene based transistors [2-4] have rapidly developed. Graphene is considered as a potential candidate for future nano-electronics owing to their remarkable structural and electronic properties. Graphane, a fully hydrogen functionalized graphene, behaves as a wide gap semiconductor or an insulator [5, 6], which is different from graphene. It means that the electronic band gap of graphene can be tuned by chemical decoration. These results are significant for the application of graphene in nano-electronics. However, the identification of the hydrogenated graphene is still an open question because it is hardly to be observed directly in experiments. In present work, using first-principles calculations, we investigate the structures and the surface work functions of five possible hydrogen decorated graphene [6-8].

Models and Method

The five kinds of graphanes considered in this work are shown in Fig.1. The chair and stirrup structures are shown in Fig.1(a) and Fig.1(b), respectively. There are three kinds of boat structures including boat1, boat2 and boat3 as shown in Fig.1(c), Fig.1(d) and Fig.1(e),

respectively. To obtain the work functions of these graphanes, the Vienna Ab initio Simulation Package (VASP) [9] is adopted to optimize their structures and perform total energy calculations. The kinetic cutoff energy is set to be 500 eV and the Brillouin zone (BZ) is sampled using a 15x15x1 Gamma-centered Monkhorst-Pack grid in our calculations. All systems are fully optimized up to the residual force on every atom is less than 0.01 eV/Å through the conjugate-gradient algorithm.

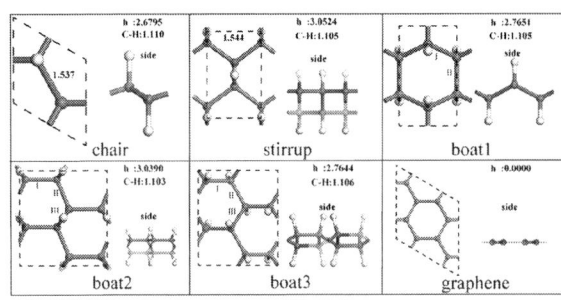

Fig.1: The top and side view of the hydrogenated graphene and the pristine graphene. White is the H atom and gray is the C atom. h is the thickness of the structure. C-H is the bond length between C atom and H atom.

Results and Discussion

The optimized configurations of the five types of hydrogenated graphene and their structural parameters are shown in Fig.1 and Tab.1. We find that the hydrogenated structures, comparing with the sp^2 pristine graphene, become sp^3 hybridization forming a zigzag configuration. All the Gibbs energies of the five configurations are listed in Tab.1. The results of Gibbs energies indicate that the chair structure is the most stable one among the five decorated configurations. After the functionalization, boat2 and boat3 configurations have three types of C-C bond, while boat1 has two types, as shown in Table.1 and Fig.1. The thicknesses (h) of the decorated structures are different, as shown in Tab.1. The chair structure is thinner than the other structures and stirrup is the thickest one. This can be used to identify those two configurations by atomic force microscope (AFM). Comparing with the other boat structures, the h of boat2 is about 0.27Å thicker than boat1 and boat3 structures. This can be distinguished by AFM.

Table.1 Summary of work function and structure properties of the systems studied. Gibbs energy (GE), Work function (WF), Length of the C-C Bond inside the graphene (dc-c).

	Chair	Stirrup	Boat1	Boat2	Boat3	Graphene
GE/ev	-0.103	-0.076	-0.052	-0.039	-0.028	0
WF/ev	3.916	4.407	4.417	4.457	4.543	4.410
dc-c/Å	1.537	1.544	I:1.537 II:1.570	I:1.573 II:1.548 III:1.542	I:1.548 II:1.539 III:1.562	1.420

Then we calculate the work functions of all the structures, as shown in Fig.2. Work function is usually defined as the minimum energy needed to take an electron away from the Fermi level to the vacuum level [10]. Work function can be measured accurately by the Kelvin probe force microscopy [11]. The average planar electrostatic potentials (APEPs, Fermi level is set to be reference) of all the systems are shown in Fig.2. The calculated work functions of all systems including pristine graphene are listed in Tab. I. All the structures are symmetrical in the orientation of z axis and the surface work functions are equal for both sides of the system, as shown in Fig.2. We can get the information that the work function varies with the different configurations. For example, the work function of boat1 is only about 4.417eV, whereas the work function of boat3 is up to 4.543eV. This can be used to distinguish those two structures by using scan tunneling microscope (STM).

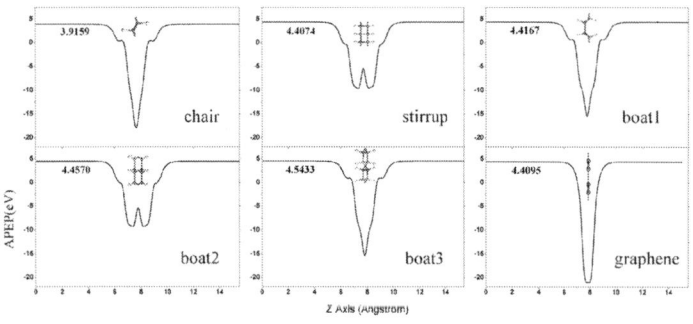

Figure.2 The work functions of the graphene and the hydrogenated graphene.

Conclusion

In summary, we have studied the structures and work functions of the hydrogenated graphene by first-principles calculations. The difference of the work function and the thickness of the structure, which induced by the different hydrogenated configurations, may provide a criterion for identifying these configurations.

Acknowledgment

This work is supported bythe National Natural Science Foundation of China (Grant Nos. 10874143), the Program for New Century Excellent Talents in University (Grant No. NCET-10-0169), and the Scientific Research Fund of Hunan Provincial Education Department (Grant Nos. 10K065).

(1) K. S. Novoselov, A. K. Geim, S. V. Morozov, D. Jiang, Y. Zhang, S. V. Dubonos, I. V. Grigorieva, A. A. Firsov, *Science* 2004 306, 666.
(2) B. Obradovic, R. Kotlyar, F. Heinz, P. Matagne, T.Rakshit, M. D. Giles, M. A. Stettler, *Appl. Phys. Lett.* 2006 88, 142102.
(3) D. Gunlycke, D. A. Areshkin, Junwen Li, J. W.Mintmire, C. T.White, *Nano Lett.* 2007 7, 3608.
(4) G. C. Liang, N. Neophytou, D. E. Nikonov, M. S. Lundstrom, *IEEE Trans. Electron DeVices* 2007 54, 677.
(5) J. C. Meyer, A. K. Geim, M. I. Katsnelson, K. S. novoselov, T. J. Booth, S. Roth, *Nature* 2007 446,60.
(6) J. O. Sofo, A. S. Chhaudhari, G. D. Barber, *Phys. Rev. B* 2007 75, 153401.
(7) O. Leenaerts, H. Peelaers, A. D.Hernández-Nieves, B. Partoens, F. M. Peeters, *Phys, Rev. B* 2010 82, 195436.
(8) Xiao-Dong Wen, Louis Hand, Vanessa Labet, Tao Yang, Roald Hoffmann, N. W. Ashcroft, Artem R. Oganov, Andriy O. Lyakhov, *PANS* 2011 108, 6833.
(9) G. Kresse, J. Furthmüller, *Phys. Rev. B* 1996 54, 11169.
(10) N. D. Lang, *Phys.Rev.B* 1971 4, 4234.
(11) M. Nonnenmacher, M. P. OBoyle, and H. K. Wickramasinghe, *Appl. Phys. Lett.* 1991 58, 2921.

Structures, stability and electronic properties of two- or four-segment BN/C nanotubes

Chaoyu He, C. X. Zhang, H. P. Xiao, L. Z. Sun[*] and J. X. Zhong[*]

Laboratory for Quantum Engineering and Micro-Nano Energy Technology, Xiangtan University, Xiangtan 411105, China

[*]lzsun@xtu.edu.cn and jxzhong@xtu.edu.cn

Abstract: The structures, stability and electronic properties of some novel two- or four-segments BN/C nanotubes are systematically investigated using the density functional theory based first-principle calculations. Our calculations reveal that the structures, stability and electronic properties of these hybridized nanotubes are dependent on their diameters, compositions and hybridizing manners.

Kewwords: Boron nitride nanotubes; Carbon nanotubes; electronic properties.

Introduction

The quasi one dimensional (1D) carbon nanotubes (CNTs) [1, 2] can be metal or insulator, depending on their chirality and diameters [3]. And the boron nitride nanotubes (BNNTs) [4, 5] always behave as wide gap insulators [6] independent of their chirality and diameters. Recently, some novel two-segment BN/C hybridized nanotubes (BNCNTs) with tunable band gaps are proposed by Aijun Du et. al [7]. Their ab initio molecular dynamics simulations reveal that armchair single wall BNCNTs can be spontaneously formed via the interaction of the two segments of ZBNNR and ZGNR at room temperature. Another work of molecular dynamics simulations and total energy calculations [8] indicated that these two-segment BNCNTs hold stability comparable to CNTs and may stable well at room temperature if their diameters are larger than 0.4 nm. Inspired by the hybridizing ideas in these researches, some four-segment armchair single walled BNC-BNCNTs and BNC-NBCNTs with diameter larger than 0.8 nm are proposed and their structures, stability and electronic properties are investigated. Our calculations reveal that the structures, stability and electronic properties of these hybridized nanotubes are dependent not only on their diameters, compositions but especially on their hybridizing manners.

Models and Method

As show in the left panel in Fig.1, the nanotubes considered in our calculations are the pure CNTs, the two-segment BNCNTs, the four-segment BNC-BNCNTs and BNC-NBCNTs as well as the pure BNNTs with 12, 16 and 20 zigzag chains. We adopt VASP [10] to investigate their stability and electronic properties. The kinetic cutoff energy is set to be 420 eV and the Brillouin zone (BZ) is sampled using a 1x1x11 Gamma-centered Monkhorst-Pack grid in our calculations. All systems are fully optimized up to the residual force on every atom is less than 0.01 eV/Å through the conjugate-gradient algorithm.

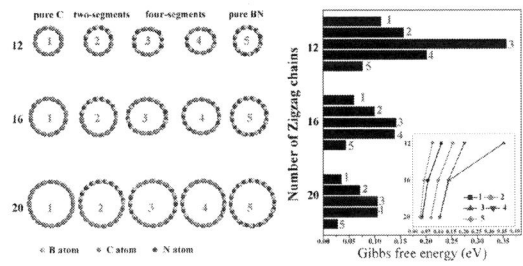

Fig.1: Shown in left are the optimized structures of the pure CNTs (1), two-segments BNCNTs (2), four-segment BNCB-NCNTs (3), BNC-BNCNTs (4) and pure BNNTs (5) with 12 (top), 16 (center) and 20 (bottom) zigzag chains, respectively; In right are their corresponding Gibbs free energies.

Results and Discussion

As shown in Fig. 1, all the fully optimized nanotubes are still cylinders except for the four-segment ones containing zigzag chains less than 20, who are staved

978-1-4673-1602-6/12 $31.00 © 2012 IEEE   83

due to the unbalanced strains in the narrow segments of ZBNNR and ZGNR. Two equivalent (inequivalent) ZGNR segments in BNC-BNCNTs (BNC-BNCNTs) are strained by unbalanced (balanced) external forces at their two sides and their curvatures are produced. The calculated Gibbs free energies of these nanotubes are shown in Fig.1. An obvious diameter-dependent stability in these tubes can be found in the dotted lines inserted in Fig. 1. From the histogram, we can see that the pure CNTs and pure BNNTs are more favorable that the multi-segment ones. Both the four-segment BNC-BNCNTs and BNC-NBCNTs are less stable than the two-segment BNCNTs due to their double unstable B-C and C-N interfaces, respectively. These results indicate that the ZBNNR segments and ZGNR segments prefer to separate from each other to form pure BNNTs and CNTs. In views of the forming of the multi-segment tubes from hybridizing the segments of ZBNNR and ZGNR [7, 8] and the successful synthesizing of the mixed B-C-N single walled nanotubes [9], it is necessary and interesting to investigate the electronic properties of these multi-segment tubes, especially of the four-segment ones.

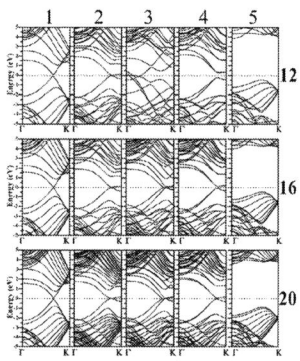

Fig.2: Band structures for the pure CNTs (1), two-segment BNCNTs (2), four-segment BNCB-NCNTs (3), BNC-BNCNTs (4) and pure BNNTs (5) with 12 (top), 16 (center) and 20 (bottom) zigzag chains, respectively.

The calculated band structures of the pure CNTs, two-segments BNCNTs, four-segment BNCB-NCNTs, BNC-BNCNTs and pure BNNTs with 12, 16 and 20 zigzag chains are shown in Fig. 2. We can see that all the pure CNTs are narrow band gap semiconductors and all the pure BNNTs are wide band gap insulators. All two-segment tubes are semimetals with tiny band gaps. Interestingly, the electronic properties of the four-segment tubes are dependent on their hybridizing manners. All the BNC-BNCNTs are semiconductors with narrow band gaps decreasing as the diameters increases, but all the four-segment BNC-NBCNTs are metals. These differences in the electronic properties are mainly derived from the different doping effects of ZBNNR segments on ZGNR segments in these two types of four-

segment tubes. In the BNC-BNCNTs, the two segments of ZGNR locating at the same chemical environment are equivalent. Both two ZGNR segments containing both B-C and C-N interfaces are doped by the interfacial B and N atoms with both n-type and p-type, respectively, and behave as semiconductors. In the BNC-NBCNTs, the two segments of ZGNR locating at very different chemical environment are inequivalent. The one with only B-C interface is doped with only n-type and another one with only C-N interface is doped or p-type, resulting in a metallic properties in the BNC-NBCNTs.

Conclusion

Using the first-principles calculations, some four-segments BNC-BNCNTs and BNC-NBCNTs have been proposed and investigated. Our results indicate that these tubes can be metals or semiconductors depending on their hybridizing manners.

Acknowledgment

This work is supported by the National Natural Science Foundation of China (Grant Nos. 11074211, 10874143, 11074213), the Program for New Century Excellent Talents in University (Grant No. NCET-10-0169).

(1) Iijima. S. *Nature* 1991 *354*, 56.
(2) Ebbesen, T. W.; Ajayan,P.M. *Nature* 1992 *358*, 220.
(3) Odom, T. W.; Huang, J. L.; Kim, P.; Lieber, C. M. *J. Phys. Chem. B* 2000 *104*. 2794.
(4) Rubio, A.; Corkill, J. L.; Cohen, M. L. *Phys. Rev. B* 1994 *49*, R5081.
(5) Chopra, N. G.; Luyken, R. L.; Cherrey, K.; Crespi, V. H.; Cohen M. L.; Louie, S. G.; Zettl, A. *Science* 1995 *269*, 966.
(6) Blasé, X.; Rubio, A.; Louie, S. G.; Cohen, M. L. *Phys. ReV. B* 1995 *51*, 6868.
(7) Du, A. J.; Chen, Y.; Zhu, Z. H.; Lu, G. Q.; Smith, S. C.; *J. Am. Chem. Soc.* 2009 *131*, 1682.
(8) Zhang, Z. Y.; Zhang, Z. H.; Guo, W. L.; *J. Phys. Chem. C* 2009 *113*, 13108.
(9) Wang, W. L.; Bai, X. D.; Liu, K. H.; Xu, Z.; Golberg, D.; Bando, Y.; Wang, E. G. J. Am. Chem. Soc.2006 128, 6530.
(10) Kresse, G.; Furthmüller, J. *Phys. Rev. B* 1996 *54*, 11169-11186. Kresse, G.; Furthmüller, J. *Comput. Mater. Sci.* 1996 6, 15-50.

Non Linear Piezoelectricity in ZincBlende GaAs and InAs Semiconductors

G. Tse, J. Pal, R. Garg, V. Haxha and M.A. Migliorato

School of Electrical and Electronic Engineering, University of Manchester, United Kingdom
Joydeep.pal@postgrad.manchester.ac.uk

*Abstract-*We investigate the strain dependence of piezoelectric effect, both linear and non linear, in zincblende GaAs and InAs semiconductors. We expanded the polarization in terms of the ionic and dipole charges, internal displacement and the exploited the ab-initio Density Functional Theory (DFT) to evaluate the dependence of all quantities on the strain tensor. By this detailed study of the non linear piezoelectric effect, we report that even third order effects are significant.

I. INTRODUCTION

It is well known that the piezoelectric effect in bulk III-V semiconductors arises from lack of inversion symmetry along particular crystallographic directions[1,2] and is found in devices as diverse as light-emitting diodes (LEDs), lasers, power electronics, transducers and micropositioners [3]. In III-V semiconductors and their nanostructures, strain with a component along the polar axis of the crystal leads to the generation of the electric dipoles. In epitaxially grown zincblende (ZB) materials such dipoles are linked to off-diagonal strain tensor components and manifest themselves as a piezoelectric field that exists in nanostructures grown on [111] oriented substrates. A similar effect is also noticeable in wurtzite (WZ) semiconductors along the polar axis of [0001] orientation and diagonal strain components are able to generate dipoles. Though non linear effect in the strain have long been recognized as important in the calculation of piezoelectric fields in ZB II-VI [4], ZB III-V [5,6,7] and WZ III-N [8] semiconductors, the two most widely used methods in the calculation of such non linearities, namely the linear response method [9] and Harrison's method[10], appear to produce different results. The predictions from those two methods have not been compared before mainly because the work using Harrison's method [5,6] has always been presented for subsets of the full strain tensor. In this work, we have tried to address this issue and report the dependence of the piezoelectric coefficients on a full set of diagonal components of the strain tensor.

II. NUMERICAL RESULTS AND DISCUSSION

In the framework of Harrison's theory and the scheme previously established by Migliorato et al [5] we use ab initio density functional theory in the local density approximation to extract the elastic and dielectric properties of single crystals under various degrees of pressure in the 3 main crystallographic directions. The results properly combined in a formulation that takes into account the dipole formation between cations and anions, together with the modification of the dipoles along all bonds in the tetrahedron of the zincblende structure, leads self consistently to the identification of strain induced polarization coefficients. As already noticed in our previous work [5,6], our data shows strong strain dependence and hence non linear piezoelectric effect, with a marked lack of inversion antisymmetry $\varepsilon_i \leftrightarrow -\varepsilon_i$ suggesting the existence of at least quadratic terms in the strain. To further prove this point we obtain the non linear piezoelectric coefficients by fitting a third order polynomial to the DFT calculated after we impose conditions on the coefficients based on the cubic symmetry of the crystal. The recent work by Beya-Wakata et al [9] presented a comprehensive study based on DFPT and linear response theory of non linear polarization in various III-V zincblende semiconductors. The method used differs from ours as it relies entirely on ab-initio calculation without any fitting parameters, which are instead used in our model (Z_H*). One of the major differences between the two models is that while the non linear response method introduces the small dependence on the quadratic shear strain, it neglects higher order contributions such as the quadratic and cubic terms in the diagonal strain components that our model instead obtains via a fitting procedure of the DFT data. The two models discussed in this paper obviously produce very different results. We pick two critical cases for comparison: InAs pseudomorphically strained on GaAs (001) ($\varepsilon_1 = \varepsilon_2 = $ -0.7, $\varepsilon_3 = $ +0.7, with a further shear given by $\gamma = 0.02$) or GaAs (111) ($\varepsilon_1 = \varepsilon_2 = \varepsilon_3 = $ -0.223, $\gamma = 0.268$). The Beya-Wakata et al [9] model predict a polarization in the growth direction equal to +0.069 C/m2 for the (001) and +0.808 C/m2 for the (111) case. We note that the linear term alone would in both cases predict a negative polarization. Our model too predicts a positive sign of the polarization, but of a much smaller magnitude: +0.002 C/m2 for the (001) and +0.201 C/m2 for the (111) case. This is a crucial difference and should be observable in experiment. The present work showcases the importance of the non-linear terms, namely second and third order, for zincblende GaAs and InAs in calculating the piezoelectric field.

ACKNOWLEDGMENT

The authors thank Dave Powell for the parameterization routines used in the work.

978-1-4673-1602-6/12 $31.00 © 2012 IEEE

REFERENCES

[1] R. M. Martin, Piezoelectricity, Phys. Rev. B **5**,1607(1972)

[2] W. G. Cady, Piezoelectricity (McGraw-Hill, New York, 1946)

[3] S. Nakamura, G. Fasol, The Blue Laser Diode: GaN Based Light Emitters and Lasers (Springer-Verlag, Berlin, 1997)

[4] L. C. Lew Yan Voon and M. Willatzen, Electromechanical phenomena in semiconductor nanostructures, Journal of Applied Physics **109**, 031101 (2011)

[5] M. A. Migliorato, D. Powell, A. G. Cullis, T. Hammerschmidt and G. P. Srivastava, Composition and strain dependence of the piezoelectric coefficients in $In_xGa_{1-x}As$ alloys, Phys. Rev. B **74**, 245332 (2006)

[6] R. Garg, A. Hüe, V. Haxha, M.A. Migliorato, T. Hammerschmidt, G. P. Srivastava: Tunability of the piezoelectric fields in strained III-V semiconductors, Appl. Phys. Lett. **95**, 041912-1–3 (2009)

[7] G. Bester, X. Wu, D. Vanderbilt, and A. Zunger, Importance of Second-Order Piezoelectric Effects in Zinc-Blende Semiconductors, Phys. Rev. Lett. **96**, 187602 (2006)

[8] J. Pal, G. Tse, V. Haxha, M. A. Migliorato and S. Tomic´, Second-order piezoelectricity in wurtzite III-N semiconductors, Physical Review B **84**, 085211 (2011).

[9] A. Beya-Wakata, P-Y Prodhomme and G. Bester, First- and second-order piezoelectricity in III-V semiconductors,Phys. Rev. B **84**, 195207 (2011)

[10] W. A. Harrison, Electronic Structure and Properties of Solids, Dover Publications Inc., New York, (1989).

Design of Silicon Photonic Crystal Integrated Optical Devices

Zhi-Yuan Li, Lin Gan, and Chen Wang

Laboratory of Optical Physics, Institute of Physics, Chinese Academy of Sciences
P. O. Box 603, Beijing 100190, China
Email address: lizy@aphy.iphy.ac.cn

Abstract—**Photonic crystal has attracted extensive interest in the past 25 years due to its great power to mold the flow of light in micrometer/nanometer scale and promising aspects in building all-optical integrated devices and circuits. In this talk we present our recent efforts of design, fabrication, and characterization of integrated optical elements and devices in infrared silicon two-dimensional photonic crystal slabs. These devices operate either on band gap confinement or on band dispersion control. We focus on topics such as the broad-band wide-angle self-collimation effect, on-chip optical diodes and isolators, new cavities without apparent confinement barriers, and polymer-silicon hybrid nonlinear photonic crystal.**

Keywords-silicon photonic crystal; optical integration; self-collimation; optical diodes; cavities with confinement barrier; ploymer-silicon hybrid nonlinear photonic crystal

Photonic crystal has offered a powerful means to mold the flow of light and manipulate light-matter interaction at subwavelength scale. The fundamental goal of photonic crystal is to realize ultrasmall integrated optical circuits on the basis of different defects introduced within photonic band gaps. Theoretical analysis and numerical simulation are very important to design and realize high-performance photonic crystal integrated optical devices.

Silicon has a large refraction index and low loss in infrared wavelengths, which makes it an important optical material that has been widely used for integrated photonics applications in the near infrared regime around 1550 nm [1-6]. In this talk we review some of our recent theoretical and experimental works on the design, realization, and characterization of infrared two-dimensional (2D) air-bridged silicon photonic crystal slab devices that are based on both the band gap and the band structure engineering. In particular, we will focus on several works that we have carried out in the past two years.

Firstly we discuss on-chip wavelength-scale optical diodes and isolators based on photonic crystal heterojunctions with directional bandgap mismatch [7]. Optical isolation is a long pursued object with fundamental difficulty in integrated photonics. As a step towards this goal, we demonstrate the design, fabrication, and characterization of on-chip wavelength-scale optical diodes that are made from the heterojunction between two different silicon 2D square-lattice

photonic crystal slabs with directional bandgap mismatch and different mode transitions. The measured transmission spectra show considerable unidirectional transmission behavior, in good agreement with numerical simulations. The experimental realization of on-chip optical diodes with wavelength-scale size using all-dielectric, passive, and linear silicon photonic crystal structures may help to construct on-chip optical logical devices without nonlinearity or magnetism for application in on-chip optics communication and signal processing, and would open up a road towards photonic computers.

Secondly we discuss a photonic crystal that exhibits self-collimation effect in a wide angle range and with a large bandwidth [8]. We have designed, fabricated, and characterized a silicon PC structure that exhibits broadband large-angle self-collimation effect of transverse-magnetic (TE) modes at wavelengths around 1550 nm. Experimentally the collimation effect is clearly observed for TE modes with different incident angles at a broad wavelength range by recording the ray trace of light scattering off the sample. Our structure can find potential applications in many areas such as beam combiner and solar energy collector.

Thirdly we design and fabricate cavities without an apparent confinement barrier by combining two incommensurate photonic crystal superlattice waveguides in 2D photonic crystal slab [9]. A resonant mode with a high quality factor shows up in the pass band of waveguides. It has nearly no influence on the propagation of waveguide mode and can be directly coupled with the waveguide mode. The experimental measurement confirms the theoretical prediction of extraordinary coexistence of localized cavity mode and continuous waveguide mode with high coupling efficiency in the same frequency and space regime. Due to the extraordinary co-existence of localized cavity mode and continuous waveguide mode in both spatial and spectrum regions, the non-barrier cavity opens up a new avenue of cavity design and may find application in integrated optical devices and solid state lasers.

Finally we present a versatile technique based on nano-imprint lithography to fabricate high-quality semiconductor-polymer compound nonlinear photonic crystal (NPC) slabs [10]. It has been shown in our previous works that polymer materials such as polystyrene has a far larger Kerr nonlinearity and much faster optical response speed (down to several femtoseconds) compared with silicon [11-15]. So it is expected that the hybrid photonic crystal structures can

978-1-4673-1602-6/12 $31.00 © 2012 IEEE

incorporate both advantages of ultrafast and low power nonlinear optical effects. We have found that the approach allows one to infiltrate uniformly polystyrene materials that possess large Kerr nonlinearity and ultrafast nonlinear response into the cylindrical air holes with diameter of hundred nanometers that are perforated in silicon membranes. Both the structural characterization via the cross-sectional scanning electron microscopy images and the optical characterization via the transmission spectrum measurement undoubtedly show that the fabricated compound NPC samples have uniform and dense polymer infiltration and are of high quality in optical properties. The compound NPC samples exhibit sharp transmission band edges and nondegraded high quality factor of microcavities compared with those in the bare silicon photonic crystal. The versatile method can be expanded to make general semiconductor-polymer hybrid optical nanostructures, and thus it may pave the way for reliable and efficient fabrication of ultrafast and ultralow power all-optical tunable integrated photonic devices and circuits

Our works presented in this review show that photonic crystals have a strong power of controlling propagation of light at micrometer/nanometer scale and possess a great potential of applications in integrated photonic circuits. New concepts are very important to bring closer the fundamental goal of ultrasmall optical integration in the framework of silicon photonic crystal. Design is always the fresh blood of the science of photonic crystal.

This work was supported by the National Basic Research Foundation of China under Grant Nos. 2011CB922002 and the Knowledge Innovation Program of the Chinese Academy of Sciences (No. Y1V2013L11).

REFERENCES

[1] L. Gan, C. Z. Zhou, C. Wang, R. J. Liu, D. Z. Zhang, and Z. Y. Li, "Two-dimensional air-bridged silicon photonic crystal slab devices", Physica Status Solidi A **207**, 2715-2725 (2010).

[2] Y. Liu, F. Qin, F. Zhou, Q. B. Meng, D. Z. Zhang, and Z. Y. Li, "Ultrafast all-optical switching in Kerr nonlinear photonic crystals", Frontiers of Physics in China **5**, 220-244 (2010).

[3] Y. Z. Liu, R. J. Liu, C. Z. Zhou, D. Z. Zhang, and Z. Y. Li, "$\Gamma-M$ waveguides in two-dimensional triangular-lattice photonic crystal slabs", Optics Express **16**, 21483-21491 (2008).

[4] Y. Z. Liu, R. J. Liu, S. Feng, C. Ren, H. F. Yang, D. Z. Zhang, and Z. Y. Li, "Multi-Channel filters via $\Gamma-K$ and $\Gamma-M$ Waveguide coupling in two-dimensionaltTriangular-lattice photonic crystal slabs", Appl. Phys. Lett. **93**, 241107 (2008).

[5] C. Z. Zhou, Y. Z. Liu, and Z. Y. Li, "90° waveguide bend in two-dimensional triangular lattice silicon photonic crystal slabs", Chin. Phys. Lett. **27**, 084203 (2010).

[6] L. Gan, Y. Z. Liu, J. Y. Li, Z. B. Zhang, D. Z. Zhang, and Z. Y. Li, "Ray trace visualization of negative refraction of light in two-dimensional air-bridged silicon photonic crystal slabs at 1.55 um", Optics Express **17**, 9962-9970 (2009).

[7] C. Wang, C. Z. Zhou, and Z. Y. Li, "On-chip optical diode based on silicon photonic crystal heterojunctions", Optics Express **19**, 26948 (2011).

[8] L. Gan, F. Qin, and Z. Y. Li, "Broadband large-angle self-collimation in two-dimensional silicon photonic crystal", to appear in Opt. Lett. (2012).

[9] C. Wang and Z. Y. Li, "Cavities without confinement barriers in incommensurate photonic crystal superlattices", submitted to EPL (2012).

[10] F. Qin, Z. M. Meng, X. L.Zhong, Y. Liu, and Z.Y. Li, "Fabrication of semiconductor-polymer compound nonlinear photonic crystal slab with highly uniform infiltration based on nano-imprint lithography technique", to appear in Optics Express (2012).

[11] Y. Liu, F. Qin, Z. Y. Wei, Q. B. Meng, D. Z. Zhang, and Z. Y. Li, "10 femtosecond ultrafast all-optical switching in polystyrene nonlinear photonic crystals", Appl. Phys. Lett. **95**, 131116 (2009).

[12] F. Qin, Y. Liu, and Z. Y. Li, "Optical switching in hybrid semiconductor nonlinear photonic crystal slabs with Kerr materials", J. Opt. A **12**, 035209 (2010).

[13] F. Qin, Y. Liu, Z. M. Meng, and Z. Y. Li, "Design of Kerr-effect sensitive microcavity in nonlinear photonic crystal slabs for all-optical switching", J. Appl. Phys. **108**, 053108 (2010).

[14] Z. M. Meng, F. Qin, Y. Liu, and Z.Y. Li, "High-Q microcavities in low-index one-dimensional photonic crystal slabs based on modal gap confinement", J. Appl. Phys. **109**, 043107 (2011).

[15] Y. Liu, F. Qin, Z. M. Meng, F. Zhou, Q. H. Mao, and Z.Y. Li, "All-optical logical gates based on two-dimensional nonlinear photonic crystal slabs", Optics Express **19**, 1945 (2011).

Nonlinear optics in photonic crystal nanostructures

(Invited Paper)

Chad Husko

Centre for Ultrahigh bandwidth Devices for Optical Systems (CUDOS), Institute of Photonics and Optical Science (IPOS),
School of Physics, University of Sydney, Sydney NSW 2006 Australia, E-mail: chad.husko@sydney.edu.au

Abstract—We examine nonlinear Kerr effects in slow-light photonic crystals in the presence of multi-photon absorption and free-carriers. We derive analytic formulations for self-phase modulation limited by three-photon absorption. These observations are confirmed with a modified nonlinear Schrödinger equation (NLSE). Experimental verifications of several nonlinear processes are presented to support the modeling.

I. INTRODUCTION

Slow-light enhanced nonlinearities have been studied extensively in semiconductor photonic crystals as key elements to future photonic technologies [1]. A variety of nonlinear processes such as temporal soliton compression [2], four-wave mixing [3], Raman scattering [4], third-harmonic generation [5], and self-phase modulation (SPM) [6], [7] have been demonstrated in photonic crystal waveguides (PhCWG). A key challenge in these works is multi-photon absorption mechanisms such as two-photon absorption (TPA) and three-photon absorption (ThPA) which restrict the desirable Kerr effect. Additionally, TPA and ThPA generate free carriers that induce both free-carrier absorption (FCA) and free-carrier dispersion (FCD). This latter effect is particularly detrimental to the propagating pulse shape. Here we present an analytic formulation of slow-light SPM for materials limited by ThPA as a leading term and compare it to TPA-restricted materials [8]. We derive critical intensity thresholds, I_c, at which FCD degrades the pulse propagation. Though the analysis here focuses on the 1.55 μm wavelength range, the results in this work are applicable to any material waveguide system limited by three-photon absorption. In particular, several groups have recently initiated systematic investigation of nonlinear optics in silicon near 2 μm where TPA is drastically reduced [9], [10].

II. NONLINEAR WAVE PROPAGATION IN SLOW-LIGHT SEMICONDUCTORS

The propagation of picosecond optical pulses in a waveguide with suppressed TPA and negligible group velocity dispersion is governed by [7], [11], [12]:

$$\frac{\partial E}{\partial z} = -\frac{\alpha}{2}E + ik_0 n_2 |E|^2 E - \frac{\alpha_3}{2}|E|^4 E + \left(ik_o\frac{dn}{dN} - \frac{\sigma}{2}\right)N_c E \tag{1}$$

where E is the electric field envelope (Intensity $I_o = |E|^2$), α the linear loss, $k_0 = 2\pi/\lambda$, n_2 the optical Kerr coefficient, α_3 the ThPA coefficient, dn/dN the index change per carrier density, σ the FCA coefficient, N_c the number of carriers, and z the distance along the waveguide of length, L. The

carrier equation: $\frac{\partial N_c(z,t)}{\partial t} = \frac{\alpha_3}{3\hbar\omega}|E(z,t)|^6 - \frac{N_c(z,t)}{\tau_c}$, describes free-carriers generated by ThPA, as well as recombination with lifetime, τ_c. We also extend this formalism to include slow group velocity enhancement of the optical field. Though slow-light effects physically affect the field intensity, here we attach the scalings to the coefficients for notational simplicity: [Kerr] $n_{2eff} = n_2(n_g/n_0)^2$; [TPA] $\alpha_{2eff} = \alpha_2(n_g/n_0)^2$ [1]; [ThPA] $\alpha_{3eff} = \alpha_3(n_g/n_0)^3(1/A_{5eff})^2$ [7], with A_{5eff} the 5th-order area, α_2 the two-photon absorption coefficient, and n_g the slow-light group index.

In Fig. 1(a) we plot phase shift, ϕ, as a function of I_0 for representative materials, GaInP (ThPA) [2], [7], and silicon (TPA) [6], with $n_g = 20$, $L = 1$ mm, and $\alpha = 10$ dB/cm, achievable in PhCWGs at present. The solid lines indicate the analytic formulation of ϕ according to the pertinent formulas in [8], [11], while the straight (dashed) lines serve as a reference in the absence of nonlinear loss. The ThPA-limited material GaInP demonstrates larger phase shift compared to TPA-limited Si. We also comment on the energy implications of these slow-light structures [13] .

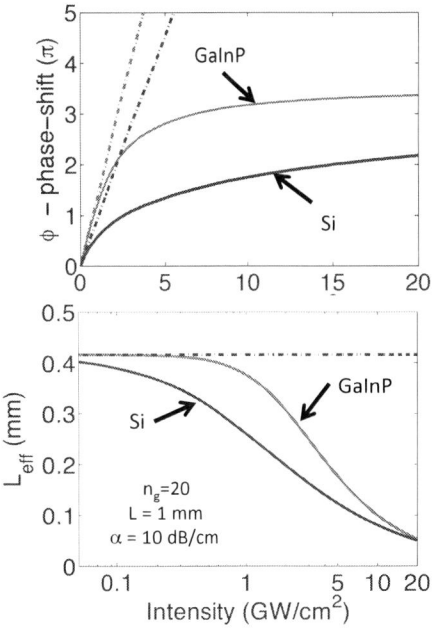

Fig. 1. (a) Phase shift, ϕ, as a function of I_o for (i) GaInP and (ii) Si. The straight dashed lines are α only, while the curves shows ϕ impacted by TPA or ThPA. (b) Effective lengths, $L_{eff}(TPA)$[Si] and $L_{eff}(ThPA)$[GaInP] vs. I_o with $L_{eff}(lin.)$ shown (dashed black line) as reference.

978-1-4673-1602-6/12 $31.00 © 2012 IEEE

In conventional optical fibers with small nonlinearities, $L_{eff}(lin.) = (1 - exp[-\alpha L])/\alpha$, e.g. no intensity dependance. In the case of materials with nonlinear absorption, different definitions of effective length should be used: $L_{eff}(TPA)$[11] or ThPA, $L_{eff}(ThPA)$, readily obtainable from re-arranging the phase equation in [8]. In Fig. 1(b), we show the corresponding L_{eff} versus I_o. While $L_{eff}(TPA)$ deviates almost immediately, note that $L_{eff}(ThPA) \approx L_{eff}(lin.)$ up to about $I_0 = 0.3$ GW/cm^2 for n_g=20, and is still 90% of $L_{eff}(lin.)$ at 1 GW/cm^2, indicating that nonlinear losses are weak under these conditions. This is in sharp contrast to the TPA material, which falls off immediately. While multi-photon absorption restricts SPM, and other nonlinear effects in general, the far greater impediment is free-carrier effects.

We plot the temporal pulse properties in Fig. 2(a). While $\phi = 1.5\pi$ is observed with only a slight onset of blue-shift for GaInP, the pulse undergoes a dramatic blue-shift in Si due to a greater number of free-carriers generated at this power level. The spectral properties in Fig. 2(b) show slight asymmetry in GaInP, as expected above a critical intensity I_c [8]. A much smaller phase shift with a strong trailing blue component are apparent in the case of Si.

We describe modeling and experimental verification of nonlinear soliton compression in PhCWGs with a modified nonlinear Schrödinger equation (NLSE) [2]. Fig. 3 shows the modeled and experimental pulse temporal compression for an example data set. Importantly, the model leads to simultaneous agreement with both the autocorrelation and spectra with no degrees of freedom. The insets of Fig. 3 show the compression factor $\chi_c = T_0/T_{comp}$ and the compression quality factor $Q_c = P_{peak}/\chi_c$, where P_{peak} is the output peak intensity normalized to the input peak intensity.

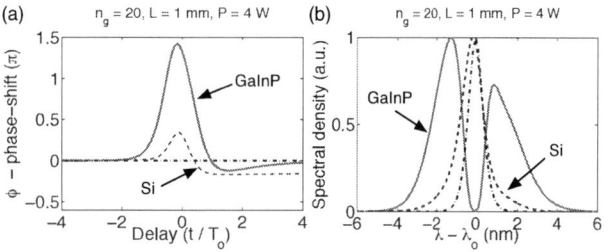

Fig. 2. (a) Phase shift (π) vs. delay for P= 4 W at n_g=20 for GaInP, solid, and Si dashed (b) Spectra corresponding to panel (a).

III. CONCLUSION

We presented analytic and numerical analysis of nonlinear Kerr effects in slow-light photonic crystals with multi-photon absorption and free-carriers. The results are modelled with a modified nonlinear Schrödinger equation (NLSE) and confirmed experimentally.

ACKNOWLEDGMENT

The authors would like to thank B. J. Eggleton, C.-W. Wong, A. De Rossi, S. Combri, P. Colman, Q. V. Tran. Current and past funding for the results include the National Science

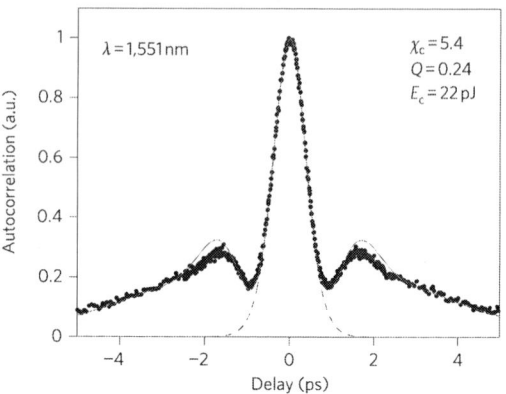

Fig. 3. Calculated (red solid line) and measured autocorrelation traces (black dots) autocorrelation traces at 1551 nm with a coupled pulse energy of 22 pJ together with a hyperbolic secant fit (blue dashed line).

Foundation, the European Seventh Framework Program, and the Australian Research Council Centres of Excellence and Discovery Early Career Researcher Award (DECRA) Program.

REFERENCES

[1] T. Baba, "Slow light in photonic crystals," *Nature Photonics*, vol. 2, p. 465, 2008.

[2] P. Colman, C. Husko, S. Combrié, I. Sagnes, C. W. Wong, and A. De Rossi, "Observation of soliton pulse compression in photonic crystal waveguides," *Nature Photon.*, vol. 4, p. 862, 2010.

[3] J. F. McMillan, M. Yu, D.-L. Kwong, and C. W. Wong, "Observation of four-wave mixing in slow-light silicon photonic crystal waveguides," *Opt. Express.*, vol. 18, pp. 15 484–15 497, 2010.

[4] H. Oda, K. Inoue, A. Yamanaka, N. Ikeda, Y. Sugimoto, and K. Asakawa, "Light amplification by stimulated Raman scattering in AlGaAs-based photonic-crystal line-defect waveguides," *Appl. Phys. Lett.*, vol. 93, no. 5, p. 051114, 2009.

[5] B. Corcoran, C. Monat, C. Grillet, D. J. Moss, B. J. Eggleton, T. P. White, L. O'Faolain, and T. F. Krauss, "Green light emission in silicon through slow-light enhanced third-harmonic generation in photonic-crystal waveguides," *Nature Photon.*, vol. 3, p. 206, 2009.

[6] C. Monat, B. Corcoran, M. Ebnali-Heidari, C. Grillet, B. Eggleton, T. White, L. O'Faolain, and T. F. Krauss, "Slow light enhancement of nonlinear effects in silicon engineered photonic crystal waveguides," *Opt. Express*, vol. 17, p. 2944, 2009.

[7] C. Husko, S. Combrié, Q. Tran, F. Raineri, C. W. Wong, and A. De Rossi, "Non-trivial scaling of self-phase modulation and three-photon absorption in iii-v photonic crystal waveguides," *Opt. Express*, vol. 17, p. 22442, 2009.

[8] C. Husko, P. Colman, S. Combrié, A. De Rossi, and C. W. Wong, "Effect of multiphoton absorption and free carriers in slow-light photonic crystal waveguides," *Optics Letters*, vol. 36, no. 12, pp. 2239–2241, 2011.

[9] A. D. Bristow, N. Rotenberg, and H. van Driel, "Two-photon absorption and Kerr coefficients of silicon for 850–2200 nm," *Appl. Phys. Lett.*, vol. 90, p. 191104, 2007.

[10] X. Liu, R. M. Osgood, Y. A. Vlasov, and W. M. J. Green, "Mid-infrared optical parametric amplifier using silicon nanophotonic waveguides," *Nature Photon.*, vol. 4, no. 8, pp. 557–560, 2010.

[11] L. Yin and G. Agrawal, "Impact of two-photon absorption on self-phase modulation in silicon waveguides," *Opt. Lett.*, vol. 32, p. 2031, 2007.

[12] N. C. Panoiu, J. F. McMillan, and C. W. Wong, "Theoretical analysis of pulse dynamics in silicon photonic crystal wire waveguides," *Selected Topics in Quantum Electronics, IEEE Journal of*, vol. 16, no. 1, pp. 257–266, 2010.

[13] C. Husko and B. J. Eggleton, "Energy efficient all-optical signal processing: a comparison of slow-light photonic crystals and nanowires ," in *Conference on Lasers and Electro-Optics (CLEO)*. OSA, 2012.

Optical design of a qubit embedded in photonic crystals for rotation gate operations

Hiroyuki Nihei[1] and Atsushi Okamoto[2]

[1]Health Sciences University of Hokkaido, Hokkaido 061-0293, Japan; nihei@hoku-iryo-u.ac.jp
[2]Hokkaido University, Kita-ku, Sapporo, Hokkaido 060-0814, Japan; ao@optnet.ist.hokudai.ac.jp

Abstract- **We have optimized the optical design parameters of a qubit composed of excited states of an atom embedded in photonic crystals for operating rotation gates, using the coherent control of spontaneous emission from the atom.**

I. INTRODUCTION

Very small defects (such as quantum dots or impurity atoms) embedded in photonic crystals can provide a promising building block of optical solid-state devices for quantum information processing [1]. Near the embedded atom, light is confined due to the presence of a photonic bandgap (PBG), which leads to the formation of a photon-atom bound state with nonzero steady-state atomic populations on excited states [2]. This may provide the basis of a qubit encoding optical and quantum information. Furthermore, robust mechanisms for processing optical and quantum information have been proposed, such as the population switching of an atom (or a quantum dot) embedded in photonic crystals [3]. So far, we have also demonstrated the coherent control of spontaneous emission from such an embedded atom with nonzero steady-state atomic populations [4]. Using the coherent control, we have also clarified a method of confining light in a single mode near the embedded atom, which is useful for a qubit [5]. A recent major theoretical challenge in this field is the investigation of carrying out a quantum logic gate operation to a qubit, especially a single-qubit rotation gate, which is one of the universal quantum logic gates and a key operation for quantum computing.

In this paper, to carry out near-complete rotation gate operation, we clarify an optical design of a solid-state qubit composed of excited states of an impurity atom embedded in photonic crystals, using the coherent control of spontaneous emission from the embedded atom, where we optimize the optical design parameters such as control laser strength and the transition frequency of the embedded atom.

II. QUANTUM DOT IN PHOTONIC CRYSTALS

We let $|0\rangle$ and $|1\rangle$ denote two excited states of an impurity three-level atom embedded in photonic crystals with a high-pass 3D PBG. It is assumed that the transition frequency between $|1\rangle$ and the ground state $|g\rangle$ (ω_{1g}) is far inside the PBG and that the transition frequency between $|0\rangle$ and $|g\rangle$ (ω_{0g}) is near the band edge frequency ω_C, where the detuning is denoted as $\delta = \omega_{0g} - \omega_C$. The coherent superposition of the far-inside bit $|1\rangle$ and the edge-side bit

$$|\psi(t)\rangle = a_0(t)|0\rangle + a_1(t)|1\rangle \qquad (1)$$

can act as a qubit with a complex amplitude $a_k(t)$ ($k=0, 1$), which provides the atomic population $n_k(t)=|a_k(t)|^2$ (namely, the probability of finding the atom in level $|k\rangle$). The time evaluation of the qubit can be shown by the Bloch sphere representation, as shown in Fig. 1. Here, the position of the qubit is determined by the norm $N(t)=n_0(t)+n_1(t)$ (the length from the origin that is given by the sum of the atomic populations on excited states), the ratio $\tan\theta(t)=|a_0(t)|/|a_1(t)|$, and the phase difference $\phi(t)=Arg(a_0(t)/a_1(t))$.

At the initial time ($t=0$), using an ultrashort pumping laser pulse, the atom is excited ($N(0)=1$), where the initial phases $\theta(0)$ and $\phi(0)$ are determined by the pulse area and phase of the laser pulses. For $t>0$, a coherent laser is used for coupling the bits $|0\rangle$ and $|1\rangle$, where the strength of the control laser is characterized by the Rabi frequency Ω. Here, the complete rotation gate operation requires qubit evaluation on the circle C in Fig. 1(a), which is determined by $N(t)=1$ and $\tan\theta(t)=1$.

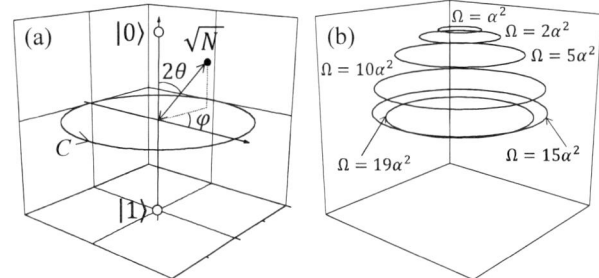

Fig. 1. Bloch sphere representation:
(a) the position of a qubit and (b) the qubit evaluation paths.

Fig. 1(b) shows the qubit evaluation paths in the Bloch sphere for various Rabi frequencies Ω and $\delta = -20\alpha^2$ with a scaled parameter α [3], where we apply the condition for making confined light composed of a single localized mode, which is fulfilled by preparing the initial atomic state to satisfy the conditions for the initial phases $\theta(0)$ and $\phi(0)$ that have been reported in Ref. [5]. Under these conditions, in addition to the nonzero steady-state norm, the population ratio $\tan\theta(t)$ remains constant, leading to the circular rotation, due to the lack of dynamical transfer of populations between the two bits $|0\rangle$ and $|1\rangle$, because Rabi splitting disappears with the destructive quantum interference. Furthermore, we find that the qubit evaluation paths approach the circle C with increasing Ω. However, to further increase Rabi frequency ($\Omega=19\alpha^2$), the norm is reduced, leading to the loss of the quantum information. So, we note that there are optimum values of the optical design parameters (the Rabi frequency Ω and the detuning δ characterizing control laser strength and the transition frequency of the atom, respectively) for rotation gate operations ($N(t)=1$ and $\tan\theta(t)=1$).

III. OPTIMIZATION OF OPTICAL DESIGN PARAMETERS

First, we consider the dependence of the norm on the optical design parameters. In Fig. 2, we plot the steady-state norm $N_S=\lim N(t)$ ($t\to\infty$) as a function of the Rabi frequency for various values of the detuning δ. Fig. 2 shows that the norm N_S is increased by increasing the detuning $|\delta|$, namely, by moving the transition frequency ω_{0g} far inside the PBG, accompanied by the enhancement of the light confinement effect. On the other hand, by increasing the large Rabi frequency, the norm N_S is decreased, and then the norm N_S decays to zero at the cutoff $\Omega_C(\delta)$, which is given by $\Omega_C(\delta)\cong|\delta|$. In the region $\Omega_C(\delta)<\Omega$, due to the Stark shift caused by the control laser, the edge-side bit $|0\rangle$ (ω_{0g}) moves out of the PBG, and then in addition to $n_0(t)$, the atomic population of the far-inside bit $|1\rangle$ ($n_1(t)$) completely decays to the ground state $|g\rangle$ through the $|0\rangle-|1\rangle$ channel coupled by the control laser. Therefore, we note that the condition $\Omega<\Omega_C(\delta)$ is required to maintain the quantum information.

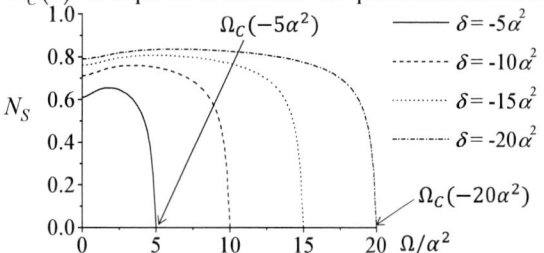

Fig. 2. Steady-state norm as a function of the Rabi frequency.

In Fig. 3, we plot the steady-state population ratio $\tan\theta_S=\lim\tan\theta(t)$ ($t\to\infty$). The steady-state population of the bit $|1\rangle$ (n_{1S}) is nearly constant with varying the detuning δ, because the bit $|1\rangle$ is far inside the PBG, whereas that of the edge-side bit $|0\rangle$ (n_{0S}) increases with increasing the detuning $|\delta|$, which leads to a decrease in the ratio $\tan\theta_S$ (but $\tan\theta_S>0$ is maintained). On the other hand, the ratio $\tan\theta_S$ increases with increasing the Rabi frequency, and then the ratio approaches 1. This is because the increase in the Rabi frequency facilitates the population transfer between the bits $|0\rangle$ and $|1\rangle$, which equalizes their populations. At $\Omega=\Omega_C(\delta)$, the ratio is 1. Therefore, we note that the ratio satisfies $0<\tan\theta_S<1$ in the region where the quantum information is maintained ($\Omega<\Omega_C(\delta)$).

Consequently, we find that both the norm and the ratio satisfy $0<(N_S$ and $\tan\theta_S)<1$, so that the optimization of the optical design parameters (the Rabi frequency and the detuning) for nearly complete rotation gate operations ($N_S\cong1$ and $\tan\theta_S\cong1$) is given by a pair of Ω and δ that maximizes $P=N_S+\tan\theta_S$. Fig. 4 shows a contour plot of P for $\Omega<\Omega_C(\delta)$ and $-20\alpha^2<\delta<0$. This figure indicates

that P has a ridge fitting to $\Omega_F(\delta)$ with $R^2=0.99$ from our numerical calculations. This line $\Omega_F(\delta)$ provides the optimum Rabi frequencies (control laser strengths). As a result, we find that, for the nearly complete rotation gate operations, the transition frequency lying far inside the PBG is an appreciable choice in suppressing the loss of quantum information, while the control laser strength should be managed as $\Omega_F(\delta)$ to enhance the population transfer in the two bits for $\tan\theta_S\cong1$ and to suppress the Stark shift for $N_S\cong1$.

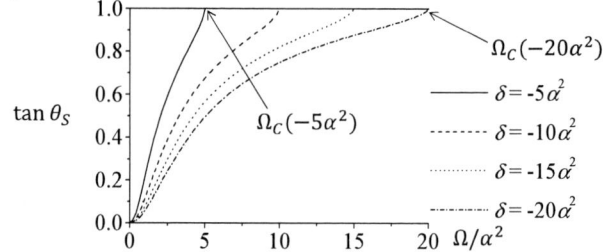

Fig. 3. Steady-state population ratio as a function of the Rabi frequency.

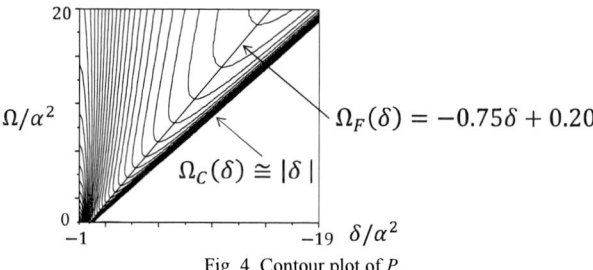

Fig. 4. Contour plot of P.

IV. CONCLUSION

We have optimized the atomic transition frequency and control laser strength for operating nearly complete rotation gates for a qubit composed of excited states of an atom embedded in photonic crystals, using the coherent control of spontaneous emission from the embedded atom. The result of this study is useful for determining the designs of future quantum logic gates based on solid-state photonic crystal systems.

REFERENCES

[1] S. John, "Light control at will," *Nature*, vol. 460, p. 337, 2009.

[2] M. Woldeyohannes and S. John, "Coherent control of spontaneous emission near a photonic band edge: A qubit for quantum computation," *Phys. Rev. A*, vol. 60, p. 5046, 1999.

[3] H. Takeda and S. John, "Self-consistent Maxwell-Bloch theory of quantum-dot-population switching in photonic crystals," *Phys. Rev. A*, 053811, 2011.

[4] H. Nihei and A. Okamoto, "Coherent control of excited atomic states inside a three-dimensional photonic bandgap", *J. Mod. Opt.*, vol. 55, pp. 2391-2399, 2008.

[5] H. Nihei and A. Okamoto, "Confined light composed of a single localized mode inside photonic crystals for a qubit", *J. Opt. and Quant. Electron.*, in press.

InGaN Nanorod LEDs: A Performance Assessment

(Invited Paper)

Bernd Witzigmann*, Marcus Deppner* and Friedhard Römer*
*Computational Electronics and Photonics Group, University of Kassel,
Wilhelmshöher Allee 71, D-34121 Kassel, Germany

Abstract—**Light Emitting Diodes (LEDs) have become an attractive concept as efficient light sources. In this contribution, the electro-optical performance of III-nitride based core-shell nanorod LEDs is investigated by detailed simulation. In contrast to their planar counterparts, they possess increased active region area on small footprint, non-polar active regions with c-plane vertical growth, and form horizontal two-dimensional active photonic crystals for the optical extraction process.**

I. INTRODUCTION

III-nitride based light emitting diodes (LEDs) have become powerful and efficient light sources for various lighting applications, with internal quantum efficiencies larger than 80%. Despite their success, they are still plagued by relatively high chip cost and a drop in efficiency at high current operation [1]. Although the root cause of efficiency droop hasn't been clarified yet, it is believed to be an intrinsic electronic effect of the active region [2].

One approach to solve these issues is the use of nanorod LEDs with the pn-junction arranged in vertical direction along the rod. These core-shell type devices can be grown on silicon substrates, and due to their large aspect ratio, can achieve a pn-junction area which is a multiple of the respective substrate area [3]. This allows LED operation at lower current densities, where the efficiency is high, and the droop effect does not occur. In this contribution, a comprehensive analysis of the core-shell nanorod device concept is presented, based on simulation. Both optical and electronic aspects are discussed.

II. DESIGN BASICS

Figure 1 shows the basic arrangement of a core-shell nanorod, with the c-axis of the wurtzite crystal aligned in vertical direction. In some implementations, the pn-junction and active region extends over the sidewalls as well as on the top in horizontal direction (as shown). Optionally, the top active region can be removed or avoided so that only the sidewall active region is present.

While in a planar LED the active area is identical to the wafer area, the core-shell nanorod LED features an active area increase (in radial approximation) of

$$m = \frac{A_{rod}}{A_{planar}} = f * \left(\frac{d}{D}\right)^2 * \left(1 + \frac{4h}{d}\right) \quad (1)$$

with $f = \frac{D^2}{a^2}$ being the area fill factor, and the other variables described in fig. 1.

Fig. 2 gives the active area enhancement for some exemplary cases of nanorod dimensions which are realistic to

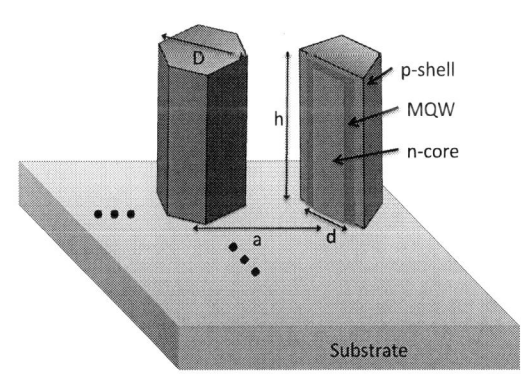

Fig. 1: Device geometry of a core-shell III-nitride nanowire. The multiple quantum-well (MQW) region includes an electron blocking layer (EBL).

achieve with available technology. An enhancement of approx. 10 is feasible, which translates into a reduced local current density if the LED is operated at the same current. The internal efficiency curve shows less droop at lower current.

III. SIMULATION RESULTS

A. Current Injection Efficiency

Homogeneous carrier injection in the active region of the high aspect ratio nanorods is a crucial requirement in order to fully benefit from the area enhancement of the rod

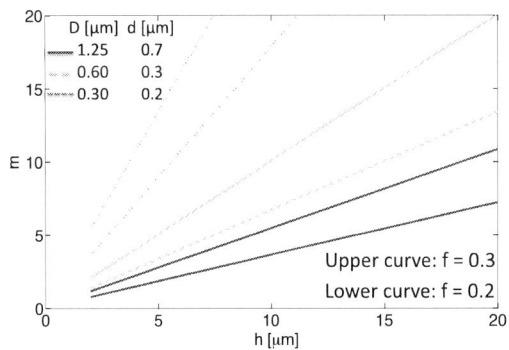

Fig. 2: Active area enhancement of core-shell nanorod LED compared to planar LED for different fill-factors f and rod height h.

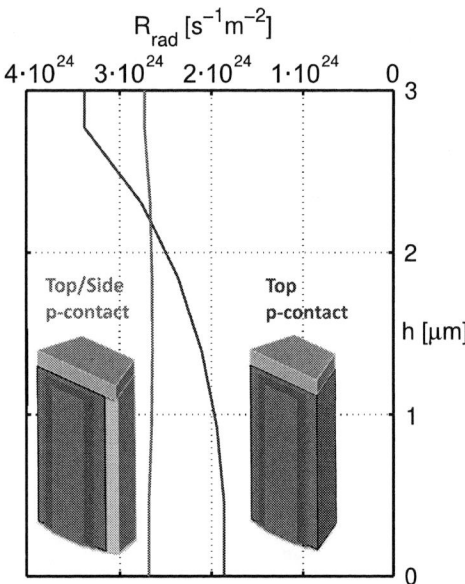

Fig. 3: Radiative recombination rate in MQW core-shell nanorod LED at a typical operating current density (I= 3.5×10^{-6}A).

arrangement. The main limiting factor is the thin p-doped shell, as the hole mobility in GaN is low. If the rod shell is contacted on top, and on bottom for the core, the active carrier density, and therefore radiative recombination rate, decreases towards the bottom of the rod . In contrast, a transparent p-contact that covers the entire sidewall gives optimum injection efficiency. Using microscopic carrier transport models, this problem can be studied in detail [4]. Fig. 3 shows a plot of the radiative recombination rate in the vertical cross-section through the quantum-well, at a typical LED operating current. The simulation is a 2-dimensional drift-diffusion model with inclusion of quantum transport [5]. The rod dimensions are $D = 1.25\mu$m, $d = 0.7\mu$m, and $h = 3\mu$m.

The n-contact is on the bottom of the rod, and two scenarios for the top contact are investigated. The red curve corresponds to the case where the entire p-shell is contacted by a transparent material; the blue curve shows an LED with a top contact only. Clearly, the thin p-shell creates a hole injection bottleneck, so that the radiative recombination decreases by almost 50%.

B. Optical Properties

Several specifics of the nanorod geometry need to be considered for the optical analysis. First, the vertical MQW active region is grown on a non-polar crystal orientation. Therefore, the active region consists of non-polar QWs without the interface piezo-polarization. This leads to a polarized optical emission, and to a higher re-absorption compared to c-plane quantum-wells [4]. The polarization of the optical emission results in a predominantly vertically directed emission (approx. 70% of emitted power in the dipole oriented in horizontal direction), which is beneficial for device operation.

The optical extraction efficiency depends on the electromagnetic properties of the nanorod array. A single nanorod can act as dielectric waveguide, and a bottom reflector increases the vertical outcoupling efficiency. As the rods can be arranged in a periodic pattern, photonic crystal effects can be exploited in the design [6]. The in-plane radiation can be suppressed by a photonic band gap, which, in the case of low internal efficiency cases, can improve the external efficiency. However, photonic band gaps also can suppress radiation with non-zero vertical wave vector, which contributes to the output power. Detailed electromagnetic simulations have been performed in order to analyze this non-trivial problem [7], which have shown operation outside of the band gap shows highest extraction efficiency. Also, transparent electrical contacts on the sidewalls are required, which are optical absorbers and limit the extraction efficiency.

IV. CONCLUSION

Nanorod LEDs offer conceptual advantages over traditional in-plane designs, such as enhanced active pn-area on a given substrate real estate, substrate material flexibility, non-polar active regions, and intrinsic photonic crystal filters. In order to benefit from these features, careful choice of the rod geometry and arrangement need to be done, which can be supported by detailed simulation. As a result, LEDs with reduced droop in efficiency can be designed.

ACKNOWLEDGMENT

We would like to thank EU project SMASH (FP7-228999-2) for the support of this work.

REFERENCES

[1] M. R. Krames, O. B. Shchekin, R. Mueller-mach, G. O. Mueller, L. Zhou, G. Harbers, and M. G. Craford, "Status and Future of High-Power Light-Emitting Diodes for Solid-State Lighting," *IEEE Journal of Display Technology*, vol. 3, no. 2, pp. 160–175, 2007.
[2] J. Piprek, "Efficiency droop in nitride-based light-emitting diodes," *physica status solidi (a)*, vol. 207, no. 10, pp. 2217–2225, 2010.
[3] S. Li and A. Waag, "GaN based nanorods for solid state lighting," *Journal of Applied Physics*, vol. 111, no. 7, p. 071101, 2012. [Online]. Available: http://link.aip.org/link/?JAP/111/071101/1
[4] F. Römer, M. Deppner, Z. Andreev, C. Kölper, M. Sabathil, M. Straßburg, J. Ledig, S. Li, A. Waag, and B. Witzigmann, "Luminescence and Efficiency Optimization of InGaN/GaN Core-Shell Nanowire LEDs by Numerical Modelling," *SPIE Proceedings, Photonics West*, 2012.
[5] S. Steiger, R. Veprek, and B. Witzigmann, "Unified simulation of transport and luminescence in optoelectronic nanostructures," in *Journal of Computational Electronics*, 2008, pp. 509–520.
[6] S. Fan, P. R. Villeneuve, J. D. Joannopoulos, and E. Schubert, "High Extraction Efficiency of Spontaneous Emission from Slabs of Photonic Crystals," *Physical Review Letters*, vol. 78, no. 17, pp. 3294–3297, 1997.
[7] C. Kölper, M. Sabathil, B. Witzigmann, F. Römer, W. Bergbauer, and S. M., "Optical properties of individual GaN nanorods for light emitting diodes: Influence of geometry, materials and facets," *SPIE Proceedings, Photonics West*, 2011.

Influence of polar surface properties on InGaN/GaN core-shell nanorod LED properties

M. Auf der Maur, F. Sacconi and A. Di Carlo

Dept. of Electronic Engineering
University of Rome "Tor Vergata", Italy
Email: auf.der.maur@ing.uniroma2.it

Abstract—**InGaN/GaN nanorod core-shell LEDs have shown to be very promising candidates for high efficiency lighting devices. Such nanorods can be grown in different ways, leading to different device geometry and in particular to different structures near the polar Ga- and N-face nanorod surfaces. In this work the influence of the properties of the polar surfaces on the electrical device behaviour is studied qualitatively based on a semiclassical simulation model.**

I. INTRODUCTION

InGaN/GaN core-shell nanorod LEDs have many advantages over planar devices [1]. First, the effective light emitting area of nanorods can be considerably higher than in planar LEDs. Second, their 3-dimensional nature leads to an efficient elastic strain relaxation and to nearly defect-free structures. Third, most of the quantum well (QW) area is provided by the lateral non-polar wells. Therefore, the problems connected to the built-in polarization fields in nitride LEDs grown along the c-axis are mostly eliminated without the need of expensive non-polar GaN substrates.

Core-shell nanorods can be grown in different ways, leading to different device structures. Numerical simulations are a valuable tool for the detailed understanding of the electrical and optical behaviour of these structures and for the identification of the most efficient ones. Several simulation studies on core-shell LEDs have been presented recently [2], [3], [4], [5], focusing on a detailed description of electroluminescence, the contributions from polar and non-polar wells or on device optimisation. In this work we propose a study on the qualitative influence of the properties of the polar nanorod surfaces and nanorod termination.

Device behaviour can be influenced by three different properties connected to the polar nanorod surfaces. First, in particular in the case of free polar surfaces, strong local strain relaxation leads to a band gap lowering and thus potentially to increased carrier densities near the surfaces. Second, the discontinuity in the polarization field at the surface tends to induce electron and hole accumulation layers at the N- and Ga-face surfaces, respectively. And third, there can be recombination centers at the surfaces leading to non-radiative losses. These effects could be important in nanorod devices especially in the case of limited aspect ratio.

The simulations are based on continuous linear elasticity theory and the drift-diffusion model.

II. SIMULATION RESULTS

Simulations have been performed on the four structures shown in Fig. 1. An $In_{0.2}Ga_{0.8}N$/GaN MQW stack with 3

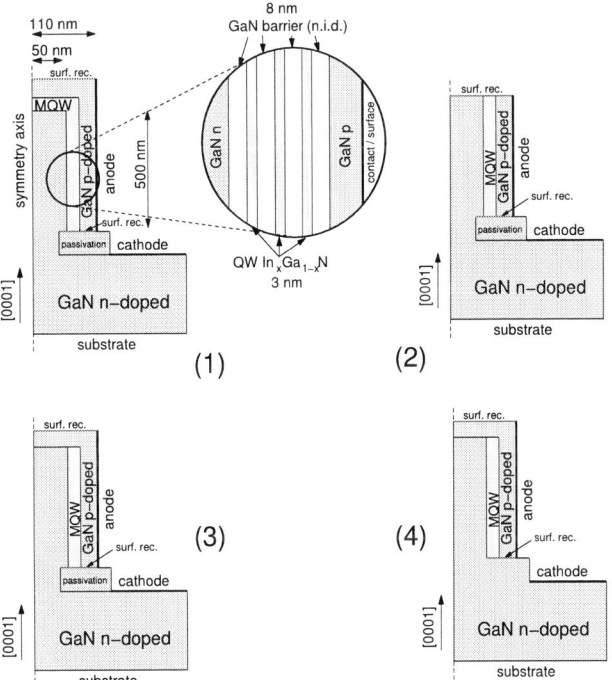

Fig. 1: The simulated InGaN/GaN MQW nanorod LEDs. Circular symmetry has been assumed.

QWs is grown on a 100 nm diameter n-doped GaN core, followed by a p-doped contact layer. In the first structure the InGaN QWs are grown all around the GaN core, such that the device includes polar c-plane QWs at the top end. The second structure assumes that these polar QWs are removed or avoided, such that the QWs extend to the top polar surface. The third structure is based on the second one, assuming however that the p-doped GaN shell is grown also on the top of the nanorod. These first three structures additionally assume a passivation layer on top of the GaN substrate so that the shell is isolated from the n-doped substrate. The fourth structure is based on the third, without the passivation layer. Additionally, we performed a set of simulations on the same devices including a surface recombination of Shockley-Read-Hall (SRH) type at the polar nanorod surfaces with surface recombination velocity of 10^4 cm/s, as indicated in Fig. 1. For all simulations we assumed the anode on the nanorod sidewall.

In the drift-diffusion simulation we consider radiative, SRH and Auger recombinations, with $B = 5 \times 10^{-11}$ cm^3/s, $\tau_{\mathrm{SRH}} = 20$ ns (inside the active intrinsic region, $\tau_{\mathrm{SRH}} \ll 1$ ns in the highly doped regions) and $C = 3 \times 10^{-31}$ cm^6/s, respectively. We assume a cylindrically symmetric device and simulate therefore a 2D slice.

Fig. 2 shows the simulated IV characteristics of the different devices. The first three devices present only slightly differing characateristics. The fourth one shows lower current first and a higher one after turn on. The former is due to the fact that the nanorod grown directly on GaN does not present an electron accumulation layer at the bottom nanorod interface due to the absence of a polarization field discontinuity on most of the interface. The higher turn on current is explained by the parasitic pn-junction at the bottom interface.

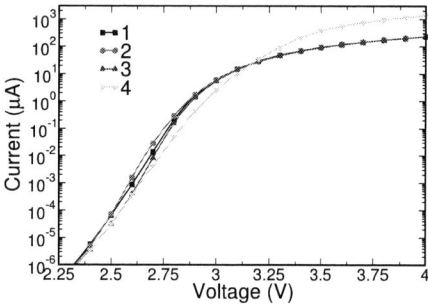

Fig. 2: The simulated IV characteristics for the different nanorods.

Fig. 3 shows the internal quantum efficiency (IQE) calculated classically for the four devices with and without surface recombination. It can be observed that device 2 where the QWs extend to the top polar nanorod surface has clearly the lowest IQE. Moreover, devices 2 and 4 show considerable sensitivity to surface recombination, contrary to the other two. In device 2 this is due to strong strain relaxation at the top nanorod surface, leading to increased carrier densities and thus increased recombination rates. For device 4 the reason is the absence of polarization induced band bending at the bottom interface, which causes an increased hole density and thus increased recombination.

Fig. 4 shows the electron density and current flow lines in the lower part of the nanorod for devices 3 and 4, showing the electron accumulation in device 3 due to the discontinuity in polarization between InGaN/GaN and SiN. This leads to an occupation of the QWs from this interface and a strong axial current component inside the wells. In device 4, electrons enter the MQW stack radially through the potential barrier formed by the leftmost intrinsic GaN barrier, leading to the slightly lower current observed in Fig. 2. In this case, however, the injection into the QWs is more homogeneous

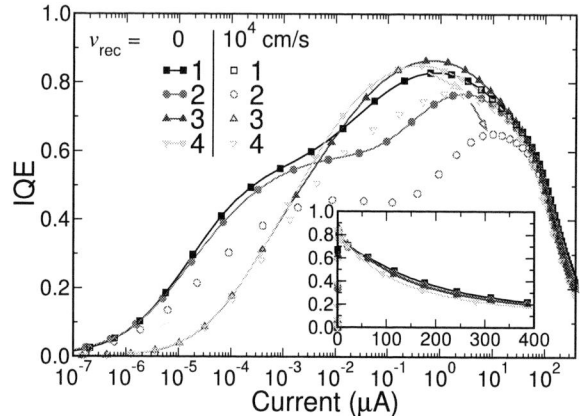

Fig. 3: The simulated IQE for the different nanorods, with and without surface recombination. The inset shows the IQE in linear scale.

along the nanorod, whereas for device 3 one might expect an inhomogeneous emission especially for high aspect ratios. On the other hand, injection from the interfaces increases homogeneity across the QWs, but leads to high local current densities.

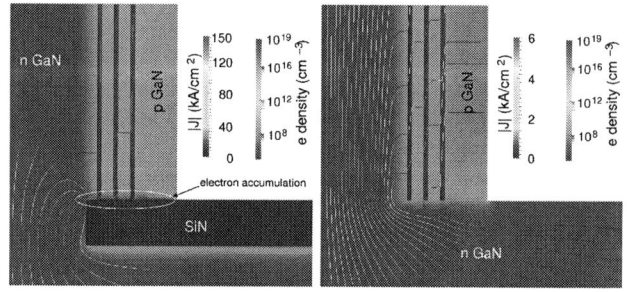

Fig. 4: Electron density and current flow lines in the lower part of the nanorod for (a) device 3 and (b) device 4.

ACKNOWLEDGMENT

The authors thank the EU Project SMASH FP7-228999-2 for support.

REFERENCES

[1] S. Li and A. Waag, "Gan based nanorods for solid state lighting," *J. Appl. Phys.*, vol. 111, p. 071101, 2012.

[2] M. Deppner, F. Römer, B. Witzigmann, J. Ledig, R. Neumann, A. Waag, W. Bergbauer, and M. Strassburg, "Computational study of carrier injection in iii-nitride core-shell nanowire-leds," in *Semiconductor Conference Dresden (SCD)*, sept. 2011, pp. 1 –4.

[3] C. Mazuir and W. V. Schoenfeld, "Modeling of nitride based core/multishell nanowire light emitting diodes," *Journal of Nanophotonics*, vol. 1, no. 4, p. 0135503, 2007.

[4] B. Connors, M. Povolotskyi, R. Hicks, and B. Klein, "Simulation and design of core-shell gan nanowire leds," in *Proc. SPIE*, vol. 7597, 2010, p. 75970B.

[5] F. Römer, M. Deppner, Z. Andreev, C. Kölper, M. Sabathil, M. Strassburg, J. Ledig, S. Li, A. Waag, and B. Witzigmann, "Luminescence and efficiency optimization of ingan/gan core-shell nanowire leds by numerical modelling," in *Proc. SPIE*, vol. 8255, 2012, p. 82550H.

Excitonic Properties of GaN/AlN Quantum Dot Single Photon Sources

Stanko Tomić[a] and Nenad Vukmirović[b]

[a] *Joule Physics Laboratory, School of Computing, Science and Engineering, University of Salford, UK*
[c] *Scientific Computing Laboratory, Institute of Physics Belgrade, University of Belgrade, Serbia*

Abstract—**Excitons and biexcitons in GaN/AlN quantum dots (QD) were investigated with special emphasis on the use of these QDs for single photon source applications. The theoretical methodology for the calculation of single-particle states was based on 8-band strain-dependent envelope function Hamiltonian, with the effects of spin-orbit interaction, crystal-field splitting, piezoelectric and spontaneous polarization taken into account. Exciton and biexciton states were found using the configuration interaction method. Optimal QD heights for their use in single-photon emitters were determined for various diameter to height ratios. The competition between strong confinement in GaN QDs and internal electric field, generally reported in wurtzite GaN, was also discussed, as well as its effect on appearance of bound biexcitons.**

I. INTRODUCTION

Modern optoelectronic devices like triggered single-photon sources ("photon on demand") are highly desired for applications in quantum cryptography and quantum information processing [1]. GaN/AlN quantum dots [2], [3] offer certain advantages for realization of single photon sources. Larger band offsets and effective masses lead to strong quantum confinement effects, which should enable the operation of single photon sources (SPS) at higher temperatures. Several single III-nitride quantum dot spectroscopy experiments were therefore performed, [4], [5], [6] which indeed led to the realization of a GaN/AlN single photon source operating at 200 K. [7] For SPS applications it is desirable to have as large as possible the value of biexcitonic shift defined as the difference between the energy of the transition line from the biexciton to exciton state and the energy of the exciton transition line, $B_{XX} = E_{XX} - 2E_X$. This is required to enable good spectral separation of the two lines. [7] It is known [3] that the built-in electric field acts to localize the electrons at the top of the dot and holes at the bottom of the dot. Consequently, the interaction between two excitons forming a biexciton is mostly determined by repulsive electron-electron and hole-hole interactions which are stronger than the attractive interaction between spatially separated electron and hole. [6] For quantum dots with larger heights the biexciton is therefore certainly unbound and biexcitonic shift increases as the height increases due to a decreasing attractive part of the interaction. From that perspective, it is desirable to have a large QD height. On the other hand, one should also have the optical transition matrix element of the exciton transition as large as possible. [8] For large QDs, this element is small due to spatial separation of electron and hole wavefunctions

and it is therefore desirable to have a small QD height from this perspective. This discussion therefore indicates that the appropriate QD geometry for single photon source applications should be determined as a compromise between the two opposite requirements, which requires detailed knowledge of the excitonic properties of these dots.

II. THEORETICAL METHOD

After the single particle states were found, using the multi-band **k·p** Hamiltonian, the (bi)exciton states were obtained using the configuration interaction (CI) method. [9], i.e. by direct diagonalization of the Hamiltonian

$$
\hat{H} = \sum \varepsilon_i \hat{e}_i^+ \hat{e}_i - \sum \varepsilon_i \hat{h}_i^+ \hat{h}_i + \frac{1}{2} \sum V_{iljk} \hat{e}_i^+ \hat{e}_j^+ \hat{e}_k \hat{e}_l
$$
$$
+ \frac{1}{2} \sum V_{iljk} \hat{h}_i^+ \hat{h}_j^+ \hat{h}_k \hat{h}_l - \sum (V_{iljk} - V_{ikjl}) \hat{e}_i^+ \hat{h}_j^+ \hat{h}_k \hat{e}_l \quad (1)
$$

where $\hat{e}$ ($\hat{e}^+$) are electron annihilation (creation) operators, $\hat{h}$ ($\hat{h}^+$) the same operators for holes, and ε_i the single-particle energies. The summation over each index takes place over electron or hole states only, depending whether that index corresponds to electron or hole operator. Coulomb integrals, V_{ijkl}, required for the diagonalization of the CI Hamiltonian were evaluated in reciprocal space and then corrected using the Makov-Payne method [10], [9] by adding the first few terms (monopole, dipole, and quadrupole) in the multipole expansion to compensate for the effect of the mirror charges induced by periodic boundary conditions. These read as

$$
V_{ijkl} = V_{ijkl}(\Omega_c) - \frac{e^2}{4\pi\varepsilon}[q_{ij}q_{kl}a_{\mathrm{mad}} + \frac{4\pi}{3\Omega_c}\boldsymbol{d}_{ij}\cdot\boldsymbol{d}_{kl}
$$
$$
- \frac{2\pi}{3\Omega_c}(q_{ij}Q_{kl} + q_{kl}Q_{ij})], \quad (2)
$$

where $V_{ijkl}(\Omega_c)$ is uncorrected Coulomb integral calculated on supercell Ω_c, and $q_{ij}(\Omega_c) = \delta_{ij}$, $\boldsymbol{d}_{ij}(\Omega_c)$, and $Q_{ij}(\Omega_c)$ are the monopole, dipole and quadrupole corrections respectively, that acquires analytic form in the PW representation. The Madelung term in Eq. (2), a_{mad}, is defined via Ewald sums in real and inverse space, and self-interaction correction. Depending on order of indices in Eq. (2) those integrals represent direct Coulomb integrals $J_{ab} = V_{aabb}$ or exchange Coulomb integrals $K_{ab} = V_{abab}$. An efficient and accurate method to evaluated these expressions in reciprocal space was described in Ref. [9]. Additionally, symmetry considerations imply that only Coulomb integrals V_{ijkl} whose wavefunctions satisfy the conservation of the total quasi-angular momentum,

$\{m_j+m_l \equiv m_i+m_k(\mathrm{mod}6)\}$ are nonzero. These are therefore the only ones that need to be evaluated, which reduces the number of integrals that need to be calculated by a factor of six. The whole methodology presented here was implemented in the kppw code. [11]

III. RESULTS AND DISCUSSION

The calculations of the single particle electron and hole states and of the excitonic structure have been performed for a set of QDs satisfying the following conditions. The QD height was varied in the range $h = 1.5$nm to $h = 5$nm with a step of 0.5nm. The diameter to height ratio D/h was varied from 4 to 10 with a step of 1, and dots with the diameter larger than 30 nm were discarded. The truncated hexagonal pyramid base angle of $\alpha = 30^{\circ}$ was assumed and the wetting layer width of 0.5185 nm. All CI calculations used the basis set consisting of $N_e = 8$ electron and $N_h = 14$ hole states. For all D/h ratios we get an expected result that the exciton energy E_X and biexcitons E_{XX} decrease as the quantum dot height is increased [12]. We have observed that for maximizing the biexciton shift, B_{XX}, smaller values of D/h ratio are required, also in agreement with the experimental findings of Ref. [7]. Within the fixed value of D/h, bigger dots tend to have larger values of the B_{XX} due to reduction of the attractive part of the Coulomb interaction. Unfortunately, the trends in dipole matrix elements, required for the bright emission from the SPS, are opposite to the trends in bi-exciton shifts. Therefore, a compromise between these trends has to be made to find the optimal quantum dot geometry. To achieve this, we define the optimization function as [13], [12]:

$$\Xi = (E_{XX} - 2E_X) \cdot \ln(p_X^{(x)}/p_X^{(0)}). \quad (3)$$

$p_X^{(x)}$ is the value of the $x-$component of the dipole matrix element of the exciton transition, $p_X^{(0)}$ is equal to $10^{-4}p_X^{(x),\mathrm{max}}$ and $p_X^{(x),\mathrm{max}}$ is the maximal value of $p_X^{(x)}$ for all quantum dots considered. While the choice of $p_X^{(0)}$ is somewhat arbitrary, we find that the positions of maxima of the optimization function are weakly dependent on its value, when it is changed within reasonable limits. The dependence of the optimization function on exciton energy for different D/h ratios is presented in Fig. 1. For $D/h = 4$ and $D/h = 5$ the optimization function is nonmonotonous with a maximum at $h =2.5$ nm and $h =2.0$ nm, respectively. For larger D/h the largest value of optimization function is reached for the smallest dots among those investigated, with the height of $h =1.5$ nm. The most optimal dots emit in the range 3.2 – 3.8 eV, as can be seen in Fig. 1. Experimental results on single photon sources operating at 200 K reported in Ref. [7] show the emission energy of around 3.5 eV, and are in very good agreement with our theoretical predictions presented here. This also suggests that their QD geometry is most likely very close to an optimal one. It was reported in Ref. [7] that the estimated dimensions of the dots based on atomic force microscopy (AFM) measurements are: the height of 4 nm and the diameter of 25 nm. Our calculation for these dimensions of QDs yields an emission energy of 1.5 eV only, as well as very low values of the optimization function. However, AFM is a surface technique that measures

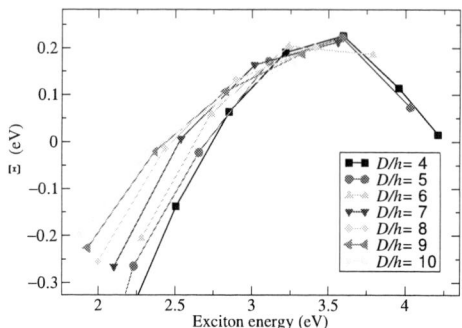

Fig. 1: The dependence of the optimization function defined as $\Xi(E_X)$ (see text for details) on exciton energy for different values of diameter to height ratios D/h.

the uncapped dots. Significant changes in the geometry of the dots after capping are possible and we believe that the dots measured in Ref. [7] actually have a much smaller height (i.e., reduced effective confinement region) than reported based on AFM measurements. We conclude with the discussion on the effect of the second order piezoelectricity [14] on the bound bi-excitons in the GaN/AlN wurtzite QD single photon sources.

REFERENCES

[1] D. Bouwmeester, A. Ekert, and A. Zeilinger, *The Physics of Quantum Information.* Springer, Berlin, 2000.

[2] B. Daudin, F. Widmann, G. Feuillet, Y. Samson, M. Arlery, and J. L. Rouviére, "Stranski-Krastanov growth mode during the molecular beam epitaxy of highly strained GaN," *Phys. Rev. B*, vol. 56, pp. R7069–R7072, 1997.

[3] A. D. Andreev and E. P. O'Reilly, "Theory of the electronic structure of GaN/AlN hexagonal quantum dots," *Phys. Rev. B*, vol. 62, pp. 15 851–15 870, 2000.

[4] S. Kako, K. Hoshino, S. Iwamoto, S. Ishida, and Y. Arakawa, "Exciton and biexciton luminescence from single hexagonal gan/aln self-assembled quantum dots," *Appl. Phys. Lett.*, vol. 85, no. 1, pp. 64–66, 2004.

[5] A. F. Jarjour, R. A. Taylor, R. A. Oliver, M. J. Kappers, C. J. Humphreys, and A. Tahraoui, "Cavity-enhanced blue single-photon emission from a single ingan/gan quantum dot," *Appl. Phys. Lett.*, vol. 91, no. 5, p. 052101, 2007.

[6] D. Simeonov, A. Dussaigne, R. Butté, and N. Grandjean, "Complex behavior of biexcitons in gan quantum dots due to a giant built-in polarization field," *Phys. Rev. B*, vol. 77, no. 7, p. 075306, 2008.

[7] S. Kako, C. Santori, K. Hoshino, S. Gotzinger, Y. Yamamoto, and Y. Arakawa, "A gallium nitride single-photon source operating at 200 k," *Nature Materials*, vol. 5, pp. 887–892, 2006.

[8] C. Santori, S. Gotzinger, Y. Yamamoto, S. Kako, K. Hoshino, and Y. Arakawa, "Photon correlation studies of single gan quantum dots," *Appl. Phys. Lett.*, vol. 87, no. 5, p. 051916, 2005.

[9] N. Vukmirović and S. Tomić, "Plane wave methodology for single quantum dot electronic structure calculations," *J. Appl. Phys.*, vol. 103, no. 10, p. 103718, 2008.

[10] G. Makov and M. C. Payne, "Periodic boundary conditions in ab initio calculations," *Phys. Rev. B*, vol. 51, no. 7, pp. 4014–4022, Feb 1995.

[11] S. Tomić, A. G. Sunderland, and I. J. Bush, "Parallel multi-band k center dot p code for electronic structure of zinc blend semiconductor quantum dots," *J. Mater. Chem.*, vol. 16, no. 20, pp. 1963–1972, 2006.

[12] S. Tomić and N. Vukmirović, "Excitonic and biexcitonic properties of single gan quantum dots modeled by 8-band $\mathbf{k} \cdot \mathbf{p}$ theory and configuration-interaction method," *Phys. Rev. B*, vol. 79, p. 245330, Jun 2009.

[13] S. D. Rinaldis, I. D'Amico, and F. Rossi, "Exciton–exciton interaction engineering in coupled gan quantum dots," *Applied Physics Letters*, vol. 81, no. 22, pp. 4236–4238, 2002.

[14] J. Pal, G. Tse, V. Haxha, M. A. Migliorato, and S. Tomić, "Second-order piezoelectricity in wurtzite iii-n semiconductors," *Phys. Rev. B*, vol. 84, p. 085211, Aug 2011.

Strain-induced modulation of mechanical properties and electronic structure of edge-modification graphene nanoribbons

Cheng Zhang[1,2], Mingsen Deng[1,2,3] and Shaohong Cai[1,2,3]

[1]Guizhou Provincial Key Laboratory of Computational Nano-material Science, Institute of
Applied Physics, Guizhou Normal College, Guiyang, 550018, China
[2]Department of Physics, Guizhou University, Guiyang, 550010, China and
[3]Guizhou Key Laboratory of Economic Systems Simulation, Guizhou University of Finance and Economics, Guiyang, 550004, China
Email: deng@iap.gzhnc.edu.cn

Abstract—**The mechanical properties and electronic structure of graphene nanoribbons(GNRs) which modified atoms or molecular groups on the zigzag edges can be tuned under the uniaxial tensile strain for application on the electronic devices. We study the elastic and plastic deformation of GNRs. The modified zigzag edges play a key role in tense force, elastic constant, critical point (the critical point of elastic and plastic deformation) and the energy gaps. In particular, it is shown that the energy gap of edge-modification GNRs can be strongly modified under uniaxial tensile strain. This way, offer new opportunities to electronic transport and force-electronic devices for next-generation electronics.**

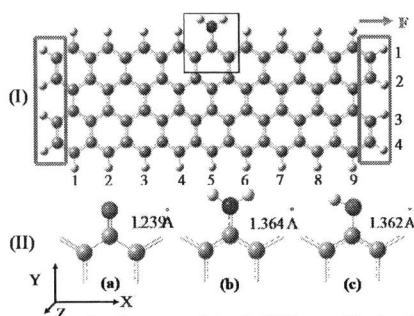

Fig. 1. (I) A schematic diagram of 4×9 GNR modified with NH_2 at the 5 site at zigzag edge; (II) Local structure of edge-modification GNRs. The zigzag edge modified with O, NH_2 and OH in (a), (b) and (c),respectively.

I. INTRODUCTION

Due to the graphene nanoribbons (GNRs) have capabilities to continuously control the band gap, it can be expected the candidates for future electronic and optoelectronic devices[1]. However, GNRs can not be produced in the same condition, a variety of lithographic fabrication (e.g oxygen plasma) are modified with oxygen at edge. In addition, diverse atoms or molecular groups may be introduced in the manufacturing process (e.g NH_2 and OH). Some theoretical and experimental studies have proved that local and uniform strain can tune the electronic properties and transport characteristics of GNRs[2], [3]. Therefore, it is vital to study the mechanical properties and electronic structure of edge-modification GNRs by the uniaxial tensile strain.

In order to describe the pristine GNRs, we discusse the mechanical properties and electronic structure of GNRs which are modified with H, O, OH and NH_2 at zigzag edges (because the zigzag interface is relatively more prone to atoms or molecular groups adsorbed). We investigate geometric and electronic properties of GNRs by the uniaxial tensile strain. We find that tension force F_T, elastic constant C and critical point (the critical point of elastic and plastic deformation) of GNRs are depend on the modified edge. Furthermore, the energy gaps of GNRs change dramatically under uniaxial strains.

II. METHODS

The electronic structure calculations are performed using the density-functional theory (DFT) by PAW potentials[4]. The exchange and correlation effects of the electrons has been approximated by Generalized Gradient Approximation (GGA), as implemented in Vienna Ab initio Simulation Package (VASP)[5]. The geometry are optimized with the conjugate-gradient algorithm method. All structures are treated within supercell geometry using the periodic boundary conditions. The size of supercells are carefully tested to maintain a sufficiently large separation between adjacent nanoribbons (>12Å form nanoribbon to nanoribbon). The basis set with kinetic-energy cutoff of 400 eV has been used, and Brillouin zone have been done using a $0 \times 0 \times 0 \Gamma$-centered k-point grid for the big molecular GNRs. A representative nanoribbon (the length of 2.2 nm and the width of 0.7 nm) is shown in FIG. 1, which the edges are modified with NH_2. Following the previous convention, the GNRs are defined according the number of dangling bonds on the armchair edges M and the number of dangling bonds on the zigzag edges N. The deformation are added along the zigzag edge (X direction) in the 4×9 GNRs.

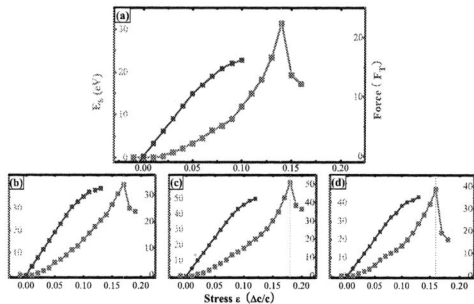

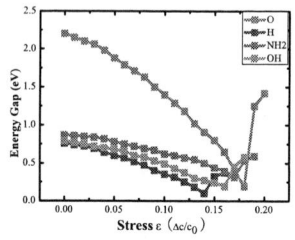

Fig. 2. Stain response of GNRs modified with H, NH$_2$, O and OH in (a), (b), (c) and (d), respectively; The F_T are showing indicating the elastic regime, while the E_S are showing in the whole process.

Fig. 3. The Strain-induced modulation of energy gaps of edge-modification GNRs.

III. RESULTS

All the geometric of GNRs are fully relaxed by minimizing the total energy. As a benchmark, the structures of GNRs are calculated under zero strain. The final structures of all atoms in GNRs are in a plane, and the lengths of optimized C-C bond are between 1.370Å and 1.450Å. The funny thing is that the X-axial direction bonds shorter than those other direction and the central bonds longer than the edge ones. The C-H bonds on the edges are approximately 1.090Å. The local structures of GNRs modified with O NH$_2$ and OH are displayed in FIG. 1. The modified angle of H-O-C is 110° and other modify angle at the zigzag are approximately 120° in GNRs. Computing by the same method, the change density analysis reveals that the total charge density inside the radius of O and N in GNRs are bigger than those in O atom and N atom. The results show that the honey structure of GNR is more likely losing electrons compared with O, NH$_2$ and OH.

We next investigate the mechanical properties, critical point and the energy gaps of GNRs. Elastic and plastic deformations are given along the X-axis as shown in FIG. 1. The tense forces and elastic constants are revealed form the strain energy $E_S=E_T(\varepsilon)-E_T(\varepsilon=0)$: the total energy with the strain ε minus the total energy at zero strain ($\varepsilon=\Delta c/c_0$, c_0 being the length of GNRs at the zero strain, Δc being the length of the strain minus the length of zero strain). In this work, in order to stretch of the GNRs, we increase the optimized length in increments of $\Delta\varepsilon=0.01$ and by uniformly expanding the structure obtained from previous optimization. Then keep the left and right side fixed. Under the elastic regime, strain energy, tension force, force constant and elastic constants can be expressed as: $F_T=-\partial E_S(\epsilon)/\partial C$, $C=(1/V_0)(\partial^2 E_S/\partial\epsilon^2)$. Here V_0 is the volume of the equilibrium supercell.

The strain energy and tense force of GNRs are shown in FIG. 2. Form the result we can obtain that the covalent bond parallel to the X-axis are stretched longer than those in other direction. As a result, the elastic constants of GNRs modified with H, NH$_2$, O and OH are 278, 256, 354, and 293 N/m, respectively (the report experimental value for graphene is $C=350\pm50$[6]). Because of the similar honeycomb structure, their elastic constants are liking those in graphene. However, the edge effects of GNRs, their stiffness values are smaller than those of the graphene. Apparently, for the existence of double bond between C atom and O atom, the elastic constant of GNR modified with O is bigger than those in other GNRs. This is the yielding points, where the strain energy drops

suddenly. The C atom which is at the 5 site (as shown in FIG. 1) is first lose their covalent bond at the yielding points (ϵ_y) by the uniaxial tensile strain. The yielding points of elastic and plastic deformation are expecting to depend on symmetries, such as vacancy and modified. In previous reports, the ϵ_y is also depend on the temperature[7]. Just as the increase of temperature the ϵ_y is decrease. In this work, the ϵ_y of GNRs modified with H, NH$_3$, O and OH are 0.14, 0.17, 0.18 and 0.16, respectively. The effect of vacancy is investigated by 4×9 armchair-edge GNR modified with H. Results indicate that ϵ_y in armchair-edge is bigger than those in zigzag-edge.

Depending on their crystalline defects the energy gaps of GNRs significant variations by the uniaxial tensile strain. As strain-dependent energy gap describing in Fig. 3, we find that the energy gap decreases with the increase the strain in the elastic deformation. The energy of highest occupied molecular orbital (HOMO) and the lowest unoccupied molecular orbital (LUMO) are also decrease as the increase the uniaxial strains. However, the energy of LUMO decrease speedy than those of HOMO. Energy gap increase significantly at the yielding points. Because the sp^2 C-C bond of the 5 site is broken and the modified atom (or molecule) is incorporated into the C atom. This way, be able to continuously control the band gap of modified GNRs for building the next-generation electronics.

IV. SUMMARY AND CONCLUSIONS

In conclusion, we have demonstrated that the Strain-induced modulation of mechanical properties and electronic structure of edge-modification graphene nanoribbons by first-principles calculation. We found that the final structures of all atoms in GNRs are in a plane. Furthermore, strain energy, tense force, elastic constant, critical point and energy gaps of GNRs are highly dependent on the modified atoms or molecular groups. In addition, the band gaps of GNRs can be changed significantly by the uniaxial tensile strain. The above result provide us a practical method to tune the energy gap of GNRs.

REFERENCES

[1] Jingshan Qi, Xiaofeng Qian, Liang Qi, Ji Feng, Daning Shi, and Ju Li, Nano Lett **10**, 1021(2012).
[2] Z. F. Wang, Yu Zhang and Feng Liu, Phys. Rev. B **83**, 041403(2011).
[3] Yi Wang, Rong Yang, Zhiwen Shi, Lianchang Zhang, Dongxia Shi, Enge Wang, and Guangyu Zhang, ACS Nano, **5**, 3654(2011).
[4] P. E. Blochl, Phys. Rev. B **50**, 17953 (1994).
[5] J. P. Perdew, J. A. Chevary, S. H. Vosko, K. A. Jackson, M. R. Pederson, D. J. Singh, C. Fiolhais, Phys. Rev. B **46**, 6671 (1992).
[6] C. Lee, X. Wei, J. W. Kysar, and J. Hone, Science **321**, 385 (2008).
[7] M. Topsakal and S. Ciraci, Phys. Rev. B **81** 024107(2010).

Quantum Mechanical Simulations of Nano-Structures and Nano-Devices

Xiang-Wei Jiang, Hui-Xiong Deng, Shu-Shen Li

Institute of Semiconductors, Chinese Academy of Sciences

Jun-Wei Luo

National Renewable Energy Laboratory

Lin-Wang Wang

Lawrence Berkeley National Laboratory

Abstract-We have investigated the quantum mechanical effects in quantum dots and nano size silicon MOSFETs using empirical psedupotential Hamiltonian model and linear combination of bulk band (LCBB) method. Unlike the traditional effective mass approximation and kp method, our approach uses a full zone expansion to represent the electronic state. This method provides a very fast yet accurate way to simulate million atom nano structures and nano devices even on a single processor personal computer.

I. INTRODUCTION

Traditional standard approaches to simulate the nano-structures are the effective-mass envelope-function approximation (EMA) and its multiband kp generalization, where the wave function of the nano-structures is expanded in terms of zone-center (k=0) Bloch bands of the underlying periodic solid. In the past 10 years, these two methods were also used to simulate the nanometer MOSFETs incorporated with transport models such as Non-Equilibrium Green's Function (NEGF). However, as pointed by [1], these methods fail to describe the whole band structure of the constituent materials, inter-valley coupling, as well as the atomistic features and strain effects in the nano-structures and devices, which could be important to the real physical properties. One failure example has been shown by Esseni and Palestri et al. [2], that the kp method can significantly misrepresent the electron density of states (DOS) for a silicon inversion layer, and the indirect band gap nature of bulk silicon presents real challenges for this model.

On the contrary, the LCBB method [1][2][3] is an successful alternative for the theoretical simulation of nano-structures and nano-devices. By expanding the wave function in terms of full-zone Bloch states of the constituent bulk solids, LCBB represents the multi-band and multi-valley effects as well as the atomistic nature of the nano-structures and nano-devices in a very efficient way. And the calculation error is well controlled in bellow 10 meV.

II. LCBB METHOD

LCBB method is based on empirical pseudo- potential Hamiltonian model (EPM), where 4-5 EPM parameters are fitted to reproduce the bulk experimental measurements of the band structure. The fitted EPM is then used to calculate the heterostructures of constituent materials. In LCBB framework [1], the single-particle Hamiltonian is constructed as a sum of the kinetic term, a superposition of peudopotentials for different atoms, a confinement energy term which represents barriers for buffer layers of the system, and the external potential term that can be calculated from Poisson Equation. Then the single particle electronic wave function is expanded as a linear combination of full-zone bulk Bloch states. All relevant k points are included in the expansion. Detailed description of the LCBB method can be found in Ref.[1].

III. SIMULATION RESULTS

A. Quantum Dots

Using LCBB method, we have investigated the ground exciton energy pressure coefficients (PC) of self-assembled InAs/GaAs quantum dots by calculating 21 systems with different quantum dot shape, size, and alloying profile [4]. Consistent with the experiment, we find the PC is in the range of 60-110 meV/GPa, much smaller than the bulk InAs and GaAs PCs. The calculated PCs as a function of QD zero-pressure exciton energy is plotted in Fig. 1 with the experimental measurements. We believe that it is necessary to consider the penetration of the electron state into the GaAs barrier in order to understand the PC reductions and their large variations.

We have also studied the electronic structure of colloidal quantum dot [5], by comparing the conduction band state results from different calculation methods. We found that a colloidal quantum dot can be considered as a heterostructure system composed of the inside semiconductor material and the outside vacuum.

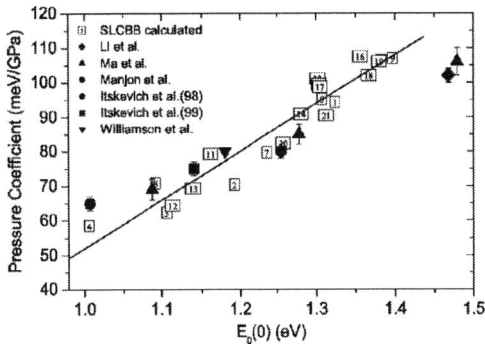

Fig. 1. The PL pressure coefficient vs PL energy

B. Nanometer Silicon MOSFETs

The ultimate downscaling of the MOSFET size today requires an accurate yet fast way to simulate the device performance before fabrication. For this motivation, we investigated the possibility of extending LCBB for device simulation [6][7][8][9][10], and we found that the same task required thousands of processors using the most widely used TB-NEGF approach can be done on a personal computer with single processor using our LCBB method without losing much accuracy. Using this quantum mechanical simulation approach based on LCBB, we have investigated the quantum mechanical effects in 25 nm MOSFET [6], including the random dopant fluctuations [8] and multi-valley coupling [7]. Based on LCBB, we have also proposed an approximated top of the barrier splitting (TBS) method to mimic the scattering states from the stationary states calculations [10].

In our simulations, we found quantum mechanical effects change the device physics in many aspects. In the simulation of 25 nm bulk MOSFET [6], we found the saturated quantum mechanical capacitance at high gate voltage is about 20% less than the classical capacitance. This is due to the fact that QM charge density is farther away from SiO2/Si interface, which is a consequence of quantum confinement effects. And the subthreshold slop is around 124 mV/dec., much larger than the theoretical limit of 60 mV/dec. This is due to the short channel effects including source to drain tunneling. In the simulation of random dopant induced fluctuation [8], we found that quantum mechanical effects increase the threshold fluctuation while decrease the threshold lowering, which is different from the results of density gradient calculations. We believe that the threshold lowering is because of the local charge density caused by the discrete dopant configuration which can be smoothed out by the quantum confinement effects, so that the quantum simulation gives a decreasing effects on threshold voltage lowering. In our investigation of multi-valley coupling [7], we found that there is very little coupling among different energy valleys and the contributions from non-X valleys are negligible. So the actual calculation can be reduced to just X valley calculation which can be further decoupled to three X valleys because the coupling among these three valleys are also negligible.

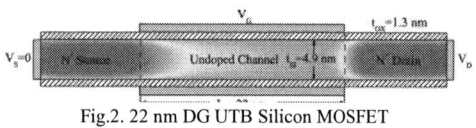

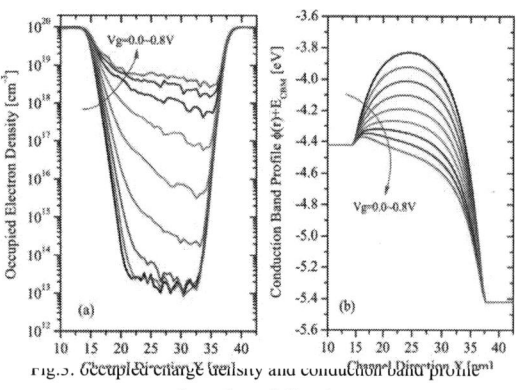

Fig.3. Occupied charge density and conduction band profile along channel direction

In our method development for efficient simulation of nano-devices, we proposed a top of the barrier splitting (TBS) method to extract scattering states information from stationary

states calculations using periodic boundary condition [10]. This method first calculates the electronic eigen states in a buffered periodic system. Then each calculated eigen state is decomposed into two parts, one is supposed from source side and another part is supposed from drain side. This decomposition is based on the physical insight of ballistic and tunneling transport and assumes the local density of states on top of the barrier is equally from source and drain. Then the source part is occupied by source side Fermi level while the drain part is occupied by drain side Fermi level to get the total occupied charge density and form the self-consistent calculation with Poisson Equation. We applied this approach to 22 nm double-gate (DG) ultra-thin body (UTB) silicon MOSFET (Fig.2). Fig.3 plots the occupied charge density and conduction band profile along channel direction. To verify our 3D LCBB-TBS model, we compared the simulated transfer characteristics with the tight-binding (TB) NEGF approach and found the results are very close, although the later does require hundreds to thousand processors to do the simulation. The simulated transfer characteristics is shown in Fig.4. The threshold voltages calculated from LCBB-TBS and TB-NEGF are 460 mV and 450 mV. The ON currents are 3265 uA/um and 3740 uA/um respectively. And the subthreshold slops calculated from both models are the same as 63 mV/dec.

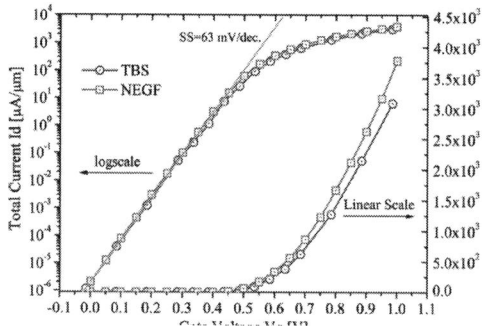

Fig.3. Simulated transfer characteristics comparing with TB NEGF method.

IV. ACKNOWLEDGMENT

This work was supported by the National Basic Research Program of China (973 Program) Grant No. G2009CB929300 and the National Natural Science Foundation of China under Grand Nos. 61106091, 60821061 and 60776061.

REFERENCES:

[1] L. Wang and A. Zunger, Phys. Rev. B 59, 15806 (1999)

[2] D. Esseni and P. Palestri, Phys. Rev. B 72, 165342 (2005)

[3] F. Chirico, A. Di Carlo, and P. Lugli, Phys. Rev. B 64, 045314 (2001)

[4] J. Luo, S. Li, J. Xia, and L. Wang, Phys. Rev. B 71, 245315 (2005)

[5] J. Luo, S. Li, J. Xia, and L. Wang, Appl. Phys. Lett. 88, 143108 (2006)

[6] J. Luo, S. Li, J. Xia, and L. Wang, Appl. Phys. Lett. 90, 143108 (2007)

[7] H. Deng, X. Jiang, J. Luo, S. Li, J. Xia, and L. Wang, J. Appl. Phys. 103, 124507 (2008)

[8] X. Jiang, H. Deng, J. Luo, S. Li, and L. Wang, IEEE Trans. Elect. Dev. 55, 1720 (2008)

[9] X. Jiang, H. Deng. S. Li, J. Luo, and L. Wang, J. Appl. Phys. 106, 084510 (2009)

[10] X. Jiang, S. Li, J. Xia, and L. Wang, J. Appl. Phys. 109, 054503 (2011)

Discretization scheme for drift-diffusion equations with strong diffusion enhancement

Thomas Koprucki and Klaus Gärtner

Weierstrass Institute for Applied Analysis and Stochastics, Mohrenstr. 39, 10117 Berlin, Germany

Email: {thomas.koprucki,klaus.gaertner}@wias-berlin.de

Abstract—**Inspired by organic semiconductor models based on hopping transport introducing Gauss-Fermi integrals a nonlinear generalization of the classical Scharfetter-Gummel scheme is derived for the distribution function $\mathcal{F}(\eta) = 1/(\exp(-\eta)+\gamma)$. This function provides an approximation of the Fermi-Dirac integrals of different order and restricted argument ranges. The scheme requires the solution of a nonlinear equation per edge and continuity equation to calculate the edge currents. In the current formula the density-dependent diffusion enhancement factor, resulting from the generalized Einstein relation, shows up as a weighting factor.**

I. Introduction

Any monotone non-Boltzmann statistics based state-equation for the carrier density of semiconductor results in a generalized Einstein relation describing the ratio of diffusion and drift current in thermodynamic equilibrium. This can be interpreted as a *diffusion enhancement* [1]. Following Scharfetter-Gummel one is interested in approximating the net electron current in order to discretize the drift-diffusion equation describing the carrier transport [2]. In the classical Scharfetter-Gummel scheme the exponential dependence of the carrier density on the chemical potential results in a current expression consisting of a weighted difference of the carrier densities. Here the usual state equation $n = N_c \mathcal{F}(\eta)$ for the carrier density in dependence on the chemical potential η, N_c denotes the density of states, is considered for the special distribution function

$$\mathcal{F}(\eta) = \frac{1}{e^{-\eta}+\gamma}, \quad 0 \le n \le \frac{N_c}{\gamma}. \quad (1)$$

This approximation can be used for the Fermi-Dirac integral of order $1/2$ with $\gamma = 0.27$ and $\eta < 1.3$ [3]. For $\gamma = 1$ it coincides with Fermi-Dirac integral of order -1 describing zero-dimensional Fermi gases, namely hopping transport between individual sites. Furthermore, it is the limit for vanishing disorder σ of the Gauss-Fermi integral [4], which is used to describe organic semiconductors [5]. The general situation is depicted in Figs. 1 and 2.

Here we present a nonlinear generalization of the Scharfetter-Gummel scheme for the approximation of the net electron current governed by the carrier density expression (1).

II. Carrier continuity equations and diffusion enhancement

The continuity equation for the electrons reads

$$\frac{\partial n}{\partial t} - \frac{1}{q}\nabla \cdot J_n = -R,$$

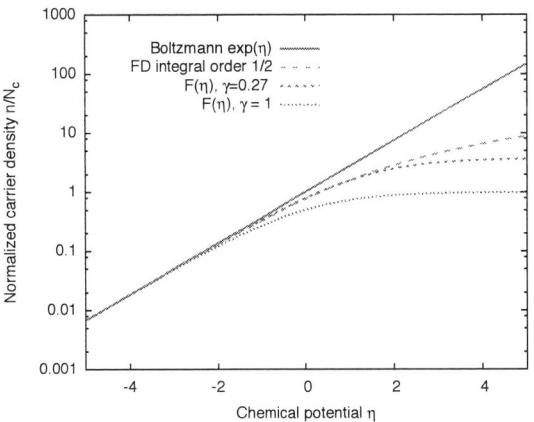

Fig. 1. Plot of distribution function $\mathcal{F}(\eta) = 1/(\exp(-\eta)+\gamma)$ in dependence of the dimensionless chemical potential η for different values of the parameter γ. In the asymptotic limit $\eta \ll -2$ a Boltzmann behavior is observed. For $\gamma = 0.27$ a good approximation of the Fermi-Dirac integral of order $1/2$ for $\eta < 1.3$ is provided, whereas the case $\gamma = 1$ corresponds to the limit of vanishing disorder of the Gauss-Fermi integral [4].

with the current expressions

$$J_n = -q\mu_n N_c \mathcal{F}(\eta)\nabla\varphi_n = -qn\mu_n\nabla\psi + qD_n\nabla n, \quad (2)$$

$$\eta = \frac{q(\psi - \varphi_n) + E_{ref} - E_c}{k_B T}, \quad (3)$$

where q denotes the elementary charge, μ_n the mobility, φ_n the quasi-Fermi potential, ψ the electrostatic potential, k_B Boltzmann's constant, T the temperature, E_{ref} a reference energy for the quasi-Fermi potential and E_c the band-edge energy. The mobility and the diffusion coefficient D_n fulfill the generalized Einstein relation

$$\frac{D_n}{\mu_n} = \frac{k_B T}{q}\frac{n}{N_c}(\mathcal{F}^{-1})'\left(\frac{n}{N_c}\right) =: \frac{k_B T}{q}g_3\left(\frac{n}{N_c}\right). \quad (4)$$

The factor g_3 in the generalized Einstein relation is describing a diffusion enhancement [1]. For our special choice of the distribution function (1) the relation becomes

$$g_3(x) = \frac{1}{1 - \gamma x}, \quad (5)$$

while the current reads

$$J_n = -qn\mu_n\nabla\psi + \mu_n k_B T\frac{1}{1 - \gamma\frac{n}{N_c}}\nabla n. \quad (6)$$

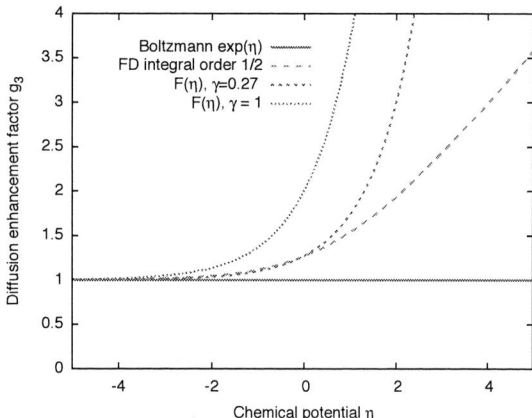

Fig. 2. Plot of diffusion enhancement factor g_3 in dependence on the dimensionless chemical potential related to the distribution function $\mathcal{F}(\eta) = (\exp(-\eta) + \gamma)^{-1}$ for different values of the parameter γ. In the asymptotic limit $\eta \ll -2$ no diffusion enhancement is observed (Boltzmann limit). Additionally, the diffusion enhancement factor g_3 related to the Fermi-Dirac integral of order 1/2 is depicted.

III. Current approximation

In the following we consider the one-dimensional case on the spatial interval $[x_a, x_b]$ and the following scaling of the equation: the potentials are given in units of the thermal voltage $U_T = \frac{k_B T}{q}$ and the current is given in units of $j_0 = q\mu_n N_c \frac{U_T}{x_b - x_a}$. The Scharfetter-Gummel discretization is derived by solving the following equation

$$\left(q\mu_n N_c \mathcal{F}\big(\eta(\varphi_n, \psi)\big)\varphi_n' \right)' = 0, \tag{7}$$

on the interval $[x_a, x_b]$ with the boundary values $\varphi_n(x_a) = \varphi_a$ and $\varphi_n(x_b) = \varphi_b$. The electrostatic potential ψ is assumed to be linearly dependent on x, the mobility μ_n is taken to be an average value on the interval $[x_a, x_b]$. First integration yields $-q\mu_n N_c \mathcal{F}\big(\eta(\varphi_n, \psi)\big)\varphi_n' = j = const$. Second integration results in an integral equation for the unknown current j

$$\int_{\eta_a}^{\eta_b} \frac{1}{\frac{j}{\mathcal{F}(\eta)} + \delta\psi} d\eta = 1. \tag{8}$$

The boundary values are

$$\eta_a = \mathcal{F}^{-1}(n_a/N_c), \eta_b = \mathcal{F}^{-1}(n_b/N_c), \tag{9}$$

and potential difference $\delta\psi$ is given by $\delta\psi = \psi_b - \psi_a$. For details of this approach see [6]. For the distribution under consideration this integral equations leads to the following nonlinear, local equation for the edge current j:

$$j = f(j) = B(\delta\psi + \gamma j)e^{\eta_b} - B(-(\delta\psi + \gamma j))e^{\eta_a}. \tag{10}$$

where $B(x) = \frac{x}{e^x - 1}$ is the Bernoulli function. This equation has a unique solution $j = j(\psi_a, \psi_b, \eta_a, \eta_b)$ due to the monotonicity of the Bernoulli function. Using the relation $\mathcal{F}^{-1}(x) = -\ln\left(\frac{1}{x} - \gamma\right)$ the current expression in terms of

densities is given by

$$j = g_3\left(\frac{n_b}{N_c}\right)B(\delta\psi + \gamma j)n_b - g_3\left(\frac{n_a}{N_c}\right)B(-\delta\psi - \gamma j)n_a. \tag{11}$$

Here, in this particular case the density-dependent diffusion enhancement factor g_3 shows up explicitly. With $\gamma = 0$ the well-known Scharfetter-Gummel expression is reproduced, while Eq. (11) is nonlinear with respect to the density and the potential difference $\delta\psi = \psi_b - \psi_a$ is modified by the local edge current and the parameter γ describing the deviation of the state-equation for the density with respect to the Boltzmann case.

The essential change compared with the classical scheme is now the solution of the nonlinear equation (11) on every edge of the spatial discretization during the assembly of each continuity equation.

IV. Conclusion

For a restricted range of arguments of the Fermi-Dirac integral of order $1/2$ a generalized, simple to implement, nonlinear Scharfetter-Gummel scheme has been derived. The effort is small compared with the introduction of an additional outer iteration. The local nonlinear equations for calculation of the edge currents can be solved due to the monotonicity properties of the Bernoulli function. The necessary conditions [7] for proving the existence of bounded steady state solutions, uniqueness of the equilibrium solution, and dissipativity are preserved, too.

Acknowledgment

The work of T. Koprucki has been supported by the Deutsche Forschungsgemeinschaft (DFG) within the collaborative research center 787 "Semiconductor Nanophotonics".

References

[1] S. L. M. van Mensfoort and R. Coehoorn, "Effect of gaussian disorder on the voltage dependence of the current density in sandwich-type devices based on organic semiconductors," *Phys. Rev. B*, vol. 78, no. 8, p. 085207, Aug 2008.

[2] D. L. Scharfetter and H. K. Gummel, "Large signal analysis of a silicon Read diode," *IEEE Trans. Electron. Dev.*, vol. 16, pp. 64–77, 1969.

[3] J. Blakemore, "The Parameters of Partially Degenerate Semiconductors," *Proc. Phys. Soc. London A*, vol. 65, pp. 460–461, 1952.

[4] G. Paasch and S. Scheinert, "Charge carrier density of organics with gaussian density of states: Analytical approximation for the gauss-fermi integral," *J. Appl. Phys.*, vol. 107, no. 10, p. 104501, 2010.

[5] R. Coehoorn, W. F. Pasveer, P. A. Bobbert, and M. A. J. Michels, "Charge-carrier concentration dependence of the hopping mobility in organic materials with gaussian disorder," *Phys. Rev. B*, vol. 72, no. 15, p. 155206, Oct 2005.

[6] R. Eymard, J. Fuhrmann, and K. Gärtner, "A finite volume scheme for nonlinear parabolic equations derived from one-dimensional local Dirichlet problems," *Numerische Mathematik*, vol. 102, pp. 463–495, 2006.

[7] K. Gärtner, "Existence of bounded discrete steady-state solutions of the van Roosbroeck system on boundary conforming delaunay grids," *SIAM Journal on Scientific Computing*, vol. 31, no. 2, pp. 1347–1362, 2009.

Beam Propagation Analysis Using Higher-Order Full-Vectorial Finite-Difference Method

Cheng-Han Du
Graduate Institute of Photonics and Optoelectronics
National Taiwan University
Taipei 10617, Taiwan

Yih-Peng Chiou
Graduate Institute of Photonics and Optoelectronics &
Department of Electrical Engineering
National Taiwan University
Taipei 10617, Taiwan
Email: ypchiou@ntu.edu.tw

Abstract—We develop a full-vectorial higher-order finite-difference formulation to model optical waveguides. Our proposed scheme yields improved convergence and field calculation results. Applications to beam propagation analysis of a photonic crystal coupler is demonstrated.

I. Introduction

Accuracy and efficiency are major concerns in numerical optical simulation of photonic devices. Finite-difference method (FDM) has been very popular among various computer-aided design (CAD) tools due to its ease of implementation. Numerical error convergence of FDM can be improved by various method, where higher-order scheme is a straight-forward solution. Moreover, proper modeling of field behavior near structure interface requires treatment with full-vectorial feature for good accuracy, while higher-order FDM requires interface continuity conditions of field and its derivatives up to corresponding differential order. This technique has been implemented in mode analysis of optical waveguides [1], where report shows that 6-, 15-, and 28-point FDM yields second-, fourth-, and sixth-order error convergence. Application in photonic crystal fiber and pipe waveguide has also been demonstrated. In this paper, a dual-core photonic crystal fiber coupler is modeled as an further demonstration of improved convergence performance and capability for complex structure simulation. We also further apply the full-vectorial higher-order FDM to beam propagation method (BPM) and visualize the coupling phenomenon of coupler. Results show that superior accuracy can be achieved with higher-order scheme even when using coarser grids. Its application in BPM also shows potential usefulness in future photonic design.

II. Formulation and Numerical Results

For z-invariant waveguide mode eigenvalue problem based on electric field, we solve the vector Helmholtz equation

$$\nabla_\perp^2 \bar{\psi} + k_0^2 n^2 \bar{\psi} = \beta^2 \bar{\psi} \qquad (1)$$

where $\bar{E} = \bar{\psi} \exp(-j\beta z)$ and β denotes the propagation constant of waveguide mode. Numerical calculation and assessment of this eigenvalue problem problem using full-vectorial higher-order finite-difference method has been demonstrated in [1]. To apply this technique to beam propagation method, we

first specify a reference refractive index $\bar{n}$. By using slowly-varying wave packet approximation, $\bar{E} = \bar{\varphi} \exp(-jk_0\bar{n}z)$, we have

$$(\nabla_\perp^2 + k_0^2 n^2)\bar{\varphi} = -\frac{\partial^2 \bar{\varphi}}{\partial z^2} + 2jk_0\bar{n}\frac{\partial \bar{\varphi}}{\partial z} + k_0^2 \bar{n}^2 \bar{\varphi}. \qquad (2)$$

We also use wide-angle BPM technique with order $(1,1)$ Padé approximant based on forward propagation operator $\Theta = (\nabla_\perp^2 - k_0^2 \bar{n}^2)/k_0^2 \bar{n}^2$ [2], which yields

$$(I + a_1^*\Theta)\varphi(z_0 + \Delta z) = (I + a_1\Theta)\varphi(z_0) \qquad (3)$$

where I is identity matrix and complex conjugate pair a_1 and a_1^* are functions of $k_0\bar{n}$.

We simulate a dual-core air-hole photonic crystal fiber coupler as assessment. Wavelength λ_0 and hole pitch Λ are 1.55 and 1.8 μm, respectively. Ratio of air hole diameter d and hole pitch is $d/\Lambda = 0.52$. Perfectly-matched layers (PML) of 20 layers are used at computation boundary. Arrangement of air holes and core and mode power distribution of even and odd modes are shown in Fig. 1. The calculated effective indices and corresponding coupling lengths using different approximation scheme and grid size $\Delta x = \Delta y = \Delta$ are listed in Table II. Calculation based on 28-point scheme yields extremely high convergence performance with variation of $\Re[n_{\text{eff}}]$ less than 10^{-8}.

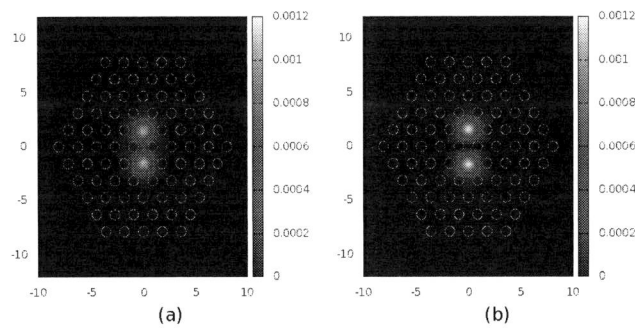

Fig. 1. Simulated E_x power distribution of (a) even and (b) odd modes.

We also apply the full-vectorial higher-order finite-difference scheme to BPM and model the same problem. An x-polarized Gaussian beam with waist of 1.0 μm is launched

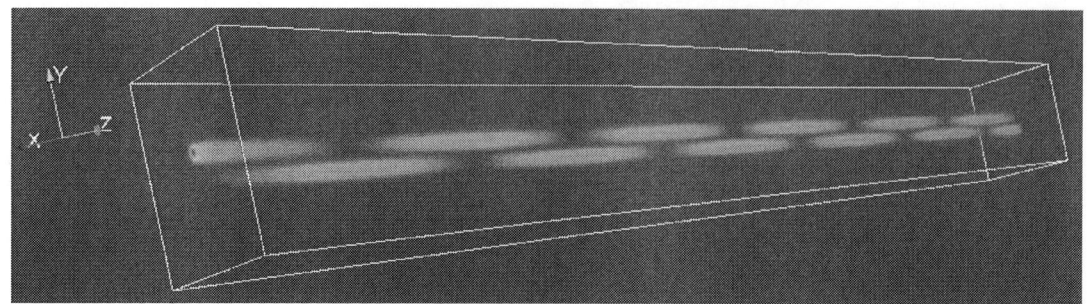

Fig. 2. 3D visualization of BPM simulation of photonic crystal fiber coupler. Computation domain is 20 μm ×24 μm ×2500 μm. Scaling along z-direction is 1 : 10.

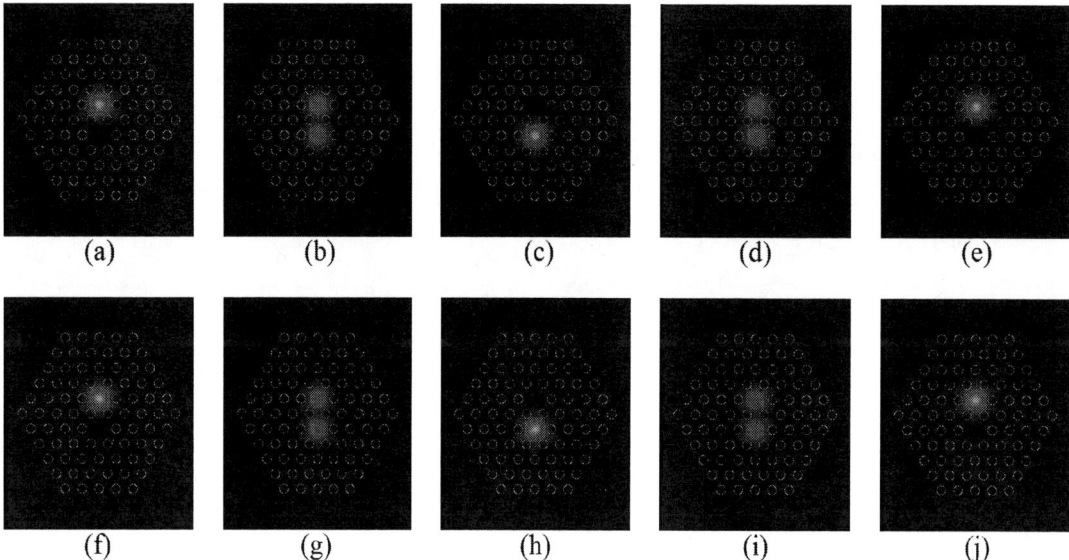

Fig. 3. E_x power distribution at $z = 457, 571, 685, 800,$ and 914 μm using (a-e) 15-point scheme with $\Delta = 0.04$ μm and (f-j) 28-point scheme with $\Delta = 0.08$ μm.

TABLE I
CALCULATED $\Re[n_{EFF}]$ OF EVEN AND ODD MODES FOR SCHEMES OF DIFFERENT ORDERS

Δ (μm)	6-point, even	15-point, even	28-point, even
0.02	1.411776100	1.411768212	1.411768194
0.04	1.411800721	1.411768276	1.411768202

Δ (μm)	6-point, odd	15-point, odd	28-point, odd
0.02	1.408383421	1.408376332	1.408376316
0.04	1.408406993	1.408376384	1.408376322

at upper core and propagates along z-direction for 2500 μm. The propagation step size $\Delta z = 1.0$ μm. Reference refractive index $\bar{n}$ is set to 1.41. From the previous mode analysis results, we have coupling length $L = \pi/(\beta_{even} - \beta_{odd}) = 228.49$ μm. The 3D visualization of $|E_x|$ distribution is shown in Fig. 3, where field coupling effect is correctly modeled. To demonstrate the calculation capability, we use 15-point scheme with $\Delta x = \Delta y = 0.04$ μm and 28-point scheme with $\Delta x = \Delta y = 0.08$ μm and plot the power distribution. Calculation results are consistent with each other and agree with [3].

III. CONCLUSION

We have successfully implemented full-vectorial higher-order finite-difference method for optical waveguide analysis. Results show good convergence performance compared with conventional method. The formulation can be applied to beam propagation method for advanced design. We will present more detailed results in the conference including the numerical property of full-vectorial beam propagation analysis.

ACKNOWLEDGMENT

This work was supported in part by the National Science Council of the Taiwan under grant NSC98-2221-E-002-170-MY3 and the Excellent Research Projects of National Taiwan University under Grant 10R80919-1. They are grateful to the National Center for High-performance Computing in Taiwan for computer time and facilities.

REFERENCES

[1] Y.-P. Chiou and C.-H. Du, "Arbitrary-Order Full-Vectorial Interface Conditions and Higher-Order Finite-Difference Analysis of Optical Waveguides," IEEE/OSA J. Lightwave Technol. , **29**, 3445–3452 (2011).

[2] Y.-P. Chiou and H. C. Chang, "An efficient wide angle beam propagation method," in *Proc. 11th Int. Conf. Integr. Opt. Opt. Fiber Commun. 23rd Eur. Conf. Opt. Commun. (IOOC-ECOC97)*, Edingburgh, U.K., **2**, pp. 220–223 (1997).

[3] K. Saitoh, Y. Sato, and M. Koshiba, "Coupling characteristics of dual-core photonic crystal fiber couplers," Opt. Express, 11, 3188–3195 (2003).

Acceleration of 3D numerical simulation of silicon solar cell using thread parallelism

B. Min, S. Suckow, U. Yusufoglu, T. M. Pletzer and H. Kurz

Institute of Semiconductor Electronics, RWTH-Aachen University
Sommerfeldstraße 24, D-52074 Aachen, Germany
E-mail: min@iht.rwth-aachen.de

Abstract— We have investigated the potential to accelerate three-dimensional numeric simulation of silicon solar cell using thread parallelism. The device simulated is a rear side passivated cell with rear point contacts (PERC). The optical and electrical behaviour of the device was simulated with Sentaurus Device (formerly dessis). We show that the simulation run-time on a four socket Opteron 6168 machine is reduced down to 6% compared to the run-time without thread parallelism. Furthermore, limits of time reduction by varying the number of threads up to 48 are studied. Thereby, the number of threads for the optimum use of the hardware resources is determined.

Keywords: simulation, PERC, silicon solar cell, high-performance computing

I. INTRODUCTION

Advanced cell concepts such as PERC, MWT (metal wrap through), EWT (emitter wrap through) or combinations of these concepts enable cell manufacturers to reduce the production cost per output power by increasing cell efficiency and reducing the cell thickness.

For this kind of complex cell concepts which include point like elements in the device structure, a true 3D device simulation is necessary, since the 2D simulation leads to errors in resistive and recombination losses [1]. Although simulations have become a very useful tool in PV research, 3D simulations are not widely carried out in comparison to 2D simulations because of time constrictions.

This work focuses on the effect of thread parallelism on the run-time of 3D device simulations of silicon solar cells. In this paper, a solar cell with rear point contact is chosen as an application example of thread parallelism. By varying the number of threads up to 48, we observed a significant reduction of simulation time.

II. APPROACH

A. Simulation model

As the simulation domain, a quarter of a symmetric element of a silicon solar cell with PERC is used. The device symmetry is illustrated in Fig. 1, where geometrical parameters are listed in Table 1.

Uniformly boron doped p-type silicon with doping of 1×10^{16} cm^{-3} ($\rho = 1.47$ Ωcm) is chosen as the substrate. The emitter of the simulated solar cells is selective, we used the

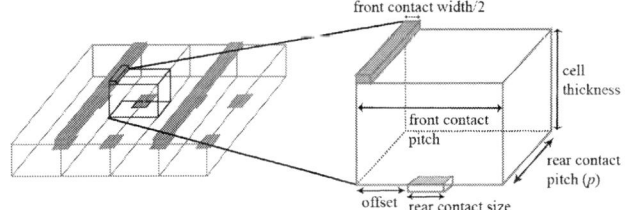

Fig. 1.   The geometrical structure of the simulation domain.

TABLE I.   THE GEOMETRICAL PARAMETERS OF THE SIMULATION DOMAIN.

Parameter	Value [µm]
cell thickness	180
front contact pitch	2
front contact width	80
front contact thickness	20
rear contact pitch (p)	350
rear contact size	85
offset	160

measured Electrochemical Capacitance-Voltage profiles of 45 /sq and 110 /sq phosphorus emitters from the work of Rudolph et al. [2]. The aluminum diffusion of the local back surface field is described by a gaussian profile with peak doping density of 2×10^{19} cm^{-3} and a junction depth of 5 µm. The minority carrier lifetime in the SRH recombination model is equal to 100 µs. The physical models adopted in these simulations include Schenk band gap narrowing [3] as well the Philips unified mobility model [4]. Fermi-Dirac statistics is adopted for the precise simulation at high dopant densities in emitter.

B. Thread parallelization

Sentaurus Device creates one process to run a simulation. The process itself can create multiple threads. Each thread occupies one processor core and is executed in parallel. The data communication between threads is established via shared memory. As solver for the computation we used the direct solver 'PARDISO', which is a high performance software package for solving large sparse symmetric or nonsymmetric systems of linear equations in parallel [5]. The software runs on a quad socket machine with four Opteron 6168 CPUs at 1.9 GHz. The main memory of 256 GB is distributed equally so that all 16 memory controllers are used.

III. RESULT

In the first part of this work the effect of the thread number variation on simulation run-time is investigated. In this case, there is only one process creating multiple threads. The number of threads varies between 1 and 48. As shown in Fig. 2 the simulation run-time drops almost linearly by increasing the number of threads up to 16, but levels off at higher numbers of threads. In comparison to the simulation run-time without thread parallelism (637 min.), 89% reduction of computation time (68 min.) was achieved at this point.

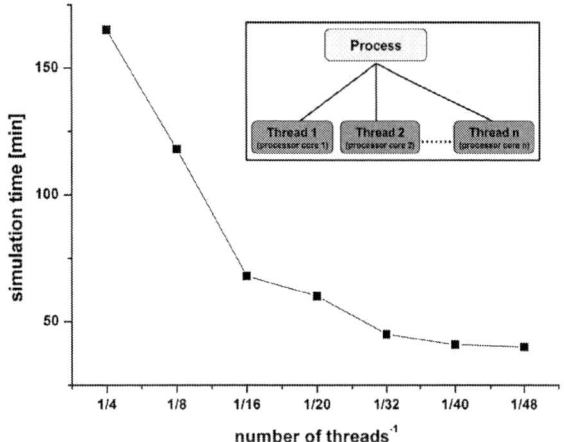

FIG. 2.   Effect of thread number variation on simulation run-time; for one process.

With the assignment of all available process cores using 48 threads, 94% reduction of computation time (39 min.) was achieved in comparison to the run-time without thread parallelism. But relative to the case of using 16 threads, a time reduction of only 57% more is achieved by use of 200% additional process cores. Therefore another way to achieve optimal throughput from the machine is followed in the second part of this work. Since the acceleration due to thread parallelism saturates, the simultaneous execution of multiple processes with multiple threads has been tried for maximum throughput.

The optimum number of processes n_{pr}^{opt} and its corresponding number of threads n_{th}^{opt} can be determined by considering the maximum number of processor cores n_{core}^{max} and the number of parameter variations which can be run in parallel n_{var} with $n_{th}^{opt} = n_{core}^{max} / n_{var}$ and $n_{pr}^{opt} = n_{var}$. As soon as this occurs, the performance will decrease dramatically.

In Fig. 3 the simulation time is plotted versus the number of simultaneously running processes. From the negligible increment of simulation time with increasing number of running processes follows that additional reduction of simulation time can be achieved by executing multiple processes. In case of $n_{var} = 3$, the run-time of 71 minutes by parallel execution of processes with 16 threads each is much less than 117 minutes by sequential execution with 48 threads. Each process can be assigned for the analysis of a specific parameter variation such as cell thickness or base resistivity.

One may argue that parallel execution produces different rounding errors that influence the number of iterations of numerical computations. The investigations performed in this study, however, show no effect on the value of any simulation results of solar cell parameters.

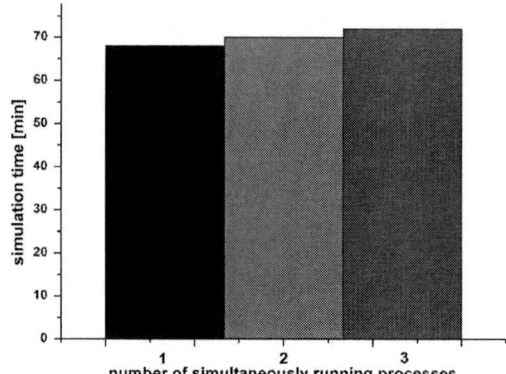

Fig. 3.   Effect of simultaneous execution of multiple processes. Each process has 16 threads.

IV. CONCLUSION

The effect of parallelization through multi-threading technique on computation speed in the device simulator Sentaurus Device by 3D simulation is significant. We demonstrated for the optimization of a PERC cell that the simulation time can be reduced to 6% using all available cores. The gain in computation speed provides a significant reduction of the "time to result". However, the scaling turned out to be less than ideal for more than 16 concurrent threads per simulation. The data accumulation in Fig. 2 provides important information on limit of maximum throughput, if several simulations need to be run and the amount of main memory is sufficient. This way three simulations with 16 threads each can be completed in 71 minutes whereas three subsequent simulations with 48 threads would require 117 minutes.

ACKNOWLEDGMENT

This work is part of the project "Kompetenzzentrum für innovative Photovoltaik-Modultechnik NRW" and has been supported by the European Union – European Regional Development Fund and by the Ministry of Economic Affairs and Energy of the State of North Rhine-Westphalia, Germany.

REFERENCES

[1] P. Altermatt, "Models for numerical device simulations of crystalline silicon solar cells – a review", Journal of Computational Electronics, 314-330, 2011.

[2] D. Rudolph et. al., "Etch back selective emitter process with single POCl₃ diffusion", 26th European Photovoltaic Solar Energy Conference and Exhibition, Hamburg, Germany, 2011.

[3] A. Schenk, J. Appl. Phys. 84, 3684, 1998.

[4] D. B. M. Klaassen, "A Unified Mobility Model for Device Simulation-I. Model Equations and Concentration Dependence", Solid.-State Electronics, vol. 35, no. 7, pp. 953-959, 1992.

[5] O. Schenk, "Scalable Parallel Sparse LU Factorization Methods on Shared Memory Multiprocessors", Series in Microelectronics, vol. 89, Konstanz, Germany: Hartung-Gorre, 2000.

Device Simulation of Intermediate Band Solar Cells

Katsuhisa Yoshida and Yoshitaka Okada
Research Center for Advanced Science and Technology,
The University of Tokyo
Tokyo, JAPAN
yoshida@mbe.rcast.u-tokyo.ac.jp

Abstract— **The intermediate band (IB) solar cells is one of the candidates to realize a higher conversion efficiency than the Shockley-Queisser limit of single-junction solar cells. By using device simulation of intermediate solar cells developed based on drift-diffusion method, we studied the fundamental properties of IB cell. Light concentration technique is very important to reduce the recombination via IB.**

Keywords-component; intermediate band solar cell; light concentration; device simulation

I. INTRODUCTION

Novel photovoltaic concepts to realize higher energy conversion efficiencies than the Shockley-Queisser's limit [1] are studied globally today. These types of solar cells are called as "third generation". The target of this generation is to utilize or reduce fundamental energy losses. The main fundamental loss mechanisms of solar cells are thermalization - energy relaxation process of carriers and transmission losses. If we use wide a bandgap semiconductor material for a single junction solar cell, the thermalization loss is reduced but the transmission loss is increased. On the other hand, use of a narrow bandgap material can reduce the transmission loss but increases thermalization loss. Thus, the optimal band gap material is decided. In the novel concepts, the key point is how to reduce or suppress these loss processes. For example, hot carrier and multi-exciton generation solar cells use a narrow band gap material with suppressing energy relaxation process of carriers in both conduction and valence bands by phonons and extracts carriers keeping a hot carrier temperature to the contacts and contribute to generate another electron-hole pairs within an energy relaxation process, respectively. In the case of multi-band or intermediate band (IB) solar cells [2], intermediate states are introduced in relatively wide band gap material to utilize lower energy photons than a host material band gap and increase total carrier generation rates in host material conduction and valence bands as shown in Fig. 1. The intermediate band solar cells are fabricated by using multi-stacked quantum dots [3, 4, 5], dilute alloy materials [6] and impurity bands [7]. The quantum dot solar cells show the concept of two-step photon absorption [5] and shows fundamental properties of the intermediate band solar cells. To improve conversion efficiency of the intermediate band solar cells, device simulations of the intermediate band solar cell can help to understand the inside physics and optimize the structure.

II. DEVICE SIMULATION OF INTERMEDIATE BAND SOLAR CELLS

We study properties of the intermediate band solar cells by using self-consistent device simulation of intermediate band solar cells in a steady state [8]. This simulation is based on the drift-diffusion method and includes carrier generation and recombination processes via the intermediate band. Electron densities in the intermediate band are decided by net transition rates via the intermediate band and affect to electrostatic potential. We assume the band as intermediate levels. Thus, the net transition rate is determined by a local balance equation,

$$G_{CI}(x) - U_{CI}(x) = G_{IV}(x) - U_{IV}(x) \qquad (1)$$

where G_{CI}, G_{IV}, U_{CI} and U_{IV} are optical generation and recombination rates between the conduction band - intermediate band, and the intermediate band and valence band, respectively. Thus, (1) describes transition rates should be balanced locally and decide the net carrier generation rate via the intermediate band. These terms in (1) have a dependence on an electron density in the conduction band, hole density in the valence band and electron density in the intermediate band. Therefore, we should solve the Poisson's equation, carrier continuity equations and the local balance equation self-consistently. We assume the carrier transport properties are well described by the drift-diffusion equation. In this calculation, we use GaAs material parameters except absorption coefficients and ideal Ohmic contacts for majority carriers and zero surface recombination velocities for minority carriers as boundary conditions. No surface recombination velocity expresses the presence of a well passivated surface or window layer.

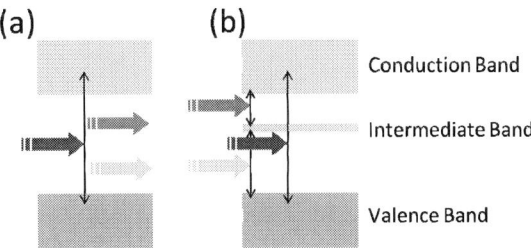

Figure 1. Schematic image of photon absorption process of the cases of single-junction solar cells (a) and intermediate band solar cells.

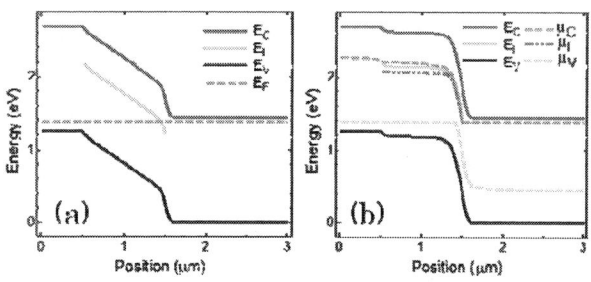

Figure 2. Calculated band diagrams in an equilibrium (a) and short-circuit condition (b), respectively. E_C and E_V are band edges of the condution and valence band. E_I is energetic position of the intermediate band. E_F is the Fermi energy. μ_C, μ_I and μ_V are quasi-Fermi energies of the conduction, intermediate and valence band, respectively

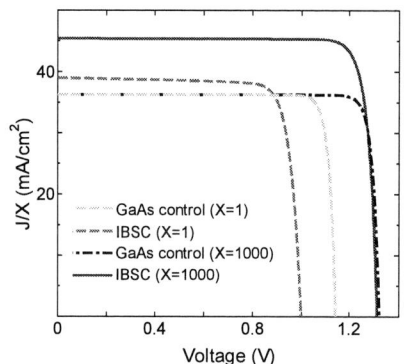

Figure 3. Curren-voltage characteristics of GaAs control cell and the intermediate band solar cell (IBSC). J and X are a current density and concentration ratio.

III. SIMULATION RESULTS

We used simplified absorption coefficients to focus on the fundamental properties of the intermediate band solar cells. The device structure consists of *p*-type top emitter layer, the intermediate band region and *n*-type bottom base layer. In the intermediate band region, the intermediate band is uniformly distributed and set 0.95eV from the band edge of the valence band. Calculated band diagrams in equilibrium and short-circuit conditions are presented in Fig. 2 (a) and (b), respectively. Electrons in the intermediate band affect the electrostatic potential profile. Under sun light illuminations, the electron density is calculated by (1) and strongly changes the band profile. Thus, the self-consistent calculation is important. In Fig. 3, current-voltage characteristics of the intermediate band solar cell and GaAs control cell which is the same structure except the intermediate band under 1 sun and 1000 suns are presented. Current densities are divided by light concentration ratio. Under 1 sun, short-circuit current of the intermediate band cell increase compared to the control cell but the open-circuit voltage and fill factor is degraded. This is the effect of the introduction of the intermediate band. The intermediate band works as carrier generation center but also recombination center. On the other hand, under 1000 suns

illumination, the short-circuit current is largely increased while the current of the control cell is the same as under 1sun condition and degradations of the open-circuit voltage and fill factor are improved. In this case, the recombination via the intermediate band is suppressed by large carrier generation rates via the band. This effect is called as photo-filling effects [9].

IV. CONCLUSION

We studied fundamental properties of the intermediate band solar cells by using the self-consistent device simulation of intermediate band solar cells. The device simulation well describes inside properties of the photovoltaic and is a great tool to optimize the structure. The intermediate band works as both carrier generation and recombination centers. To minimize the negative effects of the introduction of the intermediate band, the light concentration is very important. In quantum dot based solar cells, the absorption coefficients related to the intermediate band are small [10]. In this case, light management technique can play an important role to achieve high conversion efficiencies.

ACKNOWLEDGMENT

The authors acknowledge the funding from New Energy and Technology Development Organization (NEDO), and Ministry of Economy, Trade and Industry (METI), Japan and the discussion with Prof. N. Sano, University of Tsukuba.

REFERENCES

[1] W. Shockley and H. J. Queisser, "Detailed balance limit of effciency of p-n junction solar cells," J. Appl. Phys., vol 32, pp. 510-519, 1961.

[2] A. Luque and A. Marti, "Increasing the effciency of ideal solar cells by photon induced transition at intermeidate levels," Phys. Rev. Lett., vol. 78, pp. 5014-5017, 1997.

[3] S. M. Hubbard, C. D. Cress, C. G. Bailey, R. P. Raffaelle, S. G. Bailey and D. M. Wilt, "Effect of strain compensation on quantum dot enhanced GaAs solar cells," Appl. Phys. Lett., vol. 92, pp. 123512-123514, 2008.

[4] K. A. Sablon, J. W. Little, V. Mitin, A. Sergeev, N. Vagidov and K. Reinhardt, "Strong enhancement of solar cell efficiency due to quantum dots with built-in chrge," Nanno Lett., vol. 11, pp. 2311-2317, 2011.

[5] Y. Okada, T. Morioka, K. Yoshida, R. Oshima, Y. Shoji, T. Inoue and T. Kita, "Increase in photoncurrent by optical transitions via intermediate quantum states in direct doped InAs/GaNAs strain-compensated quantum dot solar cell," J. Appl. Phys., vol. 109, pp. 024301-024305, 2011.

[6] N. Lopez, L. A. Reichertz, K. M. Yu, K. Compman and W. Walukiewicz, "Engineering the electronic band structure for multiband solar cells," Phys. Rev. Lett., vol. 106, pp. 028701-028704, 2011.

[7] Intermediate-band photovoltaic solar cell based on ZnTe:O," Appl. Phys. Lett., vol. 95, pp. 011103-011105, 2009.

[8] K. Yoshida, Y. Okada and N. Sano, "Self-consistent simulation of intermediate band solar cells: Effect of occupation rates on device characteristics," Appl. Phys. Lett., vol. 97, pp. 133503-133505, 2010.

[9] R. Strandberg and T. W. Reenaas, "Photofilling of intermediate bands," J. Appl. Phys. vol. 105, pp. 124512-124519, 2009.

[10] S. Tomic, "Intermediate-band solar cells: influence of band formation on synamical processes in InAs/GaAs quantum dot arrays," Phys. Rev. B, vol 85, pp. 195321-195335, 2010.

Radiative and Non-radiative Processes in Intermediate Band Solar Cells

Stanko Tomić

Joule Physics Laboratory, School of Computing, Science and Engineering, University of Salford, UK

Abstract—**Intermediate band solar cells (IBSC) have emerged as an alternative design for third generation solar cells that could lead to dramatical improvements of the power conversion efficiencies. For this concept to work the intermediate band (IB) has to be located in the forbidden energy gap of the barrier material and to be separated by zero density of states from the valence and conduction band of the barrier material. We have demonstrated that a k · p multiband theory with periodic boundary conditions can easily be applied to predict electronic and absorption characteristics, as well as radiative and non-radiative carrier life-times between IB induced by semiconductor quantum dot (QD) arrays. We have identified that the most detrimental effect that might affect proper operation of the IBSC is caused by very fast, ∼ps, Auger electron cooling non-radiative process. We discuss possible QD array designs that can suppress fast Auger electron cooling.**

I. INTRODUCTION

The intermediate band solar cell (IBSC) has been proposed as a means to improve efficiency over that of a single gap solar cell. [1], [2] The IBSC comprises the so called "intermediate band material", having an electronic band (intermediate band, IB) inside what otherwise would be a conventional semiconductor bandgap, E_G. We will denote the total bandgap of the semiconductor as E_G, and its two parts, measured from the centre of the IB, as E_L and E_H. To achieve its higher efficiency potential, the IB allows absorption of below-bandgap energy photons on transitions from the valence band (VB) to the IB and from the IB to the conduction band (CB). These absorption processes induce the corresponding carrier generation rates, and these add up to the conventional carrier generation from the VB to the CB. Once carriers have been generated, they can also recombine. For preserving the output voltage of the cell (equal to the difference of electron and hole quasi-Fermi levels, $eV = E_{FC} - E_{FV}$, where e the electron charge), [3] it is necessary that quasi-Fermi level separation exists between the CB quasi-Fermi level (E_{FC}) and the IB quasi-Fermi level (E_{FI}), and also between the VB quasi-Fermi level (E_{FV}) and E_{FI}. These are increasingly difficult to achieve as the recombination rates between IB and CB and between IB and VB, involving processes other than radiative recombination, increases too.

II. THEORETICAL METHOD

To solve the multi-band system of Schrödinger equations, the **k·p** equation, for the semiconductor quantum dots (QD) array, the plane wave (PW) methodology is employed as an expansion method. [4], [5] In the PW representation the eigenvalues (E_n) and coefficients ($A_{n,\mathbf{k}}$) of the nth-eigenvector, $[\psi_n(\mathbf{r}) = \sum_{\mathbf{k}} A_{n,\mathbf{k}} e^{i\mathbf{k}\mathbf{r}}]$, are linked by the relation

$$\sum_{m,\mathbf{k}'} h_{m,n}(\mathbf{k}', \mathbf{k}) A_{n,\mathbf{k}} = E_n \sum_{\mathbf{k}} A_{n,\mathbf{k}} \tag{1}$$

where $h_{m,n}(\mathbf{k}', \mathbf{k})$ are the Fourier transform of the Hamiltonian matrix elements, and $m, n \in \{1, ..., 8\}$ are the band indexes of the 8-band **k·p** Hamiltonian. All the elements in the Hamiltonian matrix, Eq. (1), can be expressed as a linear combination of different kinetic and strain related terms and its convolution with the characteristic function of the actual QD shape, $\chi_{\text{qd}}(\mathbf{k})$. [6], [5]

The PW based **k·p** method inherently assumes *periodic Born-von Karman boundary conditions* and is particularly suited for analysis of the QD array structures. The electronic structure of such an array is characterized by a Brillouin zone (BZ) determined by the QD array dimensions. [7], [8], [9] To calculate the electronic structure the only modification to the basis set is to replace the reciprocal lattice vectors in the PW expansion with those shifted due to the QD-superlattice (QD-SL):

$$k_\nu \to k_\nu + K_\nu \tag{2}$$

where $0 \leq K_\nu \leq \pi/L_\nu$, and the L_ν are the super-lattice vectors in the $\nu = (x, y, z)$ directions. This allows sampling along the **K** points of a QD-SL Brillouin zone to be done at several points at the cost of the single QD calculation at each point **K**, avoiding laborious calculations of the large QD clusters. The QD array wave functions throughout the QD Brillouin zone become $\Psi_{n,\mathbf{K}}(\mathbf{r}) = \sum_{\mathbf{k}} A_{n,\mathbf{K}} e^{i(\mathbf{k}+\mathbf{K})\mathbf{r}}$. All the results presented here were obtained by using the kppw code. [4], [5]

III. MODEL QD ARRAY

The model QD arrays considered here consist of InAs/GaAs QDs with truncated pyramidal shape, Fig. 1. The size and shape of the QD is controlled by the pyramid base length, b, its height h, and truncation factor, t, defined as a ratio between length of the pyramid side at h, and length of the pyramid base b. The QDs are embedded in the tetragonal-like unit cell. The vertical periodicity of the QD array is controlled by $L_z = d_z + h + L_{WL}$, where d_z is the vertical separation of the QDs in subsequent layers.

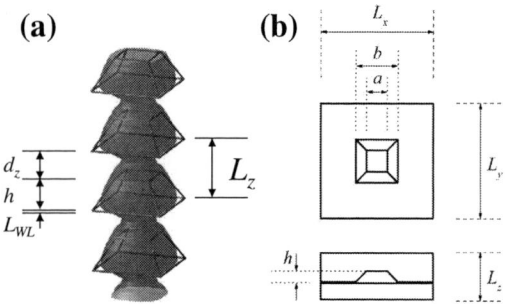

Fig. 1: Schematic of the model QD array (a), and QDs that makes array (b). QD truncation factor is defined as $t = a/b$, while all the other relevant parameters are defined in the main text

IV. RESULTS AND DISCUSSION

For typical QD array, made of truncated pyramidal InAs/GaAs QDs with $b = 10$ nm, $h = 2.5$ nm, and $t = 0.5$, and with vertical spacing between the dots in the array of $d_z = 5$ nm, we have estimated energy gaps between valence band (VB) and intermediate band (IB) of 1.2 eV and between IB and conduction band (CB) of 0.124 eV. By varying the QD sizes keeping periodicity in the z direction constant, our predictions suggest that the most promising design for an IB material that will exhibit its own quasi-Fermi level is to employ small InAs/GaAs QDs (6-10 nm QD lateral size) [8]. With appropriate design of the QD array structural parameters: (1) it is possible to achieve the regions of pure zero DOS between IB and the rest of the CB states, that is desirable for photon sorting and increased efficiency of device, and (2) it is possible to achieve the strong optically allowed excitation between IB and CB [10], [11]. Analysis of various radiative and nonradiative times, Table I, indicates that: (i) the ratio between CB to IB and IB to VB radiative times is 50, (ii) although phonon related relaxation time between CB and IB (56 ns) amounts to a half of the radiative relaxation time between CB and IB (109 ns), it is still one order of magnitude larger than the IB to VB radiative time (2.1 ns), (ii) non-radiative phonon absorption process that promotes electrons from IB to CB is very slow (650 ns) and probably would not significantly affect the transport properties of the IBSC, (iv) nonradiative Auger bi-exciton relaxation, Fig. 2, time (8.4 ns) is longer then radiative IB to VB relaxation time, indicating that this process will still be predominately radiative, (v) most detrimental effect on transport properties can originate from non- radiative Auger electron cooling process (2 ps), Fig. 2, that is three orders of magnitude faster that any other relaxation process in the IBSC [10], [11]. Special attention needs to be paid in the design of the IBSC structures in order to suppress the effects of electron cooling and to provide an increased efficiency of the IBSCs. We have shown recently that with appropriate band structure engineering, it is possible

TABLE I: Radiative, LO phonon emission/absorption, and Auger scattering times of InAs/GaAs QD array IBSC considered in the main text.

Type	minibands	Radiative (ns)	Phonons [e/a] (ns)	Auger (ns)
CB→IB	$e1 \rightarrow e0$	109	56/650	2×10^{-3}
IB→VB	$e0 \rightarrow h0$	2.1		8.4

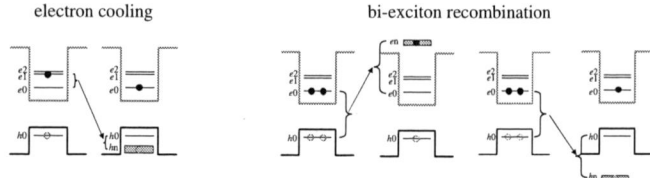

Fig. 2: Illustration of two different Auger processes: electron cooling (left) and bi-exciton recombination (right).

to place the intraband Auger electron cooling decay timescale in the ns range [12]. Such an optimised design requires a VB confinementless QD structure or Sb containing alloys in the barrier region in order to induce the type-II alignment and spatial electron/hole separation [10].

Predicted efficiency of IBSC in which nonradiative processes are suppressed and operation is dominated by radiative processes only, based on drift-diffusion model [13], and using realistic parameters obtained above, is 39% under maximal sun concentration [11]. This is increase of 56% compared to simple QD solar cell.

Our work provides a very important conceptual message – with appropriate treatment of relevant radiative and nonradiative effects, the multiband envelope function Hamiltonians with periodic boundary conditions are fully capable of capturing the all relevant effects determining operation of the QD based intermediate band solar cells.

REFERENCES

[1] A. Luque and A. Martí, "Increasing the efficiency of ideal solar cells by photon induced transitions at intermediate levels," *Phys. Rev. Lett.*, vol. 78, no. 26, pp. 5014–5017, Jun 1997.

[2] A. Luque and A. Marti, "The intermediate band solar cell: Progress toward the realization of an attractive concept," *Advanced Materials*, vol. 22, pp. 160–174, 2010.

[3] ——, "A metallic intermediate band high efficiency solar cell," *Prog. Photovoltaics*, vol. 9, no. 2, pp. 73–86, MAR-APR 2001.

[4] S. Tomić, A. G. Sunderland, and I. J. Bush, "Parallel multi-band $k \cdot p$ code for electronic structure of zinc blend semiconductor quantum dots," *J. Mater. Chem.*, vol. 16, pp. 1963–1972, 2006.

[5] N. Vukmirović and S. Tomić, "Plane wave methodology for single quantum dot electronic structure calculations," *J. Appl. Phys.*, vol. 103, no. 10, p. 103718, 2008.

[6] A. D. Andreev, J. R. Downes, D. A. Faux, and E. P. O'Reilly, "Strain distributions in quantum dots of arbitrary shape," *J. Appl. Phys.*, vol. 86, pp. 297–305, 1999.

[7] O. L. Lazarenkova and A. A. Balandin, "Miniband formation in a quantum dot crystal," *Journal of Applied Physics*, vol. 89, no. 10, pp. 5509–5515, 2001.

[8] S. Tomić, T. S. Jones, and N. M. Harrison, "Absorption characteristics of a quantum dot array induced intermediate band: Implications for solar cell design," *Appl. Phys. Lett.*, vol. 93, no. 26, p. 263105, 2008.

[9] J. W. Klos and M. Krawczyk, "Two-dimensional gaas/algaas superlattice structures for solar cell applications: Ultimate efficiency estimation," *Journal of Applied Physics*, vol. 106, no. 9, p. 093703, 2009.

[10] S. Tomić, "Intermediate-band solar cells: Influence of band formation on dynamical processes in inas/gaas quantum dot arrays," *Phys. Rev. B*, vol. 82, p. 195321, Nov 2010.

[11] S. Tomić, *in Next Generation of Photovoltaics*. (Springer, Heidleberg 2012).

[12] S. Tomić, A. Martí, E. Antolín, and A. Luque, "On inhibiting auger intraband relaxation in inas/gaas quantum dot intermediate band solar cells," *Applied Physics Letters*, vol. 99, no. 5, p. 053504, 2011.

[13] R. Strandberg and T. W. Reenaas, "Photofilling of intermediate bands," *Journal of Applied Physics*, vol. 105, no. 12, p. 124512, 2009.

Modeling of N-i-P Vs. P-i-N InGaN Solar Cells with Ultrathin GaN Interlayers for Improved Performance

Jeramy Dickerson, Konstantinos Pantzas, Tarik Moudakir, Paul L. Voss, and Abdallah Ougazzaden

Georgia Tech-CNRS, UMI 2958, Unité Mixte Internationale, F-57070 Metz, France

School of Electrical and Computer Engineering, Georgia Institute of Technology, Atlanta, GA 30332 USA

Abstract- **P-i-N structure solar cells often provide improved performance over N-i-P devices because acceptors are easier to activate when the p-type layer is close to the surface. However, for strained InGaN solar cells on GaN, the polarization-induced electric field creates a barrier for photocurrent that impedes device performance. In this paper we show that for Ga-face growth, N-i-P structures can provide improved performance because the electric field from the junction is parallel to that formed from polarization induced sheet charges. Thus the fields complement each other to assist in creating photocurrent in N-i-P devices. Additionally we simulate an N-i-P cell using the recently demonstrated insertion of ultra-thin GaN interlayers to achieve thick strained layers with high material quality.**

INTRODUCTION

The $In_xGa_{1-x}N$ system has been researched extensively for photovoltaic use after the bandgap of InN was discovered to be 0.64 eV. Furthermore this system has several beneficial qualities. It has a direct bandgap over the entire compositional range, possesses high carrier mobility and drift velocity, is highly resistant to radiation deterioration, and exhibits high optical absorption ($> 10^5$ cm^{-1}) near the bandgap.

Unfortunately this system has proven difficult to grow at high indium concentration because of phase separation [1], indium clustering [2], and cracking due to low critical thicknesses [3]. These effects quickly degrade device performance to unacceptable levels. Several groups have made significant advances despite these challenges. Both Molecular Beam Epitaxy (MBE) and Metalorganic Chemical Vapor Deposition (MOCVD) growth have resulted in single phase $In_xGa_{1-x}N$ with In content up to 40% [4-7].

Theoretical investigations on the spontaneous and piezoelectric polarization effects in the III-nitride material system has matured substantially with only around 20% difference between experiments and ab-initio predictions [8]. Despite the well-known, commercial application of these effects in high-electron mobility transistors and other devices, only a few papers mention these strong effects in solar cell devices [9, 10]. For solar cells, it beneficial to have coherently strained InGaN layers as this reduces dislocations and improves device quality. But strained InGaN on GaN substrate creates large polarization electric fields. For instance one group found the electric field to be around 2.45 MV/cm with just 18% indium [11] . As material quality and modeling are improved, it will become possible to optimize the use of polarization effects. In this paper we propose the modification of traditional P-i-N designs, showing that the recommended N-i-P configuration solar cell is a superior design over P-i-N designs for InGaN. Additionally we report simulations of

solar cells including the periodic insertion of ultrathin GaN interlayers in the InGaN absorbing region. This design has recently been shown to eliminate phase-separation in >120 nm-thick InGaN [12].

DEVICE DESIGN

The piezoelectric parameters used are listed in Table 1. The doping in p-type regions was 5e17/cm^3 while the doping in n-type regions was 1e18/cm^3. The i-region was simulated with either a 126 nm bulk $In_{0.1}Ga_{0.9}N$ layer or with 6 regions of 21 nm $In_{0.1}Ga_{0.9}N$ with 1.5 nm GaN interlayers.

RESULTS

The polarization-induced electric fields naturally formed on strained InGaN layers oppose the junction field created in typical P-i-N junctions. This effect has been noted in [9, 10] and is reproduced here in Fig. 1. The use of only 30% of the theoretical polarization sheet charge has drastic effects on the band diagram, reversing direction of the photo-generated current. It is hard to predict the exact value of the polarization field as defects, phase separation, indium clustering, doping and free carriers can screen the polarization induced sheet charges [13, 14]. Regardless, as material quality improves the effect is likely to become more substantial and could severely lower the efficiency of solar cells used in the P-i-N configuration. However, as shown in Fig. 2, switching to an N-i-P device alleviates this effect as the junction and polarization field are now in the same direction. The band diagram is only slightly modified for up to 100% of the theoretically predicted polarization. This indicates that with improved material quality, the natural polarization of the InGaN layer will assist rather then detract from the photo-generated current.

Finally, as mentioned above, the use of ultrathin GaN interlayers has been shown to prevent indium phase separation, while allowing for thicker strained InGaN layers. Fig. 3 and 4 show again that the N-i-P configuration is a superior design. Further simulations will show that for high polarization fields the predicted external quantum efficiency (EQE) of N-i-P designs should not vary substantially while the EQE of P-i-N devices is expected to approach zero.

TABLE 1
MATERIAL PARAMETERS

Parameter	Symbol	Units	GaN	$In_{0.1}Ga_{0.9}N$
Lattice Constant	A	Nm	0.3189	0.3225
Elastic Constants	c13	Gpa	106	104.6
Elastic Constants	c33	Gpa	398	380.6
Piezoelectric Tensor	e13	C/m^2	-0.527	-0.523
Piezoelectric Tensor	e33	C/m^2	0.895	0.911
Spontaneous Polarization	Psp	C/m^2	-0.034	-0.031

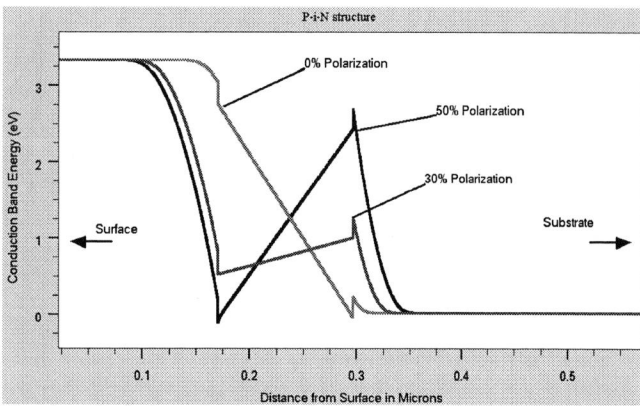

Fig. 1. Typical conduction band for a P-i-N solar cell device (0% polarization) along with photocurrent inhibiting energy band diagrams arising from polarization effects (30% and 50% shown). This and all simulations performed on Silvaco TCAD software [15].

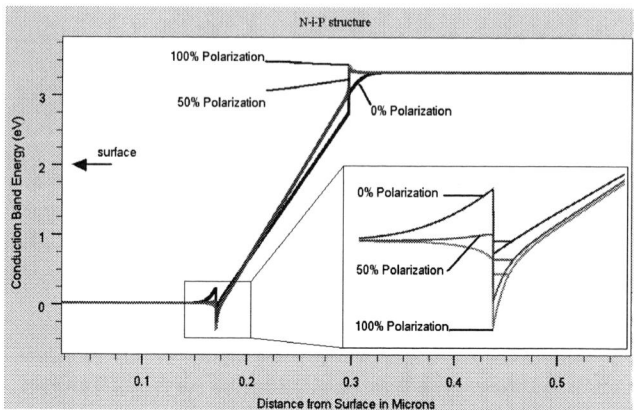

Fig. 2. Conduction band for a N-i-P solar cell device (0%, 50% and 100% polarization shown). Inset shows how the barrier at the GaN/InGaN interface is reduced and the quantum well on the InGaN side is deepened.

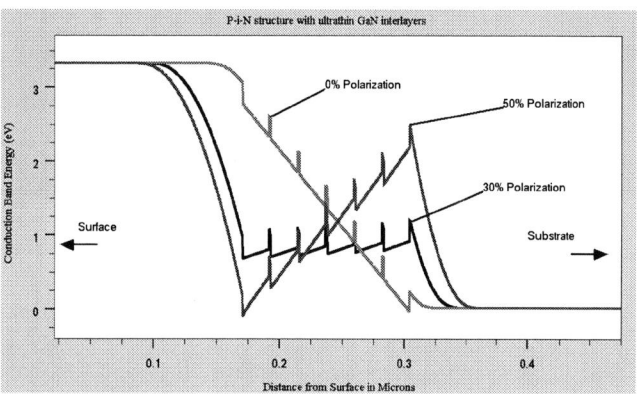

Fig. 3. Typical conduction band for a P-i-N solar cell device with ultrathin GaN layers (0%, 30% and 50% polarization shown).

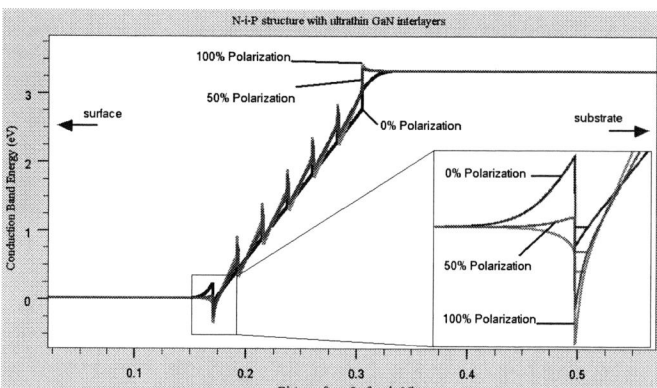

Fig. 4. Typical conduction band for a N-i-P solar cell device with ultrathin GaN layers added (0%, 50% and 100% polarization shown). Inset shows how the barrier at the GaN/InGaN interface is reduced and the quantum well on the InGaN side is deepened.

REFERENCES

[1] N. Faleev, *et al.*, "High quality InGaN for photovoltaic applications: Type and spatial distribution of crystalline defects and "phase" separation," in *Photovoltaic Specialists Conference, 2008. PVSC '08. 33rd IEEE*, 2008, pp. 1-6.

[2] I. Gorczyca, S. P. Łepkowski, T. Suski, N. E. Christensen, and A. Svane, "Influence of indium clustering on the band structure of semiconducting ternary and quaternary nitride alloys," *Physical Review B*, vol. 80, p. 075202, 2009.

[3] D. Holec, P. M. F. J. Costa, M. J. Kappers, and C. J. Humphreys, "Critical thickness calculations for InGaN/GaN," *Journal of Crystal Growth*, vol. 303, pp. 314-317, 2007.

[4] X. Chen, K. D. Matthews, D. Hao, W. J. Schaff, and L. F. Eastman, "Growth, fabrication, and characterization of InGaN solar cells," *physica status solidi (a)*, vol. 205, pp. 1103-1105, 2008.

[5] P. Misra, *et al.*, "Fabrication and characterization of 2.3eV InGaN photovoltaic devices," in *Photovoltaic Specialists Conference, 2008. PVSC '08. 33rd IEEE*, 2008, pp. 1-5.

[6] B. R. Jampana, *et al.*, "Design and Realization of Wide-Band-Gap (~2.67 eV) InGaN p-n Junction Solar Cell," *Electron Device Letters, IEEE*, vol. 31, pp. 32-34, 2010.

[7] C. J. Neufeld, *et al.*, "High quantum efficiency InGaN/GaN solar cells with 2.95 eV band gap," *Applied Physics Letters*, vol. 93, pp. 143502-3, 2008.

[8] V. Fiorentini, F. Bernardini, and O. Ambacher, "Evidence for nonlinear macroscopic polarization in III--V nitride alloy heterostructures," *Applied Physics Letters*, vol. 80, pp. 1204-1206, 2002.

[9] J. J. Wierer, A. J. Fischer, and D. D. Koleske, "The impact of piezoelectric polarization and nonradiative recombination on the performance of (0001) face GaN/InGaN photovoltaic devices," *Applied Physics Letters*, vol. 96, p. 051107, 2010.

[10] M. Lestrade, Z. Q. Li, Y. G. Xiao, and Z. M. S. Li, "Modeling of polarization effects in InGaN PIN solar cells," in *Numerical Simulation of Optoelectronic Devices (NUSOD), 2010 10th International Conference*, 2010, pp. 77-78.

[11] P. Lefebvre, *et al.*, "High internal electric field in a graded-width InGaN/GaN quantum well: Accurate determination by time-resolved photoluminescence spectroscopy," *Applied Physics Letters*, vol. 78, pp. 1252-1254, 2001.

[12] K. Pantzas, et al., "Semibulk InGaN: A Novel Approach for Thick, Single Phase, Epitaxial InGaN Layers Grown by MOVPE," *16th MOVPE Conference*, 2012, Paper FrA1-3

[13] A. D. Carlo, F. D. Sala, P. Lugli, V. Fiorentini, and F. Bernardini, "Doping screening of polarization fields in nitride heterostructures," *Applied Physics Letters*, vol. 76, pp. 3950-3952, 2000.

[14] F. D. Sala, *et al.*, "Free-carrier screening of polarization fields in wurtzite GaN/InGaN laser structures," *Applied Physics Letters*, vol. 74, pp. 2002-2004, 1999.

[15] I. Silvaco, "Silvaco Atlas User Manual, 2012 available from www.silvaco.com," ed, 2012.

Green functions for photovoltaic response of quantum wire-dot-wire junctions

A. Berbezier and F. Michelini

Institute Materials, Microelectronics and Nanosciences of Provence, Marseille, FRANCE
e-mail: aude.berbezier@im2np.fr, fabienne.michelini@im2np.fr

Abstract-To investigate the photovoltaic properties of quantum dot connected to quantum wire reservoirs we rely on numerical calculations using the non-equilibrium Green function formalism. We examine impacts of the hopping parameter that controls the dot-wire contact for a monochromatic light resonant with the isolated dot gap. Global current and power increase when the hopping decreases, in particular the short-circuit current and the open-circuit voltage.

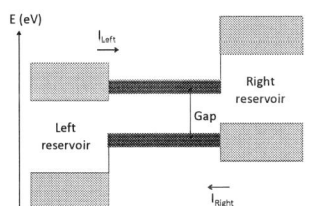

Fig1. Band diagram of the quantum wire-dot-wire junction

I. INTRODUCTION

Thanks to intelligent integration of nanoscale structures, new generation of quantum solar cells are expected to exhibit higher efficiency than conventional photovoltaic solar cells [1]. The quantum functioning of such innovative architectures imperatively needs to be understood but still very isolated works have been carried out on the subject. In this work, we rely on a mesoscopic tight-binding model to deeper understand the photovoltaic properties of quantum dots (QD) solar cells. We use the non-equilibrium Green function formalism (NEGF), perfectly adapted to studies on open interacting systems [2,3]. In order to explain quantum solar cell's behavior, a crucial point is to determine the interplay between light absorption and transport and, hence to propose new architectures.

II. MODEL

We developed a meaningful one-dimensional model that consists in a quantum dot, referred as the central nanosystem, connected to two infinite non-interacting wire reservoirs at left and right sides. The dot is given by two energy levels. So are described the wires but using a different gap value and wide bands. Two hopping parameters (here identical), one for each level, characterize contacts to left and right reservoirs (here identical). While the electron/hole selectivity is controlled by heterogeneous doping in p-n junctions, it is provided by heterogeneous working functions in the proposed cell model, which results in band offsets at each dot-reservoir contact. Band diagram is represented Fig.1.

The photovoltaic properties of the central nanosystem were calculated within the self-consistent NEGF formalism in which self-energy functions account for contact to reservoirs, exactly, and interactions, generally in perturbation. This work only included the electron-photon interaction, operating in the dot, developed in the self-consistent Born approximation.

Numerical calculations were performed on MERLIN cluster of the IM2NP, Marseille.

III. RESULTS

A. Device working

The presented study focuses on a wire/single-dot/wire unidirectional junction lighted with a single-mode monochromatic plane wave. Moreover, reference parameters of the tight-binding model have been chosen to cancel dark current and, hence, emphasize the photovoltaic conversion. Indeed, under illumination, electrons localized on down states of the central nanosystem may cross the gap and populate up ones. Radiative processes are modeled by the self-energies of the electron-photon interaction in the NEGF calculations. Effects of the self-energies of the electron-photon interaction are shown Fig.2. Due to band offsets, electrons can flow from the lowest band of the right reservoir to the highest band of the left reservoir, resulting in a net photocurrent shown Fig.3. By tuning the photon energy around the dot gap that is shifted and broadened by left and right contact self-energies, the system can evolve from a blocking state (dark current null) to a passing one (short-circuit current about 10^{-8}A) inside a resonance window of light energies which corresponds to a commutation window of the nanodevice. This resonance is

978-1-4673-1602-6/12 $31.00 © 2012 IEEE   115

shown Fig.4. Due to discrete energy levels in QDs, all responses of the nanodevice occur in narrow windows. The proposed architecture can thus be designed in order to understand and optimize how these cells operate.

B. Impact of hopping

We examine impacts of hopping to reservoirs. Contacts induce resonance shifting and broadening enhanced with h, via the amplitude of the reservoir self-energies that similarly shifts and broadens the discrete levels of the dot. In this out-of-resonance configuration, the electron-photon interaction creates a double peak in the spectral responses as shown Fig. 5 for the left current, localized inside the conduction mini-band. We observe the same for the right current, but localized inside the valence mini-band (not shown). This double peak directly arises within the self-consistent structure of interaction self-energies from the energy difference between photon energy and centered mini-band gap of the connected dot in the dark that is the resonance photon energy. This energy difference gives the energy separation between the two peak maxima, it decreases when h decreases until it cancels at resonance. When h increases, peaks decrease and broaden, as the energy levels of the connected dot. The total current produced hence decreases for all voltages when h increases (see Fig. 5). The two I-V characteristics moreover reveal that both the shunt resistance and the series resistance of the equivalent cell circuit increase when h increases. We analyze that current cancels when overlap disappears, and that the singularity point occurs when partial overlap starts, giving rise to these peculiar working behavior.

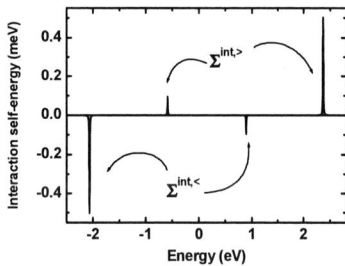

Fig2. . Real part of the self-energy of the electron-photon interaction at V_{bias}=0V. $\Sigma^{int,<}$ is plotted for the lowest level while $\Sigma^{int,>}$ is plotted for the highest level quantum

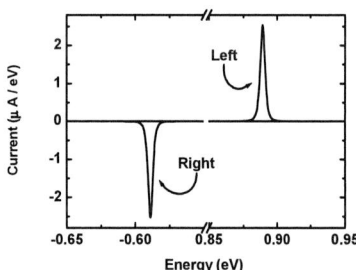

Fig3. .Left and right current spectral densities as a function of the electron energy at Vbias=0V

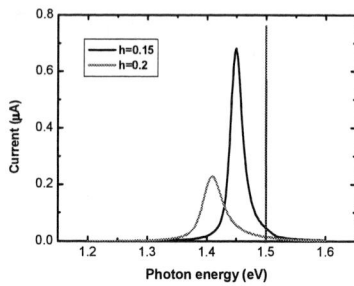

Fig4. . Left current at zero bias voltage as a function of the photon energy. The initial gap values 1.5eV (vertical line)

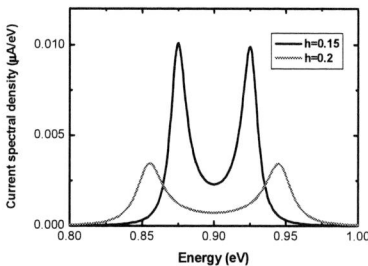

Fig5. . Left current spectral density at zero bias voltage as a function of the electron energy.

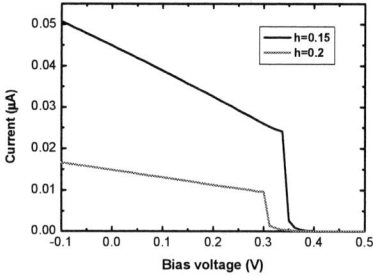

Fig6. . Left current as a function of the bias voltage.

IV. CONCLUSION

The dot-wire hopping parameter shows a strong and intricate impact on the photovoltaic cell functioning and the proposed architecture can thus be designed in order to optimize it. This model is a building block for future investigations that will include electron-electron interaction and relaxation processes.

V. REFERENCES

[1] Martin A. Green, "Third Generation Photovoltaics: Ultra-high Conversion Efficiency at Low Cost", *Prog. Photovolt: Res. Appl. 2001*, **9**: 123-135 .

[2] U. Aeberhard, "Theory and simulation of quantum photovoltaic devices based on the non-equilibrium Green's function formalism", *J Comput Electron*, 2011, 10.1007/s10825-011-0375-6.

[3] U. Aeberhard and R. H. Morf, "Microscopic nonequilibrium theory of quantum well solar cells", *Phys Rev B 77*, 2008, 125343.

Effects of sulfur incorporation into absorbers of CIGS solar cells studied by numerical analysis

Chia-Hua Huang and Hung-Lung Cheng

Department of Electrical Engineering, National Dong Hwa University,

Shou-Feng, Hualien 97401, Taiwan

Abstract— **The performance of Cu(In,Ga)Se₂ (CIGS) solar cells with the incorporation of sulfur into the surface region of the absorbers has been studied by numerical simulation. The impacts of sulfur contents and thickness of sulfurized layers in the surface region of absorbers on the performance were evaluated. The results show that the incorporation of sulfur in the CIGS films enhances the open-circuit voltage (V_{OC}), but concurrently leads to the reduction in short-circuit current density (J_{SC}). The S/(S+Se) ratios of below 0.2 could improve the cell performance for all thickness of sulfurized layers in this study. For S/(S+Se) ratios of 0.1–0.5, the thickness of 200nm was suggested to enhance the efficiency of devices.**

Keywords— **CIGS solar cells; Surface sulfurization; Thin films; Selenization**

I. INTRODUCTION

Thin-film Solar cells based on polycrystalline CIGS materials are regarded as the most promising alternative to the silicon-based solar cells due to the high conversion efficiency, low cost, large-area production, and the great potential for developing high-efficiency flexible solar cells [1]. The high efficiency of over 20% for CIGS solar cells has been demonstrated at the laboratory scale [2]. The selenization of CuGa/In metallic bi-layer precursors to prepare CIGS absorbers is a cost-effective method for the large-area production. However, it has been found that most of the Ga is accumulated near the Mo/CIGS interface. This could result in a decrease in V_{OC} due to the shrinkage of band gap near the space charge region (SCR), being ascribed to the lower Ga concentration in SCR. The surface sulfurization of CIGS absorbers has been studied to improve the performance of CIGS solar cells [3], [4], enlarging the band gap near SCR by substituting sulfur for selenium. Their results showed that the incorporation of sulfur into the surface region of CIGS absorbers did not necessarily enhance the cell performance, and even deteriorated the cell efficiencies. In order to investigate the factors impacting cell performance for the incorporation of sulfur into absorbers as well as its benefits, the effects of sulfur distribution profile, including sulfur contents and the depth of sulfur diffusion, on the performance of CIGS solar cells are evaluated with numerical simulation.

II. DEVICE MODELING

The baseline device structure of CIGS solar cells consists of Mo back contact, p-CIGS absorber, CdS buffer layer, and i-ZnO/ZnO:Al window layer. The thickness of CIGS film is

2μm and the overall Ga/(Ga+In) ratio is 0.25. The thicknesses of CdS, i-ZnO, and ZnO:Al are 60, 50, and 300nm, respectively. For the CIGS films prepared by selenization of metallic precursors, the Ga concentration of CIGS films near the CdS/CIGS interface is as low as negligible, and its maximum value is located near Mo back contact. In the simulations, we consider that the surface sulfurization of CIGS films results in the linear decrease of sulfur concentrations from the surface of absorbers. The device structure of CIGS solar cells with the sulfurized Cu(In,Ga)(S,Se)₂ (CIGSS) layer on the top of CIGS layer is shown in Fig. 1. The device simulation tool of wxAMPS [5], an update of one-dimensional solar cell simulation program AMPS-1D, was employed for the numerical study. The sulfur concentration near the surface region of CIGS films and the sulfur diffusion depth were varied. All other input parameters for the i-ZnO/ZnO:Al window layers, CdS buffer layers, and un-sulfurized CIGS absorbers were kept constant throughout all simulations as shown in Table 1. The additional deep defect levels, acting as the recombination centers, were located in the middle of band gap for all three layers.

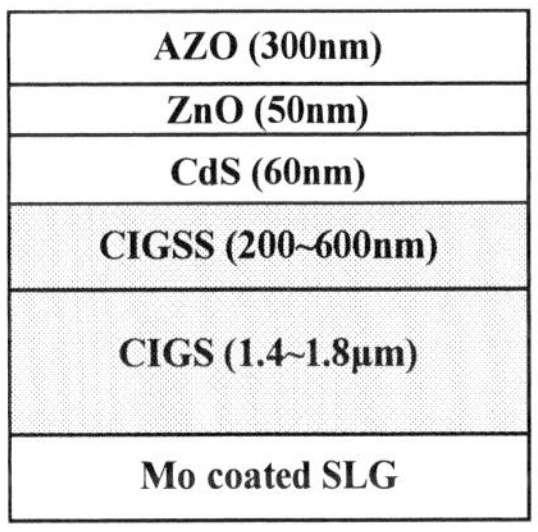

Fig.1 Device structure of CIGS solar cells for numerical simulation

TABLE I
INPUT PARAMETERS OF EACH LAYERS FOR SIMULATIONS

Layer properties	ZnO:Al	i-ZnO	CdS	CIGS
Permittivity	9	9	10	13.6
E_g (eV)	3.3	3.3	2.4	1.167
N_C (cm⁻³)	2.2×10^{18}	2.2×10^{18}	2.2×10^{18}	2.2×10^{18}
N_V (cm⁻³)	1.8×10^{19}	1.8×10^{19}	1.8×10^{19}	1.8×10^{19}
μ_e (cm²/V·s)	100	100	100	100
μ_h (cm²/V·s)	25	25	25	25
Carrier density (cm⁻³)	n: 1×10^{19}	n: 5×10^{16}	n: 5×10^{17}	p: 2×10^{16}

E_g: band gap energy; N_C: effective density of states in conduction band; N_V: effective density of states in valence band; μ_e: electron mobility; μ_h: hole mobility;

III. Results and Discussion

The simulation results of conversion efficiency, open-circuit voltage, short-circuit current density, and fill factor (F.F.) for the baseline CIGS solar cells are 17.4%, 0.566V, 40.3mA/cm², and 76.2%, respectively. With the incorporation of sulfur into the surface region of CIGS absorbers, the valence band edge shifts together with the shift in conduction band edge as the band gap increases due to the formation of CIGSS layers near the surface region of absorbers [6]. The impacts of sulfur concentration and diffusion depth on the device performance are illustrated in Fig. 2. The value of V_{OC} increases with the increase of sulfur concentration and diffusion depth, attributed to the enlargement of band gap near SCR, resulting in the reduction of recombination in SCR. At the same time, however, the value of J_{SC} decreases with the increase of sulfur concentration and diffusion depth, corresponding to the increase of minimum band gap, which results in the reduction of light absorption. When the diffusion depth of sulfur is greater than the SCR width ($\sim$0.3μm), a higher sulfur concentration will result in a notch shape in conduction band edge (shown in Fig. 3), leading to the recombination of electrons in conduction band and holes in valence band [7]. And thus, the current collection was impaired. With the thick sulfurized layer, the F.F. and the efficiency deteriorate dramatically as the S/(S+Se) ratio is more than 0.3. For thin sulfurized layer of 200nm, there are only small variations in cell performances despite the sulfur concentration.

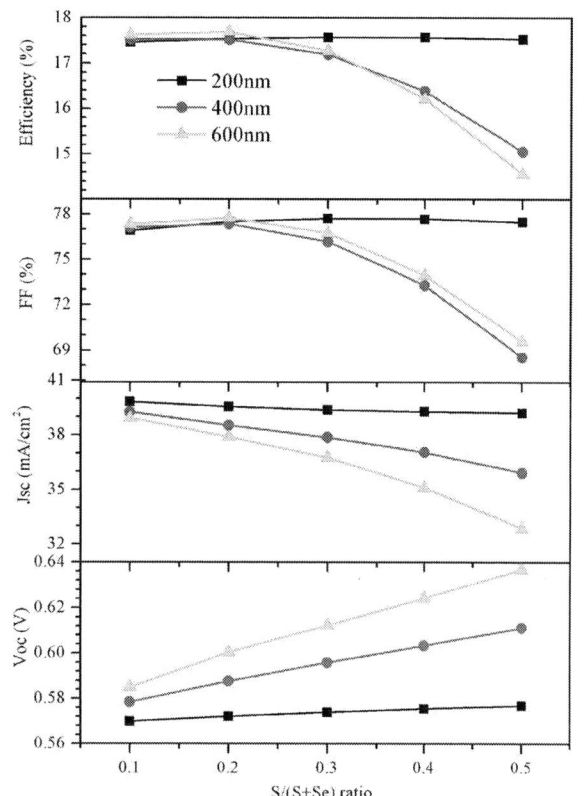

Fig.2 Performance parameters of CIGS solar cells calculated as the functions of sulfur concentration near surface region of absorbers and depth of diffusion

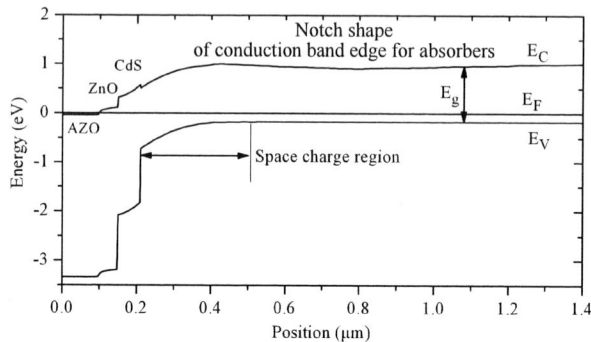

Fig. 3 Schematic energy band diagram of a CIGS solar cell with sulfurized layer thickness of 600nm and S/(S+Se) ratio of 0.5 under equilibrium condition

IV. Summary

Numerical simulations of CIGS solar cells with the surface sulfurized CIGS absorbers have been carried out using the wxAMPS simulation program. The impacts of the sulfur concentration near the surface region of absorbers and the depth of sulfur diffusion on the performance of devices were investigated. It is found that the sulfurized absorber layers can improve the cell performance as the sulfur diffusion depth is less than the SCR width. With the greater depth and high sulfur concentration, the performance of devices decreases, which could be ascribed to the low F.F. and the reduction in current collection due to the formation of notch in conduction band edge of absorbers. It is concluded that both the thickness of sulfurized layer less than SCR width and the small S/(S+Se) ratios ($\leq$0.2) are beneficial to the enhancement of cell performance.

Acknowledgment

The authors gratefully acknowledge Prof. Rockett, Yiming Liu of University of Illinois at Urbana-Champaign (UIUC) and Prof. Fonash of Pennsylvania State University for the use of wxAMPS software design in UIUC.

References

[1] F. Kessler and D. Rudmann, "Technological aspects of flexible CIGS solar cells and modules," *Solar Energy*, vol. 77, pp. 685–695, 2004.

[2] M.A. Green, K. Emery, Y. Hishikawa, W. Warta, and E.D. Dunlop, "Solar cell efficiency tables (version 39)," *Progress in Photovoltaics: Research and Applications*, vol. 20, pp. 12–20, 2012.

[3] D.Ohashi, T. Nakada, and A. Kunioka, "Improved CIGS thin-film solar cells by surface sulfurization using In₂S₃ and sulfur vapor," *Solar Energy Materials & Solar Cells*, vol. 67, pp. 261–265, 2001.

[4] U. P. Singh, "Surface sulfurization studies of thin film Cu(InGa)Se₂ solar cells," *Vacuum*, vol. 83, pp. 1344–1349, 2009.

[5] Y. Liu, Y. Sun, and A. Rockett, "A new simulation software of solar cells—wxAMPS," *Solar Energy Materials & Solar Cells*, vol. 98, pp. 124–128, 2012.

[6] M. Turcu, I. M. Kötschau, and U. Raua, "Composition dependence of defect energies and band alignments in the Cu(In₁₋ₓGaₓ)(Se₁₋ᵧSᵧ)₂ alloy system," *Journal of Applied Physics*, vol. 91, pp. 1391–1399, 2002.

[7] Adrian Chirilă et al., "Highly efficient Cu(In,Ga)Se2 solar cells grown on flexible polymer films," Nature Materials, vol. 10, pp. 857–861, 2011.

Self-Consistent Electro-Thermal-Optical Simulation of Thermal Blooming in Broad-Area Lasers

Joachim Piprek

NUSOD Institute LLC, Newark, DE 19714-7204, United States, E-mail: piprek@nusod.org

Abstract – **High-power broad-area laser diodes often suffer from a widening of the lateral far-field with increasing current, called thermal blooming. This effect is mainly caused by the non-uniform self-heating of the laser and it has been studied for several decades. For the first time, this paper presents a self-consistent electro-thermal-optical simulation of thermal blooming, including all relevant physical mechanisms. The results are in good agreement with previous measurements and reveal the blooming mechanism in detail. Common mistakes in the experimental determination of the internal temperature rise are also discussed.**

I. INTRODUCTION

High-power broad-area laser diodes often suffer from a widening of the lateral far-field with increasing current.[1,2] This effect is also referred to as thermal blooming, since self-heating is considered the main cause. The non-uniform temperature profile inside the waveguide leads to a lateral refractive index profile that enhances the index guiding of laser modes (thermal lens). Numerical simulation is a valuable tool in investigating this interaction of electronic, thermal, and optical processes, however, a comprehensive numerical analysis has not been published yet. Recent simulations combine optical mode guiding with the internal heat flux,[2] but without accurate calculation of the internal heat power distribution.

This paper present the first self-consistent electro-thermal-optical simulation of the thermal blooming effect, including the highly non-uniform heat power distribution inside the laser. A previously investigated GaAs-based broad-area quantum well (QW) laser structure with an emission wavelength near 975nm is used as an example.[2]

II. MODELS AND PARAMETERS

A customized version of the LASTIP laser simulation software is employed here,[3] which allows for the self-consistent combination of multi-mode wave guiding, drift-diffusion of electrons and holes, and heat flow in the transverse plane. The heat power distribution is calculated using the local densities of carriers and current, including Joule heat, non-radiative recombination heat, heat caused by modal absorption, as well as the Peltier/Thomson effect. Index guiding by etched trenches is considered as well as anti-index guiding by carrier-

induced index changes in the quantum well. The local refractive index $N(T)$ is calculated from the temperature distribution $T(x,y)$ using published material parameters.[4]

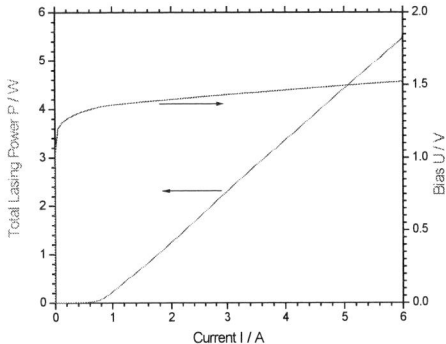

Fig. 1: Simulated laser characteristics.

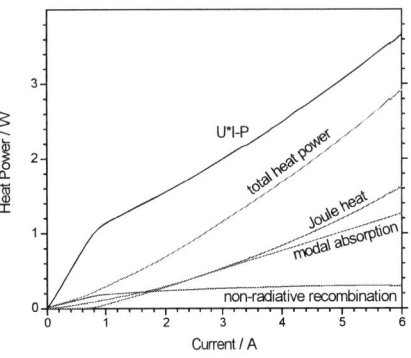

Fig. 2: Comparison of different heat sources. U*I-P gives the heat power estimated from the laser characteristics in Fig. 1.

III. SIMULATION RESULTS

The simulated light-current and current-voltage characteristics are shown in Fig. 1 and they are in excellent agreement with measurements.[2] Different heat generation mechanisms are distinguished in Fig. 2. Joule heating and modal absorption cause a similarly high contribution. Heating from non-radiative recombination hardly changes above threshold, due to the almost constant QW carrier densities. Also shown in Fig. 2 is the heat power U*I-P commonly extracted from the laser characteristics in Fig. 1,[2] which is substantially larger than the numerically calculated heat power. The main reason is the inclusion of the spontaneous photon emission as heat source in the formula U*I-P. An unknown fraction of these photons is internally absorbed and eventually generates heat, but that heat generation mainly happens far away from the active region and without much influence on its temperature. Therefore, the absorption of spontaneously emitted photons is neglected in this paper.

Since the spontaneous emission power is hard to measure, the accurate determination of the QW temperature seems to be an open problem which should be discussed at the conference. Another inaccuracy is related to the employment of a thermal resistance,[1] since this concept puts the entire heat power into the active layer, which is not correct for distributed heat sources.[5]

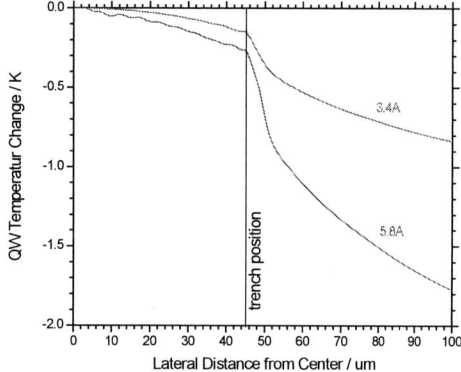

Fig. 3: Lateral QW temperature profiles relative to their peak value at two different bias points. Assuming perfect symmetry, only half the laser is considered.

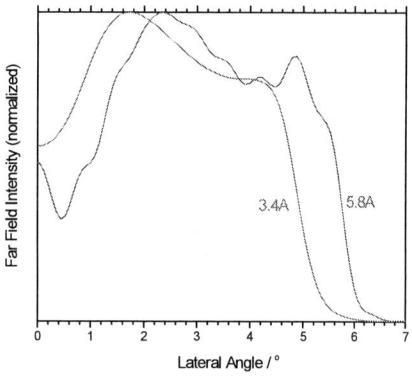

Fig. 4: Lateral far field profiles calculated at two bias points.

However, the thermal lens near the active region is not dominated by the absolute temperature but by local temperature differences. Figure 3 shows the calculated temperature profiles at two bias points. Obviously, a higher current induces a stronger temperature drop in lateral direction.

The calculated far-field profile is shown in Fig. 4 for the same bias points, clearly demonstrating the thermal blooming effect. At 3.4A current, a total of 17 lateral modes contribute to lasing. This mode number increases to 19 at 5.8A.

Higher-order modes are known to generate wider far fields. Figure 5 compares the far field of the 17th and the 18th mode. Thus, the widening of the far field can be attributed to a rising maximum mode order. The simulated near field is shown in Fig. 6. The width of the near field remains nearly constant.

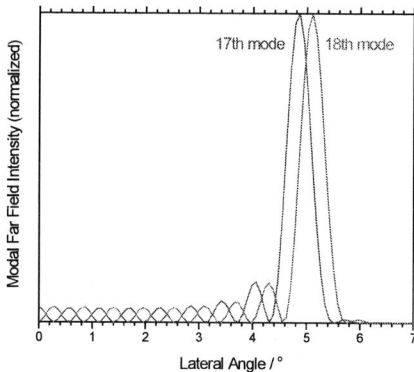

Fig. 5: Far fields of modes 17 and 18. The mode order corresponds to the total number of peaks.

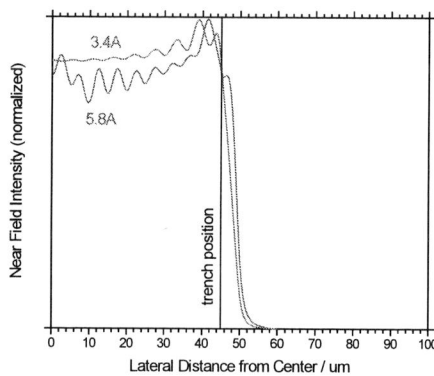

Fig. 6: Lateral near field profiles calculated at two bias points.

In summary, using self-consistent electro-thermal-optical simulation, the thermal blooming effect in high-power lasers is linked to the rising order of lateral modes due to thermal lens enhancements with higher current. This enables future laser design optimization towards smaller and more stable far fields.

ACKNOWLEDGEMENT

This work was supported by the Ferdinand Braun Institute, Berlin, Germany. Helpful discussions with H. Wenzel und P. Crump are acknowledged.

REFERENCES

[1] P. Crump et al., Sem. Science & Technol. **27** (2012) 045001
[2] H. Wenzel et al., NUSOD 2011, presentation WA2
[3] Crosslight Software, 2012 (www.crosslight.com)
[4] S. Gehrsitz et al., J. Appl. Phys. **87** (2000) 7825
[5] W. Both et al., J. Therm. Anal. **37** (1991) 61

Thermal Simulation of GaAs-Based Midinfrared Quantum Cascade Lasers

Y. B. Shi, Z. Aksamija, and I. Knezevic

Department of Electrical and Computer Engineering, University of Wisconsin
1415 Engineering Dr, Madison, WI 53706-1691, USA
Email: yshi9@wisc.edu, knezevic@engr.wisc.edu

I. INTRODUCTION

One of the limiting factors for the room-temperature continuous-wave (RT-cw) operation of quantum cascade lasers (QCLs) is the high temperature in the active region that stems from the high electrical power and poor heat extraction [1]. In order to simulate the thermal behavior of QCLs, the heat diffusion equation with appropriate source and boundary conditions needs to be solved. However, the heat generation rate of the active region under a given bias is both space- and temperature-dependent. In this paper, we present a method of extracting the heat generation rate by recording the electron-optical phonon scattering during the ensemble Monte Carlo (EMC) simulation of electron transport under different temperatures. The extracted nonlinear heat source together with appropriate thermal conductivity models enable self-consistent calculation of temperature distribution throughout QCLs. We apply the thermal model to investigate the cross-plane temperature distribution of a 9.4 μm infrared GaAs-based QCL [2]. The nonlinear effects stemming from the temperature dependence of thermal conductivity and the heat generation rate are studied.

II. THERMAL MODEL

The heat generation in the active region of QCLs originates from the optical phonons emitted due to phonon-assisted intra- and intersubband electronic transitions. These optical phonons with negligible group velocities then decay into acoustic phonons that are efficient at diffusing heat [3], [4]. The EMC simulation of electron transport provides statistic information on the scattering processes, from which the heat generation rate term Q in the heat diffusion equation (1) can be derived by Eq. (2) [5]:

$$-\nabla \cdot (\kappa \nabla T) = Q, \tag{1}$$

$$Q = \frac{n}{N_{sim} t_{sim}} \sum (\hbar\omega_{ems} - \hbar\omega_{abs}), \tag{2}$$

where n is the electron density. N_{sim} and t_{sim} are the number of particles and the simulation time, respectively. $\hbar\omega_{ems}$ and $\hbar\omega_{abs}$ are the energies of the emitted and absorbed optical phonons. Each emitted (absorbed) phonon is recorded and its energy added to (subtracted from) the sum on the right-hand side of Eq. (2). In addition, the temperature dependence of the heat generation rate can be captured by interpolating the results of a set of EMC simulation runs under different temperatures.

In QCLs, the wavelike behavior of electrons in the cross-plane direction due to confinement dictates that only a probability density of finding an electron at a given position can be known from its wave function. After the interaction with an optical phonon, an electron may end up in another subband with a completely different probability distribution. On the other hand, embedding the information on the optical phonon generation into the heat source term using Eq. (2), an exact position of each optical phonon is required. We propose a method to translate the space distribution of the optical phonons by introducing an additional random number r (uniformly distributed between [0, 1]) in the EMC simulation to determine the electron's "position" in a subband [6]. The electron in subband α is considered to be within the ith grid cell of width Δz, i.e. it is considered to be within the spatial interval $[z_i - \Delta z/2, z_i + \Delta z/2]$ if and only if

$$\int_0^{z_i - \Delta z/2} |\Psi_\alpha(z)|^2 dz < r < \int_0^{z_i + \Delta z/2} |\Psi_\alpha(z)|^2 dz. \tag{3}$$

When an electron transitions from subband α to α' by scattering with an optical phonon, two random numbers are used to find the electron's position in the initial and final subband, respectively, according to Eq. (3) and the position where the phonon is emitted or absorbed (z_{ph}) is found as their average.

We apply the thermal model to a GaAs/Al$_{0.45}$Ga$_{0.55}$As mid-infrared QCL designed for emission at 9.4 μm [2]. Temperature-dependent thermal conductivities of different layers in the device are taken into account by using an analytical model [7]. Fig. 1 shows the temperature dependence of the thermal conductivites of different layers, together with the average heat generation rate in a stage as extracted from EMC simulation. The top panel of Fig. 2 shows the subband energy levels and wavefunction moduli squared in a single stage at the threshold field (48 kV/cm) at 300 K. The three bold red lines, from top to bottom, denote the upper lasing level, the lower lasing level, and the ground level, respectively. The bottom panel shows the space distribution of the net generated optical phonons (in 30 ps) as obtained using the proposed method.

Fig. 3 shows the temperature distribution of the GaAs/Al$_{0.45}$Ga$_{0.55}$As QCL mounted epitaxial-side onto a copper heat sink at $T_0 = 300$ K calculated using (1) temperature-dependent (TD) heat generation rate and TD thermal conductivities, (2) constant active region thermal conductivity evaluated at T_0, (3) constant heat generation rate at T_0, and

(4) both constant thermal conductivities and constant heat generation rate at T_0. The results show that nonlinearity of the heat generation rate plays an important role in the accuracy of the calculated lattice temperature, while the temperature dependence of thermal conductivity of the active region should not be neglected in a rigorous calculation.

III. CONCLUSION

We presented a self-consistent thermal model for QCLs that extracts the nonuniform and temperature-dependent heat generation rate in the active region through recording the electron-optical phonon scattering events during the EMC simulation of electron transport. A GaAs/Al0.45Ga0.55As mid-IR QCL was investigated using the model. The results show that nonlinearity of the heat generation rate plays an important role in the accuracy of the calculated temperature, while the temperature-dependence of thermal conductivity of the active region cannot be neglected in a rigorous calculation.

ACKNOWLEDGMENT

This work has been supported by the Wisconsin Alumni Research Foundation and the AFOSR (grant No FA9550-09-1-0230).

REFERENCES

[1] M. Beck, D. Hofstetter, T. Aellen, J. Faist, U. Oesterle, M. Ilegems, E. Gini, and H. Melchior, "Continuous wave operation of a mid-infrared semiconductor laser at room temperature," *Science*, vol. 295, no. 5553, pp. 301–305, 2002.

[2] H. Page, C. Becker, A. Robertson, G. Glastre, V. Ortiz, and C. Sirtori, "300 K operation of a GaAs-based quantum-cascade laser at $\lambda \approx 9 \ \mu m$," *Applied Physics Letters*, vol. 78, no. 22, pp. 3529–3531, 2001.

[3] D. Vasileska, K. Raleva, and S. M. Goodnick, *Heating Effects in Nanoscale Devices, Cutting Edge Nanotechnology*, D. Vasileska, Ed. InTech, 2010.

[4] P. G. Klemens, "Anharmonic decay of optical phonons," *Phys. Rev.*, vol. 148, pp. 845–848, Aug 1966.

[5] E. Pop, S. Sinha, and K. Goodson, "Heat generation and transport in nanometer-scale transistors," *Proceedings of the IEEE*, vol. 94, no. 8, pp. 1587 –1601, Aug. 2006.

[6] Y. Shi, Z. Aksamija, and I. Knezevic, "Self-consistent thermal simulation of GaAs/Al$_{0.45}$Ga$_{0.55}$As quantum cascade lasers," *Journal of Computational Electronics*, vol. 11, pp. 144–151, 2012.

[7] S. Adachi, *GaAs and related materials: bulk semiconducting and superlattice properties*. World Scientific, 1994.

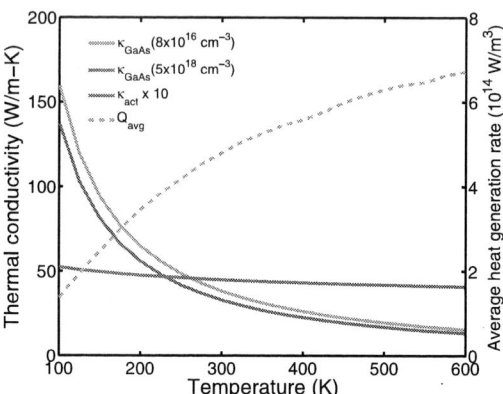

Fig. 1. Thermal conductivities of the active region and the GaAs cladding layers and the average heat generation rate in a stage as a function of temperature.

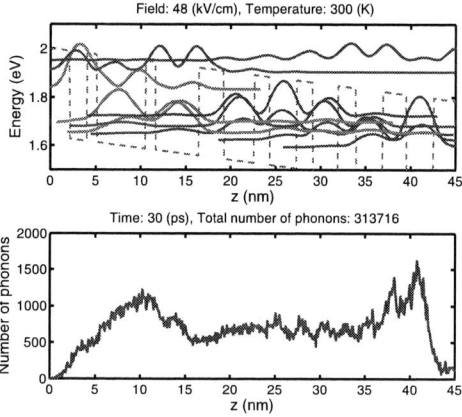

Fig. 2. A schematic conduction-band diagram of a QCL stage (top) and the real-space distribution of the generated optical phonons during the EMC simulation (bottom).

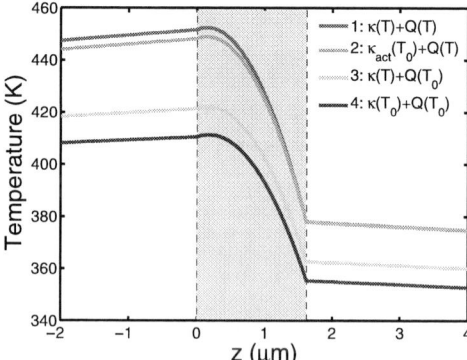

Fig. 3. Lattice temperature distribution of the QCL calculated based on (1) TD thermal conductivities and TD heat generation rate, (2) constant active region cross-plane thermal conductivity evaluated at the heat sink temperature $T_0 = 300$ K, (3) constant heat generation rate at T_0, and (4) constant thermal conductivities and heat generation rate at T_0. The shaded area marks the active region, while the white regions are the cladding layers.

On the Line Form and Natural Linewidth; Simulation and Interpretation of Experiments

M.G. Noppe

Novosibirsk State Technical University, 20 K. Marx Ave., 630092, Novosibirsk, Russia, noppe@ieee.org

Abstract - A new formula for line form inside and outside laser is derived. The linewidth is calculated on the basis of the derived formula for the line form. Our simulation of the linewidth for three Fabry-Perot lasers allows explaining all known to us experimental measurements of semiconductor laser natural linewidth.

I. INTRODUCTION

The problem of natural linewidth in semiconductor lasers is of fundamental importance both theoretically and practically (narrow linewidth lasers are in great demand in coherent optical communication systems [1]). The measurements of natural linewidth for high power lasers: GaInAs/GaInAsP (SCH), (QW), (DFB LDs) lasers [1], gain-guided V-groove laser and oxide stripe laser [2], DFB Lasers [3-5], VCSEL laser [6], **microcavity laser** [7] and Fabry-Perot lasers [8-9] have shown various functional deviations from Schawlow-Townes formula for natural linewidth Δv for one mode (the main of which is minimum of function $\Delta v = F(1/P)$, P is output power). The primary goal is to develop a theoretical model which allows explaining all known experimental measurements of semiconductor laser natural linewidth and simulation of three semiconductor lasers. Another goal is to calculate line forms versus frequency for some values of power.

II. A NEW FORMULA for the LINE FORM

Our approach to the problem is based on the following basic concepts: a laser model distributed in space; an experimentally measured dependence of the refractive index of the laser active zone on the electric field intensity E. From experiments [10] it follows that

$$n = n_0 + n_1(\omega)|E| + n_2(\omega)E^2. \quad (1)$$

The formula (1) will be used for deriving the line form and natural linewidth of the mode, basing on the representations of t creating a laser line form as a result of superposition of spontaneous radiation on laser radiation studied by a special method of the spectral theory of stochastic processes [11] which is developed in the present paper. In our paper, the formula for the line form for radiation inside the laser is derived as

$$L(\omega) = (1/2\pi) \int_{-\infty}^{\infty} \exp(-i(\omega - \omega_0)\tau -$$

$$-A_p|\tau| - B_A\tau^2 - C_A|\tau|^3)d\tau, \quad (2)$$

where $A_P = A_s^2/4E^2\Delta t$, $B_A = (\varphi A_s)^2/8$, $C_A = (\varphi A_s)^2/12\Delta t$, (3)

where A_s is a spontaneous fluctuation amplitude; Δt is the average time between two acts of emission of spontaneous fluctuations, m is the number of mode; for Fabry-Perot lasers

$$\varphi - \frac{c\,\pi m(n_1 + 2n_2|E|)}{L_z(n_0 + n_1|E| + n_2E^2)^2}, \quad (1)$$

where L_z is the length of laser, c the light velocity. A formula for laser form, differing from (2) in that $B_A = 0$, was proposed in [11]. But the calculations made in the paper have shown that B_A defining the peak fluctuations influences the natural linewidth. Hence it is impossible to neglect B_A. If $\varphi = 0$, a general expression for the line form (2) takes the form of Lorentz line, where natural linewidth is $\Delta\omega = A_s^2/2\Delta tE^2$; it is a modification of Schawlow-Townes formula [12]. The calculations made for one model of transmission coefficient have shown an insignificant change in the natural linewidth and the line center outside the laser. Therefore the line form looks like (2). Formulas for parameters (3) are derived.

III. CALCULATIONS of LINE FORM and NATURAL LINEWIDTH

The intention in this section is to simulate the natural linewidth Δv versus inverse power (1/P) for three lasers (see Fig.1, 3): possessing very small Δv (SCH laser, see Fig.3. in [1]) and possessing very large Δv (V-groove and oxide stripe lasers, see Fig.2. in [2]). Another goal is to calculate line forms versus frequency $v = (\omega - \omega_0)/2\pi$ for some values $1/P_k$ (see Fig.2). We present some calculations for Fabry-Perot (p-GaAs) - laser with

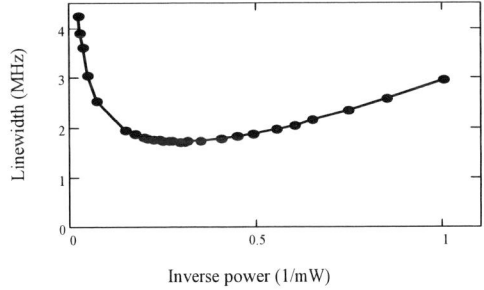

Fig.1. Calculated natural linewidth Δv versus inverse power (1/P) simulates experimental natural linewidth (SCH laser, see Fig.3 in [1])

the following model parameters: C is the acceptor with concentration $N_a=2 \cdot 10^{18} cm^{-3}$; T=293K; $(L_x:L_y:L_z)=$ (0.2µm:20µm:350µm); $R_1=0,99$; $R_2=0,01$. Selecting other parameters, we can present the results of the experimental natural linewidth simulation (see Fig.3 in [1] and Fig.2 in [2]) in Fig.1, 3 respectively. The line forms $L(v)$ versus frequency v for some values (1/P_k) from Fig.1 are transformed into one amplitude. These are presented in Fig.2.

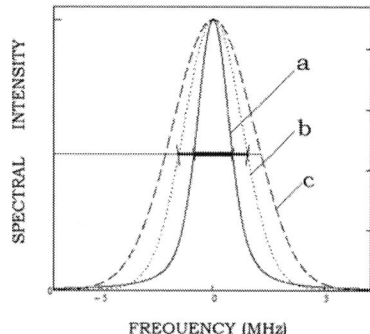

Fig.2. Calculated laser line forms versus frequency $v=(\omega-\omega_0)/2\pi$ for some values 1/P_k (from Fig.1): a) (1/P_a)=0,3(mW^{-1}); $\Delta v_a = 1,72$ (MHz);

b) (1/P_b)=0,05(mW^{-1}); $\Delta v_b = 3,05$ (MHz);

c) (1/P_c)= 0,025(mW^{-1}); $\Delta v_c = 4,24$ (MHz).

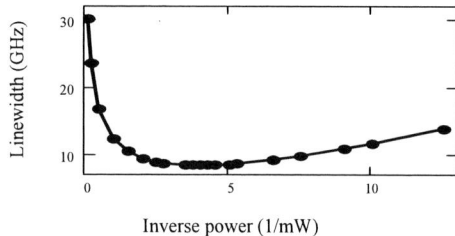

Fig.3. Calculated natural linewidth Δv versus inverse power (1/P) simulates experimental natural linewidth (see Fig.2 in [2]).

Comparison of the calculated natural linewidth (Fig.1,3) with the experimental natural linewidth (see Fig.3 in [1] and Fig.2 in [2]) shows their good fit.

IV. EXPLANATION of EXPERIMENTS

1. Thus, simulation of experimental measurements of natural linewidth for the Fabry-Perot semiconductor laser has been conducted and presented in Fig.1, 2. In view of generality of physical background for deriving (2) for different types of lasers it is possible to ascertain qualitative generality of the obtained simulation results of Fabry-Perot lasers for other types of lasers. It allows us to explain the behavior of all known experimental measurements of semiconductor laser natural linewidth which have a minimum [1-8] within the framework of our theoretical model.

2. We can explain the minimum, basing on the formulas (2) – (4) derived: the influence of phase perturbations of spontaneous radiation decreases with the growth of power (see (3) for A_p); the influence of amplitude perturbations of spontaneous radiation increases with the growth of power (see (3) for B_A); their influence is identical for some (1/P)$_{min}$.

3. Our model may serve as a basis for simulation of experimental measurements of semiconductor laser natural linewidth. The first examples of simulation Δv =F(1/P) (small Δv [1] and large Δv [2]) are presented in this paper. Comparison of the calculated natural linewidth with the experimental natural linewidth shows their close agreement.

4. Recommendations for the decrease of Δv on the basis of the developed theory and simulation conducted have been worked out. The potential for elaborating further recommendations has not been exhausted.

5. We suggest that experimenters measure $L(v)$ and compare it with the calculated $L(v)$ by way of (2).

REFERENCES

[1] S. Takano, T.Sasaki, H.Yamada, M.Kitamura, I.Mito, "Sub-MHz spectral linewidth in 1.5µm separate-confinement heterostructure (SCH), quantum-well, DFB LDs", *Elect. Let.*, vol.25, pp.356-358, 1989.

[2] W. Elsasser, E.O. Gobel, J. Kuhl. "Coherence properties of gain- and index-guided semiconductor lasers", *IEEE Journal of Quantum Electronics,* vol.19, pp.981-985, 1983.

[3] F.-J. Vermersch, V. Ligeret, S. Bansropun, M. Lecomte, O. Parillaud, M. Calligaro, M. Krakowski, G. Giuliani "High-Power Narrow Linewidth Distributed Feedback Lasers With an Aluminium-Free Active Region Emitting at 852 nm", *IEEE Phot. Tech. Lett.*, vol. **20**, pp.1145-1147, 2008.

[4] S. Spießberger, M. Schiemangk, A. Wicht, H. Wenzel,O. Brox, G.Erbert, "Narrow Linewidth DFB Lasers Emitting Near a Wavelength of 1064 nm", *J. of Light. Tech.*, vol. **28**, pp.2611-2616, 2010.

[5] Y.Inaba, M.Kito, T.Takizava, H.Nakayama et al., "Reduced Spectral Linewidth in High Output Power DFB Lasers", *Semiconductor Laser Conference*, pp.85-86, 2000. 07803-6259-4/00/$10.00©2000 IEEE

[6] T. Fordell, A.M. Lindberg, "Experiments on the linewidth-enhancement factor of vertical-cavity surface-emitting laser", *IEEE J. of Quantum Electronics*, vol.43, pp.6-15, 2007.

[7] M.Baghery, M.H. Shin, S.J Choi., J.D.O'Brien, P.D.Dapkus, "Microcavity Laser Linewidth Close to Threshold", IEEE J. of Quantum Electronics **45,** 945-949, 2009.

[8] P.G. Eliseev, "Switching lasers", In *Proc. of P.N. Lebedev Phys. Inst.*" **v**216. 3-55, 1992 (in Russian).

[9] D. Welford, A. Mooradian, "Output power and temperature dependence of the linewidth of single frequency cw (GaAl)As diode lasers", *Phys. Lett.*, v40, pp.865-867, 1982.

[10] A. Partovi, E.Garmire, "Band-edge photorefractivity in semiconductors: Theory and experiment", *J. Appl.Phys*, v.69, pp.6885-6897, 1991.

[11] A.P. Kovalevski, M.G. Noppe "On New Formulas for Line Form and Natural Linewidth in Semiconductor Lasers." In *Proc. of IX Intern. Conf. on Actual Problems of Electronic Instrument Engineering Proceedings (APEIE – 2008), NSTU, Novosibirsk,* 2008, v.5, pp.32-37. 978-1-4244-2825-0/08/$25.00©2008 IEEE

[12] L.A. Rivlin, A.T. Semenov, S.D. Jakubovich, *The dynamics and spectrums of radiation of semiconductor lasers*, Radio and Communication, Moskow., 1983, 208p. (in Russian).

Simulation of a Ridge-Type Semiconductor Laser for Separate Confinement of Horizontal Transverse Modes and Carriers

Hiroki Kato, Hazuki Yoshida, and Takahiro Numai
Graduate School of Science and Engineering, Ritsumeikan University
1-1-1 Noji-Higashi, Kusatsu, Shiga 525-8577, Japan
Phone: +81-77-561-5161 E-mail: numai@se.ritsumei.ac.jp

1. Introduction

High power 980-nm semiconductor lasers are indispensable for pumping sources of erbium doped optical fiber amplifiers [1]. Generally, 980-nm semiconductor lasers have ridge structures so as not to expose their active regions to air during their fabrication, because the active regions are easily oxidized and degraded in air. In the ridge structures, higher-order transverse modes as well as the fundamental transverse mode are confined. As a result, with an increase in injected current, higher-order transverse modes lase; kinks appear in their current versus light-output (*I-L*) curves [2]. These kinks are attributed to changes in the local gain profile and refractive index owing to spatial hole burning, the free-carrier plasma effect, and heating. To obtain high fiber-coupled optical power, semiconductor lasers with high kink levels operating in the fundamental transverse mode are required. To date, to increase kink levels, coupling of the optical field to the lossy metal layers outside the ridge [3], highly resistive regions in both sides of ridge stripe [4], and incorporation of a graded V-shape layer [5] have been demonstrated. To increase kink level and decrease the threshold current further, a ridge structure with optical antiguiding layers have been proposed [6] , [7], but the fabrication process is fairly complicated.

In this paper, a ridge-type semiconductor laser with selectively proton-implanted cladding layers is proposed to make the fabrication process more simple, increase kink level, and decrease threshold current. In this semiconductor laser, horizontal transverse modes are confined by the distribution of the effective refractive index; carrier distributions are controlled by selectively proton-implanted cladding layers. The proton-implanted cladding layers have higher refractive index than p-/n- cladding layers by appropriate annealing, because concentrations of the free carriers in the proton-implanted regions are low.

From simulations, it is found that the kink level is higher and the threshold current is lower than those of the ridge-type semiconductor lasers with optical antiguiding layers for horizontal transverse modes [6], [7]. In addition, it is revealed that they are optimal when the space between the proton-implanted regions in the p-cladding layer S_p is 1.3 μm.

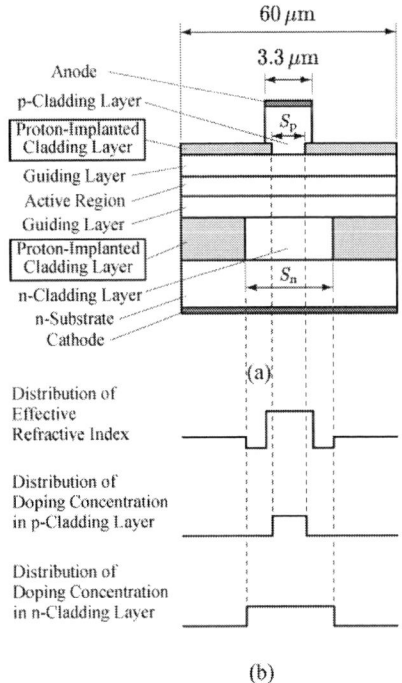

Fig. 1 (a) Schematic cross-sectional view of a proposed ridge structure with selectively proton-implanted cladding layers and (b) distributions of the effective refractive index of the semiconductor laser and doping concentration of the cladding layers. The shaded areas are undoped. Here, S_p is the space between the proton-implanted regions in the p-cladding layer; S_n is the space between the proton-implanted regions in the n-cladding layer.

2. Laser Structures and Simulations

Figure 1 (a) shows a schematic cross-sectional view of a proposed ridge structure with selectively proton-implanted cladding layers and (b) distributions of the effective refractive index of the semiconductor laser and doping concentration of the cladding layers. The shaded areas are undoped. Here, S_p is the space between the proton-implanted regions in the p-cladding layer; S_n is the space between the proton-implanted regions in the n-cladding layer. The p-cladding layer is 50 nm thick; the n-cladding layer is 1.5 μm thick. As a result, the distribution of the refractive index of the n-cladding layer contributes to the effective refractive index; the contribution of the p-cladding layer to the effective refractive index is negligible. Rectangular mesa is 1.55 μm high and 3.3 μm wide. The relatively wide mesa is selected in order to obtain high light-output power. The base is 60 μm wide, and the cavity is 1200 μm long. Reflectivities of the front and rear facets are 2 and 90%, respectively.

Layer parameters such as band gap energy, refractive index, thickness, electron effective mass, hole effective mass, and doping concentration are the same as those described in Refs. 6 and 7. Lasing characteristics are simulated by using a device simulation software, ATLAS (Silvaco).

3. Simulation Results and Discussions

Figure 2 shows a kink level as a function of the space S_p between the proton-implanted regions in the p-cladding layer. The parameter is the space S_n between the proton-implanted regions in the n-cladding layer. The kink level has a maximum value of 603 mW at S_p =1.3 μm and S_n =2.3 μm. This kink level is 1.9 times as high as that in Ref. 7.

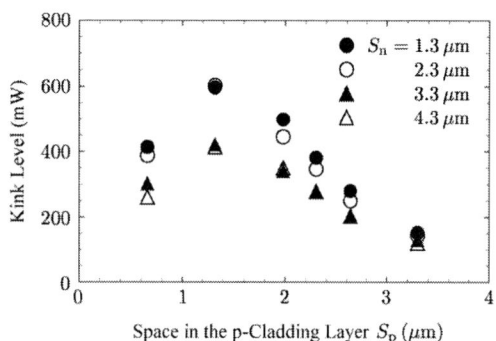

Fig.2 Kink level as a function of the space S_p between the proton-implanted regions in the p-cladding layer. The parameter is the space S_n between the proton-implanted regions in the n-cladding layer.

Figure 3 shows the threshold current for the fundamental transverse mode as a function of the space S_p between the proton-implanted regions in the p-cladding layer. The parameter is the space S_n between the proton-implanted regions in the n-cladding layer. The threshold current I_{th} has a minimum value of 46.3 mA at S_p =1.3 μm and S_n =4.3 μm. This value of the threshold current is 99 % of the previous result in Ref. 7.

There is the optimum value for S_p to obtain the lowest threshold current. The reason for this is that the lateral spreading of the carriers is suppressed most efficiently at S_p =1.3 μm. In this structure, it is considered that lateral spreading of the carriers is determined by the ambipolar lateral diffusion by the narrower space in the cladding layers S_p. The am-

bipolar lateral diffusion length L_a = 1.35 μm, which is quite similar to S_p =1.3 μm. Here, in $In_{0.2}Ga_{0.8}As$ QW active layers, it is assumed that the mobility of electrons is μ_e = 8.00×10^3 cm^2 V^{-1}s^{-1}; the mobility of holes is μ_h = 3.70×10^2 cm^2 V^{-1} s^{-1}. The lifetime τ_n of both electrons and holes is supposed to be 1.00×10^{-9} s.

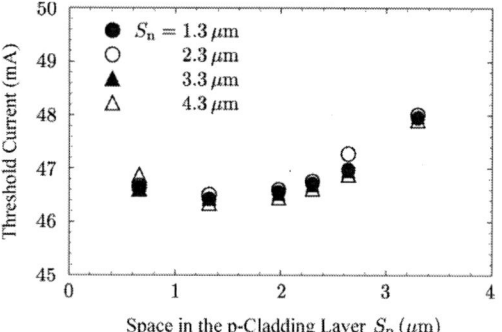

Fig.3 Threshold current I_{th} for the fundamental transverse mode as a function of the space S_p between the proton-implanted regions in the p-cladding layer. The parameter is the space S_n between the proton-implanted regions in the n-cladding layer.

4. Conclusions

To improve kink levels and decrease threshold current in ridge-type semiconductor lasers, a novel ridge structure with selectively proton-implanted cladding layers was proposed and simulated. It is found that kink level was 1.9 times as high as that in the previous result. The lowest threshold current was 99 % of the previous result.

References

[1] C. S. Harder, L. Brovelli, H. P. Meier, and A. Oosenbrug, *Proc. Optical Fiber Communication Conf.* '97, February, FC1, 1997, p.350.

[2] M. F. C. Schemmann, C. J. van der Poel, B. A. H. van Bakel, H. P. M. M. Ambrosius, A. Valster, J. A. M. van den Heijkant, and G. A. Acket, *Appl. Phys. Lett.*, vol.66, pp.920-922 (1995).

[3] M. Buda, H. H. Tan, L. Fu, L. Josyula, and C. Jagadish, *IEEE Photonics Technol. Lett.*, vol.15, **pp.**1686-1688 (2003).

[4] M. Yuda, T. Hirono, A. Kozen, and C. Amano, *IEEE J. Quantum Electron.*, vol.40, pp.1203-1207 (2004).

[5] B. Qiu, S. D. McDougall, X. Liu, G. Bacchin, and J. H. Marsh, *IEEE J. Quantum Electron.*, vol.41, pp.1124-1130 (2005).

[6] N. Shomura, M. Fujimoto, and T. Numai, *IEEE J. Quantum Electron.*, vol.44, pp.819-825 (2008).

[7] H. Yoshida and T. Numai: Jpn. J. Appl. Phys., vol.48, 082105, 2009.

978-1-4673-1602-6/12 $31.00 © 2012 IEEE

Efficient Simulation Method for DFB Lasers with Large Gain Saturation Effect

Yanping Xi[1], *Member, IEEE,* Lin Han[2], Xun Li[2], *Senior Member, IEEE*

[1]Wuhan Nantional Laboratory for Optoelectronics, Huazhong Universtiy of Science and Technology
Wuhan, Hubei, China 430074

[2]Department of Electrical & Computer Engineering, McMaster University,
Hamilton, Ontario, Canada L8S 4K1

Abstract—In this paper, we proposed a nonlinear approximation scheme to the time-dependent longitudinal carrier distribution with a view toward finding the efficient solution scheme of the previously proposed standing-wave model for simulation of DFB lasers. It shows its advantage over the existing linear approximation scheme in cases where the optical power is large enough to trigger the saturation effect. The validity of this improved scheme is demonstrated by simulation example of $\lambda/4$-shifted DFB lasers.

I. INTRODUCTION

In the description of DFB lasers, the accurate analysis of the longitudinal spatial hole burning (LSHB) effect is thought to be indispensible[1]. Several longitudinal one-dimensional models have been reported with the complex LSHB effect included[2-5]. The standing-wave model (SWM-CCM) in [5], utilizes the mode expansion technique to avoid the spatial discretization of the optical field, and consequently shows its higher efficiency in modeling DFB lasers. By linearly linking the shape of the longitudinal optical field to that of the carrier, Kinoshita[6] proposed an idea to "globally" approximate the carrier distribution with only two scalar quantities: one measures the average; the other represents the inhomogeneity of the distribution. A significant improvement of the computational efficiency of the SWM was also achieved by using this splitting idea[7]. However, the inhomogeneous part of the carrier distribution is assumed to linearly change with that of the optical field distribution in [6,7], which is not valid when the nonlinear gain saturation effect is large. To remedy this problem, we propose in this paper a nonlinear approximation to the carrier distribution for describing the large nonlinear interaction between the optical field and the carrier density distributions.

II. THEORY

The time-dependent carrier densities at different longitudinal positions are normally solved by the carrier rate equation

$$\frac{dN(z,t)}{dt} = \frac{I}{eV} - \frac{N(z,t)}{\tau_c} - \upsilon_g \Gamma g(z,t) P(z,t) \quad (1)$$

where I the injected current, V the active region volume of the laser diode, e the electron charge, τ_c effective carrier life time and $P(z,t)$ the photon density distribution. For the SWM in [6], the carrier density evolution at different position inside the laser cavity is still obtained by discretizing the cavity into a number of spatial segments. Considering that the longitudinal carrier distribution normally has an opposite pattern from that of the photon distribution, $N(z,t)$ is linearly approximated by a deviation shape function $f(z,t)$ extracted from $P(z,t)$ in [7] to reduce the complexity of the existing SWM, i.e. if we define the time-dependent average photon density along the laser cavity as

$P_{av}(t)$, the photon density distribution can be expressed by $P(z,t) = P_{av}(t)\left[1+f(z,t)\right]$, with $\int_0^L f(z,t)\,dz = 0$. The carrier distribution can then be approximated by

$$N(z,t) = N_a(t) - D(t)f(z,t). \quad (2)$$

where $N_a(t)$ and $D(t)$ measure the average and the inhomogeneity of the distribution, respectively. Equation (2) is valid by assuming the moderate optical power inside the cavity, therefore is no longer an appropriate approximation for the DFB laser if nonlinear saturation starts to kick on. In such cases, the carrier profile should be modeled by a nonlinear relationship in terms of the shape function $f(z,t)$. We thus proposed the following approximation to the carrier density distribution

$$N(z,t) = \frac{x(t)}{1+f(z,t)} - \frac{y(t)}{\left[1+f(z,t)\right]^2} + N_0. \quad (3)$$

The function $f(z,t)$ indicates the carrier non-uniformity and is determined mainly by parameters of the laser structure such as κL and facet reflectivities, and hence it is reasonable to take $f(z,t)$ to be slowly varying compared with time-dependent coefficient $x(t)$ and $y(t)$. As such, when (3) is applied, Equation (1) is reduced to

$$\frac{dx(t)}{dt} = \left(\frac{I}{eV} - \frac{N_0}{\tau_c}\right)\left(2+\frac{\theta_3}{\theta_2}\right) - \frac{x(t)}{\tau_c}$$
$$-\upsilon_g \frac{\Gamma g_N P_{av}(t)}{1+\varepsilon P_{av}(t)}\left[\left(2+\frac{\theta_3}{\theta_2}\right)x(t) - y(t)\right] \quad , \quad (4)$$

$$\frac{dy(t)}{dt} = \left(\frac{N_0}{\tau_c} - \frac{I}{eV}\right)\frac{\theta_2}{\theta_f} - \frac{y(t)}{\tau_c} + \upsilon_g \frac{\Gamma g_N P_{av}(t)}{1+\varepsilon P_{av}(t)}\left[\frac{\theta_2}{\theta_f}x(t)\right], \quad (5)$$

where $\theta_2 = \frac{1}{L}\int_0^L f^2\,dz$, $\theta_3 = \frac{1}{L}\int_0^L f^3\,dz$, $\theta_4 = \frac{1}{L}\int_0^L f^4\,dz$ and $\theta_f = \frac{1}{L}\int_0^L \frac{f}{f+1}\,dz$. Along with the governing equation for the complex amplitude of each longitudinal mode in the SWM[6], i.e.

$$\frac{d\mathbf{A}_m(t)}{dt} = p'_{mm}(t)\mathbf{A}_m(t) + \sum_{n=1(n\neq m)}^{K} p'_{mn}(t)\mathbf{A}_n(t) + \tilde{\eta}_m(t), \quad \text{m=1,2,..K,} \quad (6)$$

the DFB laser can be simulated in a selfconsistant manner. K is the number of eigenmodes used in the field expansion. More details of (6) can be found in [5,7].

III. SIMULATION RESULTS

In a $\lambda/4$-shifted DFB laser, the carrier profile becomes highly non-uniform due to the severe LSHB effects in the presence of strong coupling strength and relatively high injection level, and is thus chosen to be the simulation example.

978-1-4673-1602-6/12 $31.00 © 2012 IEEE

The SWM-CCM without the carrier approximation is treated as our benchmark and denoted as Scheme I. The SWM-CCM with linear carrier approximation [8] is denoted as Scheme II and the proposed one with nonlinear carrier approximation is designated as Scheme III.

we define the error of the carrier distribution as

$$\varepsilon_N = \left(\sqrt{\sum_{z_i \in [0,L]} \left| \tilde{N}(z_i) - \hat{N}(z_i) \right|^2} \bigg/ \sqrt{\sum_{z_i \in [0,L]} \left| \hat{N}(z_i) \right|^2} \right) \times 100\% \quad , \text{where} \quad \tilde{N}(z_i)$$

denotes the approximated carrier distribution obtained by scheme II or III, and $\hat{N}(z_i)$ represents the benchmark calculated by scheme I. Fig. 3 plots the change of ε_N with the bias level for different normalized coupling coefficients. As is expected, the simulation results show that the proposed scheme III gives a more accurate description of the carrier distribution than scheme II when the bias exceeds certain critical level, where the nonlinear saturation starts to take effect. With the increase of the coupling strength, the range of the injection level where the scheme III prevails is increasing. And the improvement of the accuracy is more pronouncing for lasers with large κL, i.e. with more severe nonlinear saturation effects.

One of the most important characteristics of the later is the output power obtained at the laser facets. Since both of the facets of the structure under investigation are AR-coated, the steady-state output power in our simulation is extracted at one end when the transient settles down. The error of the output power is defined as $\varepsilon_P = \left| \left(P_{II(\text{or }III)} - P_I \right) \big/ P_I \right| \times 100\%$, where P_I, P_{II} and P_{III} denote the output power obtained by scheme I, II and III, respectively. The change of this error with the coupling strength for a given bias is shown in Fig. 4. Similar performances are observed around $\kappa L = 1.25$, while scheme III outperforms scheme II when the normalized coupling strength is away from this value. This is due to the fact that the optical field is relatively uniform when $\kappa L = 1.25$, and becomes non-uniform enough to jeopardize the assumption of the linear approximation when the peaks of the optical field reach the saturation level for the laser with undercoupling or overcoupling.

The proposed scheme is of equal importance and can be viewed as a complementary scheme to scheme II. The detailed comparisons of computation complexity between three schemes are listed in Table I.

IV. Summary

This paper presents an approximation to the carrier distribution which is nonlinearly related to the photon density profile. The proposed scheme is demonstrated to be effective in simulating DFB lasers with large saturation effects through comparing with the benchmark scheme as well as the existing linearly approximated scheme.

References:

[1] J. E. A. Whiteaway, B. Garretty, G. H. B. Thompson, A. J. Collar, C. J. Armistead and M. J. Fice, *IEEE J. Quantum Electron.* 28 (1992) 1277.

[2] P. Vankwikelberge, G. Morthier, and R. Baets, *IEEE J. Quantum Electron.* 26 (1990) 1728.
[3] B. -S. Kim, Y. Chung, and J. -S. Lee, *IEEE J. Quantum Electron.* 36 (2000) 787.
[4] H. Wenzel, U. Bandelow, H. -J. Wünsche, and J. Rehberg, *IEEE J. Quantum Electron.* 32(1996) 69.
[5] Y. Xi, X. Li and W. -P. Huang, *IEEE J. Quantum Electron.* 44 (2008) 931.
[6] J. Kinoshita and K. Matsumoto, *IEEE J. Quantum Electron.* 24 (1988) 2160.
[7] Y. Xi, W. –P. Huang and X. Li, *IEEE/OSA J. Lightwave Tech.* 27 (2009) 3227.

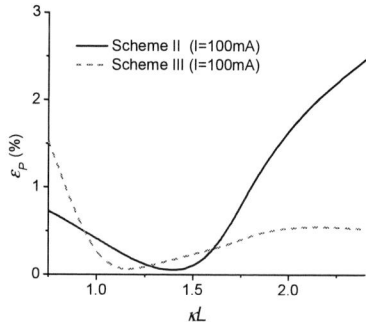

Fig.1 Variations of the error of carrier distributions with injection level for structures with different normalized coupling coefficients

Fig.2 Variations of the error of the output power with normalized coupling coefficients for scheme II and III

TABLE I

COMPARSONS OF SCHEME I, II AND III

| | Optical field | | Carrier density | | Total # of ODEs | Scope of Applications |
	Discret-ization	# of ODEs	Discret-ization	# of ODEs		
Scheme I	no	K	yes	M	$K+M$	General
Scheme II	no	K	no	2	$K+2$	Small to moderate nonlinear saturation
Scheme III	no	K	no	2	$K+2$	Moderate to large nonlinear saturation

M: The number of sub-sections divided along the laser cavity (normally in several tenths for cavity length of several hundred micrometers).

K: The number of eigenmodes used in the field expansion.

Recent progress in theory of nonlinear pulse propagation in subwavelength waveguides

(Invited Paper)

Shahraam Afshar V., Wenqi Zhang, M. A. Lohe, and Tanya M. Monro

Institute for Phtonics & Advanced Sensing, School of Chemistry & Physics, The University of Adelaide

Adelaide, SA 5005, Australia

Email: shahraam.afshar@adelaide.edu.au

Abstract—**High index subwavelength waveguides form a new platform for highly nonlinear photonic devices. This paper reviews the recent progress in the theory of nonlinear pulse propagation in these waveguides and highlights the opportunities that these waveguides have opened up in terms of active photonic devices.**

Recently there has been significant interest in design and manufacturing of high index subwavelength waveguides mainly due to their extreme nonlinearity and possible applications for all optical photonic-chip devices. Examples of these waveguides include silicon, chalcogenide, or soft glass optical waveguides, which have formed the base for three active field of studies; silicon photonics [1], chalcogenide photonics[2], and soft glass microstructured photonic devices [3].

It has recently been shown that the standard (scalar) theory of nonlinear pulse propagation (SNPP), which relies on the well-known scalar Helmholtz equation [4], can not provide accurate descriptions of nonlinear phenomena in HIS-WGs [5], [6]. We have recently reported the development of a vectorial nonlinear pulse propagation (VNPP) model that can be employed to describe the nonlinear processes in any waveguides, especially in HIS-WGs. The new VNPP indicates that the propagating modes have significant components along the direction of propagation, which causes the propagating modes to be non-transverse. Based on VNPP, new vectorially based expressions of effective nonlinear coefficient, γ and Raman gain, g_R, have been given. Based on these expressions, we predicted significantly higher values of γ [5] and g_R [6] in the HIS-WG parameter regime compared to those predicted by SNPP. We attributed these results to the large z component of the propagating modes in the subwavelength regime. Fig. 1 left shows the predictions of VNPP and SNPP for the nonlinear coefficient, γ, of a nanowire, made of chalcogenide glasses, for different core diameters. Results in Fig. 1 left demonstrate that VNPP predicts much higher values for γ, in the subwavelength regime. In an attempt to confirm these results, we have been successful in fabricating a suspended subwavelength-core fiber made of bismuth glass [7]. Using this fiber, we have not only achieved a world-record nonlinearity in microstructured optical fibers [8], γ, but also been able to confirm the prediction of VNPP model for γ of subwavelength waveguides [9], see Fig. 1 right.

The new VNPP and vectorial definition of γ, lead to a

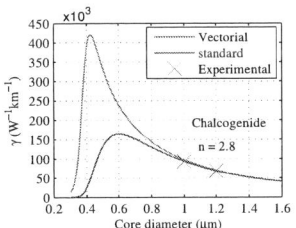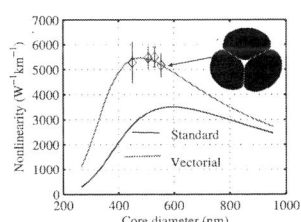

Fig. 1. (Left) γ for a chalcogenide nanowire as a function of core. (Right) γ as a function of core diameter for bismuth suspended core fibers. Experimental results are shown by diamond signs.

new regime of polarization switching which has not been observed before. According to VNPP, the interactions between polarizations of waveguide modes can be described by the following coupled nonlinear Schrödinger equations,

$$\frac{\partial A_j}{\partial z} + \sum_{n=1}^{\infty} \frac{(i)^{n-1}}{n!} \beta_{jn} \frac{\partial^n}{\partial t^n} A_j =$$
$$i(\gamma_j |A_j|^2 + \gamma_c |A_k|^2) A_j + i\gamma_c' A_j^* A_k^2 \exp(-2i\Delta\beta_{jk}) \quad (1)$$

in which $A_{j,k}$ ($j, k = 1, 2$ and $j \neq k$) are the amplitudes of the fields with two polarizations, β_{jn} are the n-th order propagation constants of the two polarizations, $\Delta\beta_{jk} = \beta_j - \beta_k$ is the linear birefringence, γ_j, γ_c and γ_c' are the effective nonlinear coefficients representing self phase modulation, cross phase modulation and coherent coupling of the two polarizations, respectively [4]. The vectorial definitions of γ_1, γ_2, γ_c, and γ_c' in VNPP model show that in general $\gamma_1 \neq \gamma_2 \neq \frac{3}{2}\gamma_c \neq 3\gamma_c'$ which is contrary to what is commonly used in SNPP. SNPP uses the approximations, $\gamma_1 = \gamma_2 = \gamma$, $\gamma_c = 2\gamma_c' = (2/3)\gamma$, which is based on the fact that (a) the waveguide material is isotropic and has only electronic-based Kerr nonlinearity, (b) the two polarized modes have same effective mode area [5]. These approximations work well with low index contrast and large dimension waveguides but are no longer appropriate for HIS-WGs. VNPP model and the resultant inequality $\gamma_1 \neq \gamma_2 \neq \frac{3}{2}\gamma_c \neq 3\gamma_c'$ indicate that optical waveguides with isotropic and electronic-based Kerr nonlinear materials can also display anisotropic nonlinearity due to the difference in the mode field distributions and the z-component (along the direction of propagation) of eigenmodes of the two polarizations.

It can be shown that Eqs. (1) lead to two classes of steady polarization states, one of which is unstable and results in polarization switching [10], [11]. We focus on this class of steady state but unstable solutions which do not exist if $\gamma_1 = \gamma_2 = 3/2\gamma_c = 3\gamma_c'$, which is the common assumption of SNPP. For these steady state and unstable solutions, polarizations with certain initial powers P_1 and P_2 and phase difference $\Delta\phi$ do not change as they propagate through a waveguide. Any small perturbations in powers or phase difference push the fields away from these steady states, however, the fields do not become chaotic, rather both the powers and phase difference are periodic functions. The period T can be expressed as a function of the waveguide parameters, initial power and phase of the input fields.

For switching solutions, the phase difference between the two polarization vectors experiences abrupt phase shifts through π as the light propagates within the waveguide. As a result, the state of polarization flips between two well-defined polarization states, where the flipping angle depends on fiber parameters and initial condition. Figure 2 shows an example of switching behavior of the polarization state for which the $a = b = 2$ with $p_{10}/(p_{10} + p_{20}) = 1/2$ and $\Delta\phi = (\phi_{10} - \phi_{20}) = 10^{-4}$. Here a and b are dimensionless parameters related to the input power and the parameters of the fiber by $a = -\Delta\beta/\gamma_c'P_0 - (\gamma_c - \gamma_2)/\gamma_c'$ and $b = (\gamma_1 + \gamma_2 - 2\gamma_c)/2\gamma_c'$ and p_{10}, ϕ_{10}, p_{20}, and ϕ_{20} are initial powers and phases of the two polarizations. This example corresponds to a linearly polarized input laser beam in which the polarization vector makes an angle of $45°$ to either of the principle axes of the waveguide. We plot $v = p1/(p_1 + p_2)$ and $\cos(\theta/2 = \Delta\phi)$ as functions of dimensionless length $\tau = 2\gamma_c'P_0z$, showing the periodicity of these functions and the switching behavior of $\cos(\theta/2 = \Delta\phi)$. Since $v_0 = 1/2$, the angular flipping of the polarization vector is $\pi/2$, because $\cos(\theta/2 = \Delta\phi)$ flips between values ± 1 as shown in the inset of Fig. 2. This in principal can lead to optical limiting or switching devices [12].

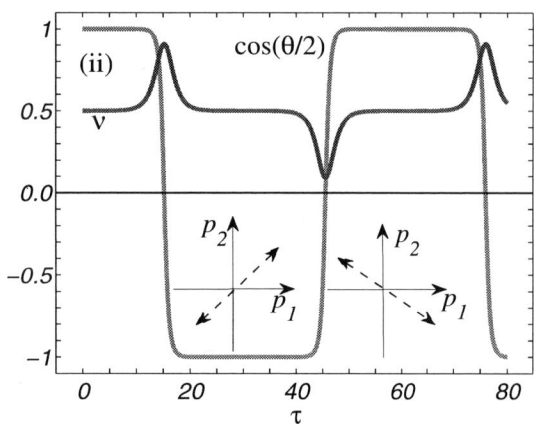

Fig. 2. (a) v and $\cos(\Delta\phi)$ as functions of normalized length τ. The polariztion state flipps by $\pi/2$ as the laser light propagates.

I. CONCLUSION

High index contrast and subwavelength dimension waveguides have opened up a new era in the field of nonlinear guided optics both in terms of fundamental theories and applications. We have presented a new vectorial nonlinear pulse propagation (VNPP) model to describe nonlinear processes in these waveguides, demonstrated significant differences between the predictions of VNPP and SNPP for Kerr nonlinear coefficient and Raman gain, γ, g_R, respectively, confirmed experimentally the prediction of VNPP for higher γ, compared to that of SNPP, and predicted a new regime of nonlinear polarization switching.

II. ACKNOWLEDGMENT

This research was supported under the Australian Research Council's Discovery Project funding scheme (project number DP110104247). Tanya M. Monro acknowledges the support of an ARC Federation Fellowship.

REFERENCES

[1] R. K. W. Lau, M. M´enard, Y. Okawachi, M. A. Foster, A. C. Turner-Foster, R. Salem, M. Lipson, and A. L. Gaeta, "Continuous-wave mid-infrared frequency conversion in silicon nanowaveguides," Opt. Lett. 36, 1263–1265, 2011
[2] B. J. Eggleton, B. Luther-Davies and K. Richardson, "Chalcogenide photonics," Nature Photonics 5, 141–148, 2011
[3] G. Qin, X. Yan, C. Kito, M. Liao, T. Suzuki, A. Mori, and Y. Ohishi, "Highly nonlinear tellurite microstructured fibers for broadband wavelength conversion and flattened supercontinuum generation," Journal of Applied Physics 107, 043108, 2010
[4] P. Agrawal, *Nonlinear Fiber Optics*, Academic press, 2007
[5] S. Afshar V. and T. M. Monro, "A full vectorial model for pulse propagation in emerging waveguides with subwavelength structures part I: Kerr nonlinearity," Opt. Express 17, 2298–2318, 2009
[6] M. D. Turner, T. M. Monro, and S. Afshar. V., "A full vectorial model for pulse propagation in emerging waveguides with subwavelength structures part II: Stimulated Raman scattering", Opt. Express 14, 11565, 2009
[7] T. M. Monro, H. Ebendorff-Heidepriem, W. Qi Zhang, and S. Afshar V., IEEE JQE, Vol 45, No 11, 2009
[8] W.Q. Zhang, S. Afshar V., H. Ebendorff-Heidepriem, T. M. Monro, "Record nonlinearity in optical fibre", Electronics Letters 44, 1453, 2008
[9] S. Afshar V., W. Zhang, H. Ebendorff-Heiperiem, and T. M. Monro, "Small core optical waveguides are more nonlinear than expected: experimental confirmation", Optics Letters, 34, 3577, 2009
[10] W. Q. Zhang, M. A. Lohe, T. M. Monro, and S. Afshar V., "Nonlinear polarization bistability in optical nanowires," Opt. Lett. 36, 588–590 (2011).
[11] S. Afshar V., M. A. Lohe, W. Q. Zhang, and T. M. Monro, "Polarization effects in a full vectorial model of pulse propagation in high index sub-wavelength waveguides", submitted to Optics Express, arXiv:1203.6167
[12] W. Q. Zhang, M. A. Lohe, T. M. Monro, and S. Afshar V., "Nonlinear self-flipping of polarization states in asymmetric waveguides", submitted to IEEE PTL, arXiv:1203.6416

Spatio-Temporal Pulse Propagation in nonlinear dispersive optical Media

Carsten Brée, Shalva Amiranashvili and Uwe Bandelow

Weierstrass Institute for Applied Analysis and Stochastics, Mohrenstr. 39, 10117 Berlin, Germany

Email: see http://www.wias-berlin.de/contact/staff

Abstract—We discuss models for the propagation of ultrashort optical pulses through nonlinear dispersive optical media. Starting from a single-mode fiber with fixed radial field structure and one propagation coordinate we turn to a full three-dimensional model for propagation of ultrashort pulses in gases.

I. INTRODUCTION

Ultrashort laser pulses have dramatically triggered both fundamental and applied science and also created new challenges from the numerical side. A straightforward solution of the underlying field and material equations becomes impractical because too different space- and time-scales are involved, and the common slowly varying envelope approximation (SVEA) is no longer valid for ultrashort pulses. Therefore new models which allow for an efficient numerical treatment have to be developed [1], [2]. We discuss several such models starting from the case of a single-mode waveguide in which the field structure in the radial direction is fixed and only one propagation coordinate is involved [3]–[7]. Thereafter we turn to the full three-dimensional modeling of propagation of ultrashort pulses in gases [8].

II. SCALAR CASE

We start with an exemplary straightforward numerical solution for an ultrashort pulse propagating in a single-mode fiber [4], Fig. 1. As shown there, the envelope structure is destroyed in the course of propagation. Another observation is that the pulse carrier frequency is shifted and therefore not well defined. Such extreme propagation regimes require more careful treatment than the traditional envelope description.

In principle, an optical pulse in a single-mode waveguide can be described by a single field component $E(\vec{r}, t)$, which, to a good approximation, is governed by a scalar wave equation

$$(\partial_z^2 + \vec{\nabla}_\perp^2)E - \frac{1}{c^2}\partial_t^2(\hat{\epsilon}E) = \mu_0\partial_t^2 P_{\mathrm{NL}}, \quad \vec{\nabla}_\perp^2 = \partial_x^2 + \partial_y^2, \quad (1)$$

where the dispersion operator $\hat{\epsilon}$ is defined in the frequency domain $(\hat{\epsilon}E)_\omega = \epsilon(\omega)E_\omega$ and P_{NL} denotes the nonlinear part of the induced polarization. We decompose the real-valued electric field $E(\vec{r}, t) = \sum_\omega E_\omega(\vec{r})e^{-i\omega t}$, $E_{-\omega} = E_\omega^*$, into the complex-valued negative- and positive-frequency parts

$$E = \sum_{\omega<0} E_\omega e^{-i\omega t} + \sum_{\omega>0} E_\omega e^{-i\omega t} = E^{(-)} + E^{(+)},$$

introduce the analytic signal $\mathcal{E}$ for the electric field $E = \mathrm{Re}[\mathcal{E}]$

$$\mathcal{E}(\vec{r}, t) = 2E^{(+)}(\vec{r}, t) = 2\sum_{\omega>0} E_\omega(\vec{r})e^{-i\omega t},$$

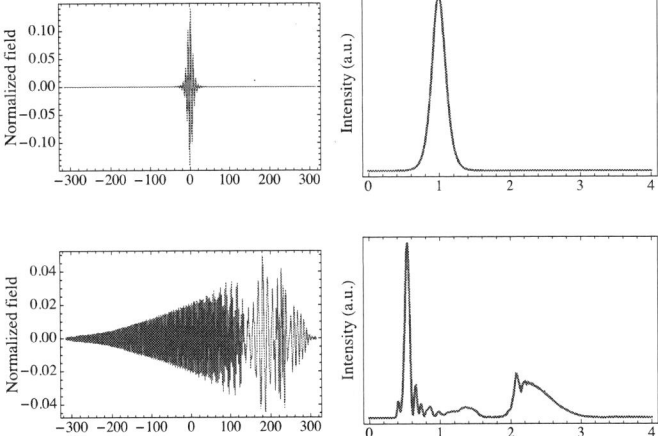

Fig. 1. Top: electric field (left) and spectrum (right) of the initial pulse. Bottom: the same after 10 ps propagation in a bulk fluoride glass. One sees that the envelope structure of the initial pulse is gradually destroyed in the course of propagation.

and obtain in the unidirectional approximation

$$\left(i\partial_z + \hat{\beta}\right)\mathcal{E} + \frac{\hat{\beta}^{-1}}{2}\vec{\nabla}_\perp^2\mathcal{E} = \frac{3\chi^{(3)}}{8c^2}\hat{\beta}^{-1}\partial_t^2\left(|\mathcal{E}|^2\mathcal{E}\right)^{(+)}. \quad (2)$$

Equation (2) is similar to the nonlinear Schrödinger equation (NSE). However, Eq. (2) is completely independent on SVEA. If the pulse can be characterized by a narrow spectrum around the carrier frequency ω_0 and the corresponding wave vector $\beta_0 = \beta(\omega_0)$, we get the equation

$$\left(i\partial_z + \hat{\beta}\right)\mathcal{E} + \frac{1}{2\beta_0}\vec{\nabla}_\perp^2\mathcal{E} + \frac{3\omega_0\chi^{(3)}}{8cn(\omega_0)}|\mathcal{E}|^2\mathcal{E} = 0,$$

which can be related to the 1D NSE for the pulse envelope ψ by a standard transformation to the pulse-comoving frame

$$\mathcal{E}(\vec{r}, t) = \mathfrak{R}(x, y)\psi(z, \tau)e^{i(\beta_0 z - \omega_0 t)}, \quad \tau = t - \beta_1 z,$$

where $\beta_1 = \beta'(\omega_0)$ is the reverse group velocity and $\mathfrak{R}(x, y)$ is the transverse mode profile [9]. A similar elimination of the radial coordinates can be appled directly to Eq. (2).

III. VECTORIAL CASE

In a homogeneous medium without (linear) waveguiding one has to account, in principle, for a fully vectorial description of the electric field $\vec{E}$, e.g., in the frequency domain by

$$[\vec{\nabla}^2 + \beta^2(\omega)]\vec{E}_\omega = \vec{S}_\omega. \quad (3)$$

The source term $\vec{S}_\omega$ is given by

$$\vec{S}_\omega = -\mu_0 \omega^2 \vec{P}_{\mathrm{NL},\omega} + i\mu_0 \omega \vec{J}_\omega + \frac{1}{\epsilon_0} \vec{\nabla}\left(\rho - \vec{\nabla}\cdot\vec{P}_\omega\right), \quad (4)$$

and takes account of the nonlinear part $\vec{P}_{\mathrm{NL},\omega}$ of the total polarization density $\vec{P}_\omega$, the existence of free carriers with density ρ and current density $\vec{J}_\omega$, respectively. The last term on the r.h.s. of Eq. (4) models vectorial effects which become important for strongly divergent beams occuring under extreme focusing conditions. Scalar approximations can be restored for many experimental situations of interest. Nonlinear self-focusing effects may increase the optical intensity to trigger photoionization, which requires to include free carrier terms. We obtain a set of coupled equations for forward and backward electric field components

$$(i\partial_z \pm |k_z|)\mathcal{E}_\omega^\pm = -\mu_0 \omega^2 \left[1 - \frac{\vec{k}\otimes\vec{k}}{k^2}\right](\vec{P}_{\mathrm{NL},\omega} + i\vec{J}_\omega/\omega). \quad (5)$$

As the operator $1 - \vec{k}\otimes\vec{k}/k^2$ projects out longitudinal field components, the evolution of the latter is governed by a source-free equation, while transverse components are governed by Eq. (5). Due to the presence of $1 - \vec{k}\otimes\vec{k}/k^2$, this bidirectional equation requires very costly numerics. The scalar, unidirectional limit $\vec{\mathcal{E}} \to \mathcal{E}$ comparable to Eq. (2) is obtained by letting

$$\frac{\vec{k}\otimes\vec{k}}{k^2}(\vec{P}_{\mathrm{NL},\omega} + i\vec{J}_\omega/\omega) \approx 0, \qquad \vec{\mathcal{E}}^- \approx 0, \quad (6)$$

i.e. by neglecting longitudinal field components and decoupling orthogonal polarization states, as well as by dropping backward propagating waves. However, while in the case Eq. (2) forward propagating field components can be identified with the positive frequency part of the electric field, this correspondence breaks down in the non-waveguiding case. It may be restored under the paraxial approximation $k_\perp \ll k_z$. In the unidirectional limit, this definition of directional fields leads to the forward Maxwell equation (FME) [10], which is successfully used in the context of femtosecond filamentation.

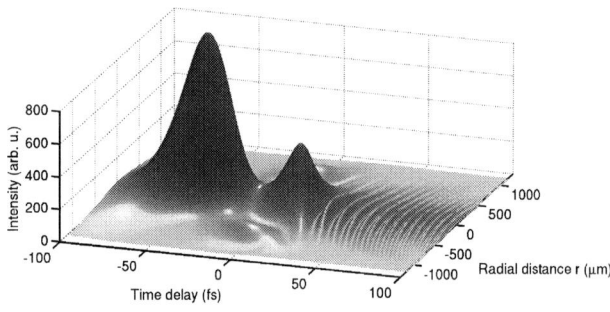

Fig. 2. Numerical simulation of optical wavebreaking in a femtosecond filament in argon.

The formation of femtosecond filaments can be observed when pulse femtosecond laser radiation is loosely focused into a dielectric medium, provided that the peak optical power P of the pulses exceeds a certain critical threshold P_{thr}. These filaments are narrow, longitudinally extended structures of dilute plasma and light. While for moderate ratios P/P_{thr} the cylindrical symmetry of the input beam can be maintained, an azimuthal modulation instability leads to multifilamentation and loss of cylindrical symmetry for higher input powers. The conservation of cylindrical symmetry leads to a strong simplification of the required numerics and holds for many interesting scenarios like pulse self-compression and harmonic generation within filaments. Furthermore, in these cases it is often justified to work within the paraxial approximation, such that the evolution of the electric field is governed by the FME. Numerically, this is solved using a pseudospectral split-step scheme, where that part of the propagation equation governed by the radial component $\Delta_r = 1/r\partial_r r\partial_r$ of the Laplacian is discretized using an implicit Crank-Nicolson scheme in the frequency domain, while the nonlinear part of the FME is evaluated in the time domain, involving the need for repeated Fourier transforms between temporal and spectral domain and accounting for aliasing errors. A characteristic radially symmetric field arising in a simulation of femtosecond filamentation is shown in Fig. (2) [11], which depicts the higher dimensional analogue of optical wavebreaking occuring during fiber propagation [9]. It is caused by modulation instability and is also known as hyperbolic shock-wave formation [12].

ACKNOWLEDGMENT

The work of Sh. Amiranashvili has been supported by the Deutsche Forschungsgemeinschaft (DFG) within the collaborative research center MATHEON under Grant D14.

REFERENCES

[1] T. Brabec and F. Krausz, "Nonlinear optical pulse propagation in the single-cycle regime," *Phys. Rev. Lett.*, vol. 78, no. 17, p. 3282, 1997.

[2] M. Kolesik and J. V. Moloney, "Nonlinear optical pulse propagation simulation: From Maxwell's to unidirectional equations," *Phys. Rev. E*, vol. 70, no. 3, p. 036604, 2004.

[3] S. Amiranashvili, U. Bandelow, and A. Mielke, "Pade approximant for refractive index and nonlocal envelope equations," *Opt. Commun.*, vol. 283, pp. 480–485, 2010.

[4] S. Amiranashvili, A. G. Vladimirov, and U. Bandelow, "A model equation for ultrashort optical pulses around the zero dispersion frequency," *Europ. Phys. J. D*, vol. 58, pp. 219–226, 2010.

[5] S. Amiranashvili and A. Demircan, "Hamiltonian structure of propagation equations for ultrashort optical pulses," *Phys. Rev. A*, vol. 82, no. 1, p. 013812, 2010.

[6] S. Amiranashvili, U. Bandelow, and A. Mielke, "Calculation of ultrashort pulse propagation based on rational approximations for medium dispersion," *Opt. Quantum Electron.*, vol. Online First, 2012.

[7] S. Amiranashvili and A. Demircan, "Ultrashort optical pulse propagation in terms of analytic signal," *Advances in Optical Technologies*, vol. 2011, p. 989515, 2011.

[8] L. Berge, S. Skupin, R. Nuter, J. Kasparian, and J. P. Wolf, "Ultrashort filaments of light in weakly ionized, optically transparent media," *Rep. Prog. Phys.*, vol. 70, p. 1633, 2007.

[9] G. P. Agrawal, *Nonlinear Fiber Optics*, 3rd ed., ser. Optics and Photonics. Academic Press, 2001.

[10] A. V. Husakou and J. Herrmann, "Supercontinuum generation of higher-order solitons by fission in photonic crystal fibers," *Phys. Rev. Lett.*, vol. 87, no. 20, p. 203901, Oct 2001.

[11] C. Brée, J. Bethge, S. Skupin, L. Bergé, A. Demircan, and G. Steinmeyer, "Cascaded self-compression of femtosecond pulses in filaments," *New J. Phys.*, vol. 12, p. 093046, 2010.

[12] L. Berge, K. Germaschewski, R. Grauer, and J. J. Rasmussen, "Hyperbolic shockwaves of the optical self-focusing with normal group-velocity dispersion," *Phys. Rev. Lett.*, vol. 89, p. 153902, 2002.

Characterization of Subwavelength Grating Waveguides with 3D Finite Element Method

Yuri H. Isayama[1], Marcos S. Gonçalves[2] *IEEE Member*, Hugo E. Hernández-Figueroa[1] *IEEE Senior Member*

1 Dept. of Microwave and Optics, School of Electrical and Computer Engineering,
University of Campinas, Campinas, Brazil. Email: {yurihi, hugo}@dmo.fee.unicamp.br
2 School of Technology, University of Campinas, Limeira, Brazil. Email: marcos@ft.unicamp.br

Abstract—A subwavelength grating waveguide was numerically analyzed by a 3D finite element method. Waveguide parameters as core height, width, duty cycle, and index contrast were varied and its effects investigated. Frequency shifts of the order of 40THz were obtained for the dispersion relation.

I. INTRODUCTION

A subwavelength grating (SWG) waveguide confines light through index-guiding, with a core composed of alternating segments of a material of high refractive index and a material with lower refractive index. Due to the small size of the grating pitch, Bragg condition is not satisfied and diffraction is frustrated [1].

The core height of a SWG waveguide has dimensions comparable to its width and, because of that, a 2D approach of the problem might be imprecise and, thus, a 3D analysis is necessary to investigate the characteristics of these waveguides. In this work, we present a 3D finite element method (FEM) approach to analyze the effects of changes in the SWG waveguide core height, width, duty cycle, and refractive index of the core over its modal behavior.

II. FORMULATION

The vector wave equation for the electric field in the frequency domain is given by

$$\nabla \times \nabla \times \mathbf{E}\left(\mathbf{r}\right) = \left(\frac{\omega}{c}\right)^2 \epsilon\left(\mathbf{r}\right) \mathbf{E}\left(\mathbf{r}\right), \tag{1}$$

where $\mathbf{r}$ is the position vector, $\epsilon(\mathbf{r})$ is the electric permittivity, $\mathbf{E}(\mathbf{r})$ is the electric field, and c is the speed of light in vacuum. Assuming a periodic structure, the electric field can be written as [2] $\mathbf{E}(\mathbf{r}) = \mathbf{u}(\mathbf{r})e^{-j\mathbf{k}\cdot\mathbf{r}}$, where $\mathbf{k}$ is the wave vector, $\mathbf{u}(\mathbf{r})$ is a periodic function defined as $\mathbf{u}\left(\mathbf{r}\right) = \mathbf{u}\left(\mathbf{r}+\mathbf{a}\right)$, and and $\mathbf{a}$ is the lattice vector. Using the given definition of $\mathbf{E}(\mathbf{r})$ in (1), applying Galerkin's Method [3], and both the Divergence and Green's theorems, we have

$$\iiint_V \Big\{ \nabla \times \mathbf{u}(\mathbf{r}) \cdot \nabla \times \mathbf{w}(\mathbf{r}) - j\mathbf{k} \times \mathbf{u}(\mathbf{r}) \cdot \nabla \times \mathbf{w}(\mathbf{r}) +$$
$$+j\mathbf{k} \times \mathbf{w}(\mathbf{r}) \cdot \nabla \times \mathbf{u}(\mathbf{r}) + \mathbf{k} \times \mathbf{u}(\mathbf{r}) \cdot \mathbf{k} \times \mathbf{w}(\mathbf{r}) \Big\} dV +$$
$$+ \oiint_S \Big\{ \mathbf{w}(\mathbf{r}) \times [j\mathbf{k} \times \mathbf{u}(\mathbf{r}) - \nabla \times \mathbf{u}(\mathbf{r})] \Big\} \cdot \mathbf{n} dS =$$
$$= \left(\frac{\omega}{c}\right)^2 \iiint_V \epsilon(\mathbf{r})\mathbf{u}(\mathbf{r}) \cdot \mathbf{w}(\mathbf{r}) dV, \tag{2}$$

where $\mathbf{w}(\mathbf{r})$ represents a proper trial function, V is the whole domain volume, and $\mathbf{n}$ is the unit normal vector with respect to the surface S of the computational domain.

For perfect electric conductor (PEC) and perfect magnetic conductor (PMC) boundary conditions, the surface integrals in (2) over the boundaries are zero. For periodic boundary conditions, the fields on two parallel surfaces at the boundaries of the domain must be the same and, since the normal vectors of two parallel surfaces have opposite directions, the surface integrals of (2) on two parallel surfaces must be zero.

Applying the FEM and considering the aforementioned boundary conditions, the following eigenvalue problem arises:

$$[K]\{u\} = \left(\frac{\omega}{c}\right)^2 [M]\{u\}. \tag{3}$$

The elementary matrices, related to the global matrices, are given by:

$$[K^e]_{m,n} = \iiint_{V_e} \Big[\nabla \times \mathbf{W}_m^e \cdot \nabla \times \mathbf{W}_n^e - j\mathbf{k} \times \mathbf{W}_m^e \cdot \nabla \times \mathbf{W}_n^e +$$
$$+j\mathbf{k} \times \mathbf{W}_m^e + \mathbf{k} \times \mathbf{W}_m^e \cdot \mathbf{k} \times \mathbf{W}_n^e \Big] dV,$$

$$[M_{m,n}^e] = \iiint_{V_e} \mathbf{W}_m^e \cdot \mathbf{W}_n^e dV.$$

where $\mathbf{W}_\xi^e$ is the Whitney basis function. The basis function associated to the edge ξ that connects the nodes I and j is given by $\mathbf{W}_\xi^e = L_I \nabla L_j - L_j \nabla L_I$, where $L_{i,j}$ are nodal basis functions associated to the nodes I and j, respectively, and V_e is the volume of an element of the discretized domain.

978-1-4673-1602-6/12 $31.00 © 2012 IEEE

III. SIMULATION RESULTS

The computational domain simulated had dimensions $x \times y \times z = 2\mu m \times 2\mu m \times 0.3\mu m$, and periodic conditions were applied at the planes z = 0 and z = 0.3 μm. A segmentation pitch of $\Lambda = 0.3$ μm, substrate height of 0.4 μm, and an upper cladding (air) height of 1.6 μm were employed. The refractive index of the subtrate and the upper cladding are, respectively, $n_{SiO_2} = 1.44$, and $n_{air} = 1.0$. The waveguide core is made of either Si ($n_{Si} = 3.476$) or Si_3N_4 ($n_{Si_3N_4} = 1.99$).

Figs. 1 and 2 show the dispersion relation of quasi-TE and quasi-TM modes, respectively, for a core width w = 300 nm, refractive index $n_{Si} = 3.476$, and different values of core height (h) and segment length (l). The duty cycle of the waveguide corresponds to the ratio l/Λ. As expected, it is observed that increasing the waveguide height or width the dispersion curves of the waveguide suffer a shift to lower frequencies and the modes become more confined to the core. Increasing the duty cycle also has the effect of shifting the dispersion curves to lower frequencies and it decreases the cutoff frequency of both quasi-TE and quasi-TM modes. Increasing w from 300 nm to 500 nm in a h = 300 nm and 50% duty cycle waveguide produces a shift of 42 THz for the quasi-TE mode and 15.7 THz for the quasi-TM mode. For a w =300 nm and l = 150 nm waveguide, increasing h from 300 nm to 600 nm produces a shift of 14THz for TE polarization and 46 THz for TM polarization.

Fig. 3 presents the dispersion relation for the quasi-TM mode (fundamental) for different core materials (Si and Si_3N_4), w = 300 nm and l = 150 nm. Reducing the refractive index of the core has the effect of moving the dispersion curves to higher frequencies. The variations of the refractive index of the waveguide core, as well as the variations of waveguide width could be analyzed by a 2D approach. However, Figs. 1 and 2 show that changes in the SWG waveguide height can result in big variations in its dispersion relation and, because the effects of waveguide height are not considered in a 2D model, in order to properly investigate the waveguide presented here, a 3D formulation is fundamental.

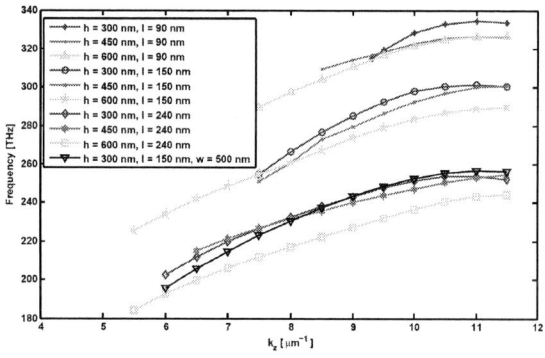

Fig. 1. Dispersion relation for the quasi-TE mode for w = 300 nm.

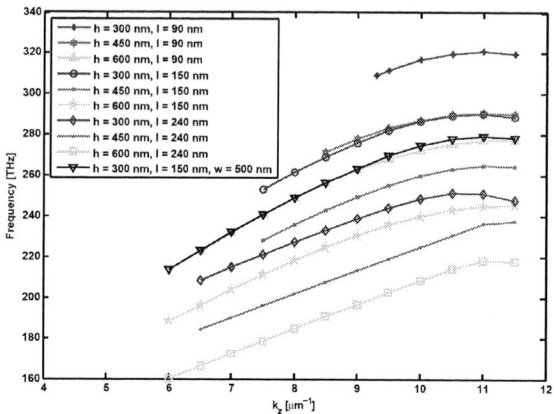

Fig. 2. Dispersion relation for the quasi-TM mode for w = 300 nm.

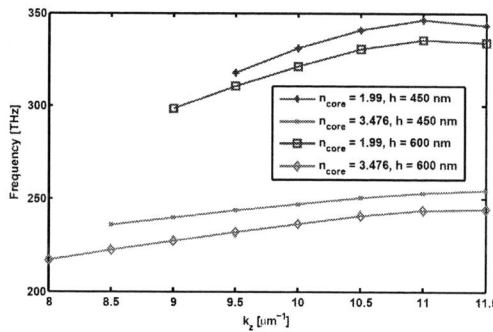

Fig. 3. Dispersion relation for the quasi-TM mode for w = 300 nm and l = 150 nm.

IV. CONCLUSION

In this paper, the modal characteristics of a SWG waveguide were analyzed with a 3D finite element method. Frequency shifts of the order of 40 THz were obtained, by manipulating the waveguide height, for TM polarization, and waveguide width, for TE polarization. In conclusion, it is possible to engineer the effective index not only by varying the duty cycle of the waveguide and its refractive index, but also by altering parameters as its height and width. Additionally, the possibility to produce large frequency shifts in the dispersion relation makes the SWG waveguide suitable for waveguide coupling, and, therefore, it represents an important building block for integrated optics.

ACKNOWLEDGMENT

This work was supported by INCT-Fotonicom, CAPES, FAPESP, and FAEPEX.

REFERENCES

[1] I. Molina-Fernandez et al., "New concepts in silicon component design using subwavelength structures," in *Proc. SPIE*, 2012, vol. 8266, p. 82660E.

[2] J. D. Joannopoulus, R. D. Meade, and J. N. Winn, *Photonics crystals: molding the flow of light*, Princeton University Press, 1995.

[3] J. M. Jin, *The finite element method in electromagnetics*, 2nd edition, Wiley, New York, 2002.

All-Optical Discrete Fourier Transform for OFDM Demultiplexing and its Sensitivity to Phase Errors

Stefan Schwarz, Christian G. Schaeffer
Chair of High-Frequency Engineering
Helmut Schmidt University, Holstenhofweg 85,
22043 Hamburg, Germany
Email: Stefan.Schwarz@hsu-hh.de

Abdul Rahim, Juergen Bruns, Klaus Petermann
Institute of High-Frequency Engineering
Berlin University of Technology, Einsteinufer 25,
10587 Berlin, Germany

Abstract—**We present the design of an optical OFDM-demultiplexer for the separation of 8 sub-channels. Using simulations, we investigate the tolerance towards phase errors in the structure which could be realized as a planar lightwave circuit (PLC).**

Keywords-Optoelectronic integrated circuits, Optical communications, Optical transforms, Multimode interference devices

I. INTRODUCTION

Latest experiments in the field of optical orthogonal frequency division multiplexing (OFDM) technology have shown the feasibility of a bandwidth-efficient and dispersion-tolerant optical transmission system at high data rates [1]. The corresponding structure for the discrete Fourier transform (DFT) or a computation efficient Fast Fourier transform (FFT) in the receiver as well as their inverse counterparts in the transmitter can be realized by various methods using optical delay interferometers (DI, [1]), arrayed-waveguide grating routers (AWGR, [2],[3]) or multimode interference (MMI) couplers [4]. In this paper, we present a new structure for an 8-channel OFDM demultiplexer which consists of 2×2 and 4×4 MMI couplers, delay lines and phase shifters. This two-stage layout with only a small number of couplers allows for compact device on a small footprint. After introducing the principles of operation, we investigate the impact of phase errors on the signal quality, Q. Using Monte-Carlo simulation, we evaluate how much phase deviation this DFT structure can tolerate without severe performance deterioration.

II. OPERATION PRINCIPLES AND SIMULATION SETUP

The separation of one OFDM super-channel into N subcarrier channels X_0 to X_7 is performed by applying a discrete Fourier transformation

$$X_m = \sum_{n=0}^{N-1} e^{-j2\pi\frac{mn}{N}} x_n, \quad m = 0,...,N-1, \quad (1)$$

where x_n represents the N equidistant samples of the incoming signal x(t) over a period of time T. Figure 1 shows the proposed design for an optical 8-point DFT circuit. The first stage consists of two symmetrical 2×2 MMI couplers and delay lines with a length difference of $4T$, while symmetrical 4×4 MMI couplers are being connected by

delay lines of length $0T...3T$ on each arm of the second stage. Additional tuning elements, denoted by φ_{Ai} and φ_{Bi}, are used to adjust the phases. However, only phase differences between the delay lines change the behavior of the filter. The filter structure shown in Fig. 1 is similar to the DFT architecture presented in [5] except that we use multi-mode instead of slab coupler and resort the delay line order.

It should be noted that, in contrast to the electrical processing, this optical DFT is computed continuously. Instead of superimposing weighted, sampled signal values, we feed the signal and its delayed copies to the last MMI couplers in the upper and lower arm which perform the addition. Following these outputs, electro-absorption optical switches can be used as optical gates to perform the needed time gating.

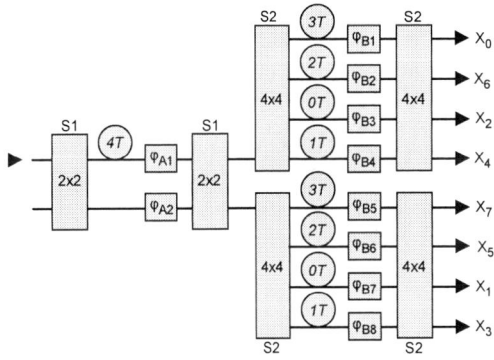

Figure 1. Structure for an 8 channel OFDM demultiplexer consisting of symmetrical 2×2 and 4×4 MMI couplers (labeled as S1 and S2), different delay lines (unit delay T is equal to the symbol period) and phase shifters.

Phase errors on the delay lines, caused by the manufacturing uncertainty or imprecise adjustment of the phase shifters result in a shift of the equidistant filter zeros. Consequently, the subcarriers are no longer orthogonal which leads to interchannel interference (ICI). In this context, a statistical analysis of the estimated filter sensitivity prior to the realization step is helpful for the design process.

We used VPItransmissionMaker^{TM} 8.7 to evaluate the performance of the proposed OFDM demultiplexer statistically. Therefore, we feed 8 QPSK modulators, each transmitting with a symbol rate of 25 GSymbols/s, with a separate wavelength obtained from an optical comb generator. Pulse shapers with a rise time of 5 ps were implemented for smoothing the modulator driving signals. Since no cyclic prefixes were used, the carrier frequencies were spaced at the reciprocal value of the symbol duration, which is 40 ps, to fulfill the OFDM condition, i.e. the orthogonality of the signals. As our investigation is focused on the sensitivity of the DFT circuit to phase errors, we assumed an ideal transmission link without dispersion or noise impairments. The receiver at each output of the demultiplexer is modeled as an ideal 90° hybrid. Furthermore, each sub-channel receiver was set to perform a perfect clock and carrier recovery at the optimal sampling time.

In a Monte-Carlo simulation with 1000 iterations, we add phase errors to each delay line of the filter. These errors were assumed to be Gaussian distributed with mean values of zero and specific standard deviations. We then evaluated 256 transmitted symbols on each channel for every iteration step to determine the signal quality Q, which was derived according to the definition in [6]. Assuming Gaussian statistics, Q-values of 15.56 dB, 12.6 dB and 9.8 dB translate to bit error rates of 10^{-9}, 10^{-5} and 10^{-3}, respectively.

III. RESULTS AND DISCUSSION

A comparison of system quality Q versus the standard deviation of the phase errors can be seen in Fig. 2. The results have been averaged over 1000 iteration runs for every channel and every step of standard deviation. Due to rise and fall time limitations in the transmitter and receiver, the orhogonality condition is not perfectly fullfilled in this setup. Even an optimal phase adjustment in the DFT structure will lead to intersymbol and interchannel interferences and hence this will lower the Q-factor to around 15 dB to 17 dB. The resulting crosstalk on each channel increases the closer it is located to the center of the super-channel. With rising deviation from ideal phase conditions, the Q-factor decreases at a rate of ~2 dB every additional 5° phase error.

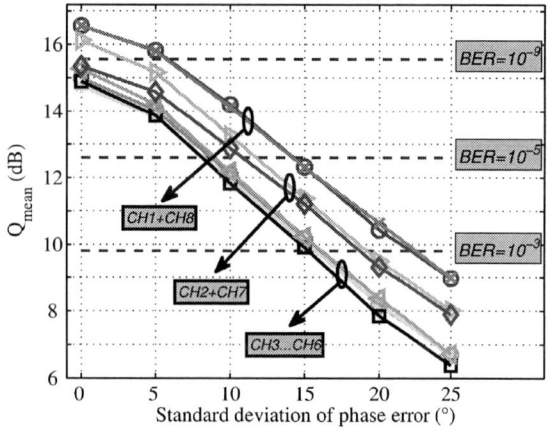

Figure 2. Mean Q versus the standard deviation of applied phase errors.

The financial support of German Research Foundation (DFG) in the framework of this project is gratefully acknowledged.

The yield curves as a function of penalty tolerance are shown in Fig 3. Following the definition in [3], a yield of 100% means that all simulation runs from one specific set of phase errors with a certain standard deviation provide a system performance, Q, which is greater than a certain chosen boundary. Since we set this threshold to $Q_{Limit} = 9.8$ dB, the value for the yield at 2 dB penalty tolerance, for example, refers to the fraction of simulation runs with an Q-value larger than 9,8 dB – 2 dB = 7,8 dB. The results obtained in this simulation show that this filter structure is resilient against phase errors caused by imprecise adjustment of the phase shifters and manufacturing uncertainties. The advantage of this approach over other coupler-based alternatives is its compact design requiring lower number of delay lines. Consequently, there are less elements in the filter structure susceptible to phase errors.

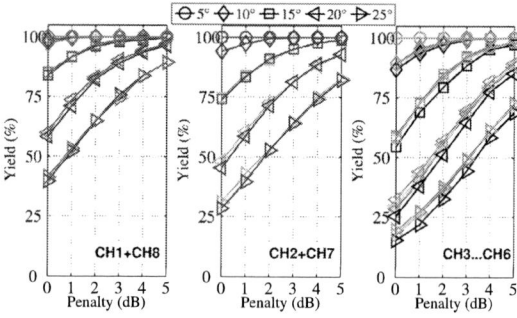

Figure 3. Yield curves versus acceptable penalty ($Q_{Limit} = 9.8$ dB).

IV. CONCLUSION

We presented a new filter layout for an 8-channel OFDM demultiplexer which could be realized in planar waveguide technology. Using a Monte-Carlo simulation, the sensitivity to different sets of delay line phase errors was analyzed. It is shown that coupler based structures for OFDM application can be used, if properly designed with higher order MMI couplers, as an alternative to AWG realizations.

REFERENCES

[1] D. Hillerkuss et al., "26 Tbit s⁻¹ line-rate super-channel transmission utilizing all-optical fast Fourier transform processing," Nature Photonics, vol. 5, pp. 364–371, May 2011.

[2] A. J. Lowery, and L. Du, "All-optical OFDM transmitter design using AWGRs and low-bandwidth modulators," Opt. Expr. 19(17), pp. 15696–15704, August 2011.

[3] S. Lim and J.-K. K. Rhee, "System tolerance of all-optical sampling OFDM using AWG discrete Fourier transform," Opt. Express 19(14), pp. 13590–13597, July 2011.

[4] K. Takiguchi et al., "Integrated-optic OFDM demultiplexer using multi-mode interference coupler-based optical DFT circuit," OFC/NFOEC'12, paper OM3J.6, Los Angeles, CA, March 2012.

[5] G. Cincotti, "Generalized fiber Fourier optics," Optics Letters 36(12), pp. 2321–2323, June 2011.

[6] A. J. Lowery et al., "Performance of Optical OFDM in Ultralong-Haul WDM Lightwave Systems," J. Lightw. Technol. 25(1), pp. 131–138, January 2007.

Determination of Resonance Frequencies in Silica Fiber using SRS Gain

Mrinal Sen[1] and Mukul Kumar Das[2], *Senior Member, IEEE*

[1,2]Dept. of Electronics Engineering, Indian School of Mines, Dhanbad, India
[1]mrinal.sen.ahm@gmail.com, [2]das_mkdas@yahoo.co.in

Abstract--- This paper presents a novel approach for the determination of resonance frequencies in silica fiber using composite susceptibility model and Stimulated Raman Scattering (SRS) gain. The Raman gain coefficient in silica fiber is calculated in terms of composite susceptibility model where resonance frequencies of material are considered. Composite susceptibility is optimized using Genetic Algorithm to match calculated gain with the experimental results and, hence eight resonance frequencies are found.

I. INTRODUCTION

Optical fiber communication link covers a massive part of the existing telecommunication network around the globe. Performance of present day Wavelength Division Multiplexing (WDM) technique in optical network is limited by narrow bandwidth of Erbium Doped Fiber Amplifiers (EDFA) [1-3]. Recently there had been a lot of interest among researchers around the world in Distributed Fiber Raman Amplifier (DFRA) due to its ultra-wide bandwidth, flexibility in operation, low noise and capacity to alleviate fiber nonlinearities which is significant at high power transmission.

This paper presents a comprehensive theoretical study of gain in optical fiber under the influence of SRS. The generation of SRS in optical fiber is accompanied by intense molecular or lattice vibrations which have a high degree of temporal and spatial coherence. These molecular vibrations modulate the incoming light beam and generate sidebands [4]. The signal or stoke waves, that are separated by an amount of the frequency of the lattice vibrations from the pump wave, are get amplified. Nonlinear susceptibility plays a major role in most of the nonlinear optical processes like SRS [5,8]. Thus, quantitative measurement of the stimulated Raman gain spectrum allows determining frequency dependent Raman susceptibility of the material. In this paper, a numerical model for composite Raman susceptibility of bulk silica material is proposed considering classical mechanics. The Raman gain is calculated and compared with the reported experimental results to find the material resonant frequencies and composite susceptibility.

II. MODEL DESCRIPTION

The energy transfer procedure from a pump photon to another photon of shifted frequency is inelastic in nature. The inelastic behavior can be explained by quantum mechanics or by classical mechanics (to a reasonable extent). In a molecular system, due to the light-particle interaction, material absorbs/delivers energy from/to the optical signal (pump signal) and produces photon of lower (Stokes) or higher (Anti-Stokes) frequency.

Intermolecular vibration plays an important on the shifting of frequency of photons. This vibration has already been described as a simple harmonic motion of single resonance frequency [5]. However, for accurate depiction of molecular vibration, instead of single resonance frequency, multiple frequencies need to be considered because molecular vibration is not isotropic in space due to polarization of light signal. Authors proposed the modified equation for j^{th} vibrational mode of molecule as

$$\frac{d^2\tilde{q}_j}{dt^2} + 2\gamma_j \frac{d\tilde{q}_j}{dt} + \omega_{vj}^2 \tilde{q}_j = \frac{\tilde{F}_j(t)}{m} \qquad (1)$$

where, $\tilde{q}_j$ is the deviation of the intermolecular distance from equilibrium under the influence of applied force, $\tilde{F}_j(t)$, ω_{vj} is resonance frequency of oscillation and γ_j is damping constant. This equation actually presents the motion of the molecular vibration under applied force that acts on the jth vibrational degree of freedom with m as reduced nuclear mass.

Now, optical polarizability of molecule depends upon the intermolecular distance as

$$\tilde{\alpha}(t) = \alpha_0 + \left(\frac{\partial\alpha}{\partial q}\right)_0 \tilde{q}(t) \qquad (2)$$

and the composite Raman susceptibility can be obtained as

$$\chi_R(\omega_S) = \sum_j \frac{(N/6m)(\partial\alpha/\partial q)_{0j}^2}{\omega_{vj}^2 - (\omega_L - \omega_S)^2 + 2i(\omega_L - \omega_S)\gamma_j} \qquad (3)$$

The Raman Gain Coefficient g_r is given by [8].

$$g_r = -\frac{12\mu_0^2\omega_L\omega_S^2}{A_{eff}^R K_L K_S} Im(\chi_R) \qquad (4)$$

where,

$$A_{eff}^R = \frac{\iint|E_{pump}(r,\varphi)|^2 r\, dr d\varphi.\iint|E_{stoke}(r,\varphi)|^2 r\, dr d\varphi}{\iint|E_{pump}(r,\varphi)|^2.|E_{stoke}(r,\varphi)|^2 r\, dr d\varphi} \qquad (5)$$

N represents the number density of molecules, $(\partial\alpha/\partial q)_{0j}$ represents the rate of change of polarizability with the deviation of intermolecular distance from equilibrium for the jth resonance, ω_L is the pump angular frequency and ω_S is the stoke signal angular frequency. Here the summation is taken over all possible contribution of resonance on Raman susceptibility.

With a typical pump frequency and typical amplitudes for Pump and Stokes power, effective area is calculated using

Eq.5 and hence, resonance frequency dependent composite susceptibility is optimized to match the normalized theoretical Raman gain with the experimental data [7]. The Optimization is done using Genetic Algorithm (GA) where we have considered the independent variables as $(\partial\alpha/\partial q)^2_{0j}$, γ_j and ω_{vj} for a range of j from 2 to 10. The fitness function has been considered as the sum of the square of the differences of simulated and measured values. In several different run we found that the fitness with eight different frequency components (i.e., j) is best among the others. In the process optimization we have observed that the variation of $(\partial\alpha/\partial q)^2_{0j}$ and γ_j can affect a little on the peaks of the composite susceptibility whereas ω_{vj} is extremely responsible for determining the peaks. With these observations we have reached to a conclusion that though it is not possible to determine the proper value of $(\partial\alpha/\partial q)^2_{0j}$ and γ_j with the available knowledge, as there may a set of different solutions be available, we can at least predict the resonance frequencies of the molecular vibrational system. The results of this methodology are presented below.

III. RESULTS AND DISCUSSIONS

Pump power of 500mW at 1450 nm and signal (Stokes) power of 1mW in the range of wavelengths are considered. The Pump is co-polarized and co-propagating with the Stokes wave. Step index single mode fiber of core diameter 12μm is taken. The result of 10 different run of GA for eight resonance frequencies with their corresponding fitness values has been presented in table I. Hence the results are produced with the eight resonance frequencies and their respective $\partial\alpha/\partial q$ and γ obtained by the GA with minimum fitness value. The comparison of theoretically calculated gain and the measured gain has been shown in Fig 1. The theoretically calculated Raman Effective Area as a function of frequency shift is shown in Fig 2. Finally, the composite susceptibility with the contribution of individual resonances has been presented in Fig 3. The resonance frequencies are found centered at 7.3594, 13.2039, 14.6445, 18.0898, 24.1391, 32.3625, 10.4297 and 14.2273 THz frequencies.

TABLE I. RESULTS OF GENETIC ALGORITHM FOR TEN DIFFERENT RUN

Fitness Value	Resonance Frequencies (THz)							
	1	2	3	4	5	6	7	8
0.0756	7.75	13.2	14.656	18.25	24.1	32.05	10.566	14.325
0.0746	7.5625	13.2	14.519	18.140	24.037	31.894	10.641	14.254
0.0628	7.4595	13.106	14.614	18.25	24.161	32.05	10.649	14.258
0.0959	7.5469	13.325	14.5	18.25	24.193	32.05	10.687	14.575
0.0699	7.25	13.137	14.671	18.187	24.037	32.3	10.626	14.262
0.0822	7.2969	13.2	14.574	18.043	24.178	31.753	10.402	14.246
0.083	7.7188	13.2	14.578	18.125	24.094	32.057	10.687	14.293
0.0592	**7.3594**	**13.203**	**14.644**	**18.089**	**24.139**	**32.362**	**10.429**	**14.227**
0.0995	8.0044	13.325	14.5	18.125	24.147	32.430	11.187	14.637
0.0729	7.5	13.227	14.554	18.117	24.209	32.55	10.648	14.2

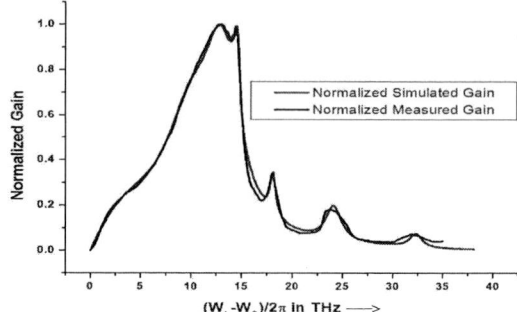

Fig.1. Raman frequency shift verses theoretically calculated and practically measured Raman Gain Plot.

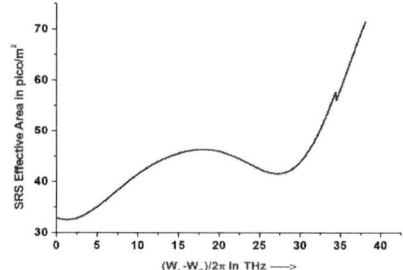

Fig.2. Raman frequency shift verses Raman Effective Area plot.

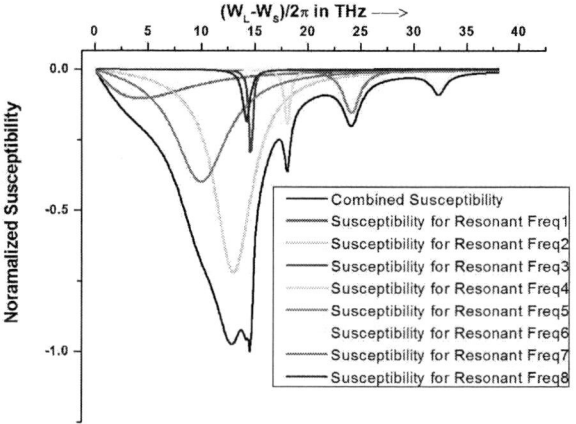

Fig.3. Raman frequency shift verses Composite Susceptibility plot. The thin line represents the individual effect of different resonances 7, 10, 13.2, 14.2, 14.5, 18, 23.6 and 31.8 THz frequencies.

REFERENCES

[1] Mohammed N. Islam (Ed.), *Raman Amplifiers for Telecommunications 1- Physical Principles*, Springer Series in optical sciences, **2004**.

[2] A. Sano et al., in *Proc. of European Conf. on Optical Comm.*, Th 4.1.1, **2006**.

[3] A. Saleh and J. Simmons, in *IEEE J. Lightwave Technol.*, Vol. 24, pp.3303-3321, **2006**

[4] Max Maier, "Applications of Stimulated Raman Scattering," in *Appl. Phys.*, 11, 209—231, **1976**.

[5] Robert W. Boyd, *Nonlinear Optics*, Second Edition, Academic Press, Elsevier Science (USA), **2003**.

[6] R.H. Stolen and E.P. Ippen, "Raman gain in glass optical waveguides," in *Appl. Phys. Lett.*, 22:6, **1973**.

[7] Jake Bromage, "Raman Amplification for Fiber Communications Systems," in *Journal of Lightwave Technology*, vol. 22, no. 1, January **2004**.

[8] A. Yariv and Pochi Yeh, *Photonics – Optical Electronics in Modern Communication*, Sixth Edition, Oxford University Press, **2006**.

NUSOD 2012 Author Index

Afshar S	129	Deng M	99	Hu WD	33
Ahmadi V	69	Deppner M	93	Hu WD	35
Aksamija Z	121	DiCarlo A	95	Hu WD	45
Amiranashvili S	131	Dickerson J	113	Hu WD	51
Auf der Maur M	95	Ding JY	55	Hu XN	11
Bandelow U	131	Ding RJ	11	Hu XN	19
Berbezier A	115	Ding RJ	19	Hua J	75
Bree C	131	Du CH	105	Huang CH	117
Bruns J	135	Ebuchi S	15	Huang L	47
Cai SH	99	Fan L	57	Huang L	59
Cao JC	9	Fukuchi Y	27	Huang LJ	3
Chen X	47	Gaertner K	103	Huang LJ	55
Chen X	59	Gan L	87	Huang YZ	5
Chen XS	1	Gao K	37	Hung SY	67
Chen XS	3	Garg R	85	Husko C	89
Chen XS	17	Gonçalves MS	133	Iiyama K	15
Chen XS	29	Gu Z	37	Isayama YH	133
Chen XS	33	Guo FM	57	Jamali M	69
Chen XS	35	Guo N	1	Jeon MY	43
Chen XS	45	Guo N	45	Jiang c	7
Chen XS	55	Guo XG	9	Jiang XF	53
Chen Z	7	Hagness SC	79	Jiang XW	101
Cheng HL	117	Han SP	43	Jiang Y	3
Cheng L	51	Hashimoto Y	15	Jiao N	81
Cheng LW	21	Haxha V	85	Kaatuzian H	73
Cheng LW	23	He C	81	Kato H	125
Cheng LW	35	He C	83	Kaya OA	41
Chiou YP	105	He L	11	Kim JH	65
Cicek A	41	He L	19	Kim N	43
Cui H	39	Hernández HE	133	Knezevic I	79
Cui H	77	Hong XK	71	Knezevic I	121
Dai Z	75	Hu WD	1	Ko H	43
Das MK	137	Hu WD	3	Kong F	31
Das N	13	Hu WD	21	Koprucki T	103
Deng HX	101	Hu WD	29	Kurz H	107

Lee HJ	65	Lu W	29	Roemer F	93
Lee S	65	Lu W	33	Ru GP	25
Lee TJ	65	Lu W	45	Ryu HC	43
Li G	15	Lu W	47	Sacconi F	95
Li G	47	Lu W	51	Schaeffer CG	135
Li G	59	Lu W	55	Schwarz S	135
Li GH	3	Lu W	59	Sen M	137
Li GH	55	Luo JW	101	Shahriari M	73
Li K	31	Luo T	37	Shao CX	3
Li L	63	Lv XM	5	Sheng Y	21
Li L	61	Lv YQ	45	Sheng Y	23
Li N	17	Maruyama T	15	Sheng Y	51
Li Q	17	Mashayekhi HR	13	Shi YB	121
Li SS	101	Masouleh FF	13	Si JJ	45
Li X	127	Michelini F	115	Suckow S	107
Li XY	33	Migliorato MA	85	Sule N	79
Li Y	11	Min B	107	Sun LZ	81
Li Y	19	Monro TM	129	Sun LZ	83
Li YY	25	Moudakir T	113	Sun X	75
Li Z	47	Ni B	47	Tang N	39
Li Z	59	Ni B	55	Tang N	77
Li ZF	17	Ni B	59	Tang Z	39
Li ZM	21	Nihei H	91	Tang Z	77
Li ZM	23	Noppe MG	123	Tomic S	97
Li ZM	25	Numai T	125	Tomic S	111
Li ZY	87	Okada Y	109	Tse G	85
Liang J	35	Okamoto A	91	Ulug B	41
Lin C	11	Osawa Y	27	Voss PL	113
Lin C	19	Oshige R	27	Vukmirovic N	97
Lin H	127	Ougazzaden A	113	Wang C	87
Lin JD	5	Pal J	85	Wang L	1
Liu HC	9	Pantzas K	113	Wang L	29
Liu YS	53	Park JW	43	Wang L	33
Lohe MA	129	Park KH	43	Wang LW	101
Lu H	71	Petermann K	135	Wang X	31
Lu W	1	Piprek J	119	Wang XD	33
Lu W	3	Pletzer TM	107	Wen J	61
Lu W	17	Rahim A	135	Wen J	63
Lu W	21	Razaghi M	69	Weng QC	61

Weng QC	63
Willis KJ	79
Witzigmann B	93
Woo DH	65
Wu X	3
Xi Y	127
Xia CS	21
Xia CS	23
Xia CS	35
Xia CS	51
Xiao HP	83
Xiong DY	61
Xiong DY	63
Xu JT	33
Yang XF	53
Yao QF	5
Ye ZH	11
Ye ZH	19
Yoshida H	125
Yoshida K	109
Yucel MB	41
Yusufoglu U	107
Zhang C	99
Zhang CX	81
Zhang CX	83
Zhang DB	71
Zhang W	129
Zhang XL	45
Zhang Z	31
Zhong JX	83
Zhou D	49
Zou LX	5

9781467316026